The Ecosystem Approach in Anthropology

From Concept to Practice

Emilio F. Moran, Editor

Ann Arbor
The University of Michigan Press

Copyright © by the University of Michigan 1990
All rights reserved
Published in the United States of America by
The University of Michigan Press
Manufactured in the United States of America

1993 1992 1991 1990   4 3 2 1

Distributed in the United Kingdom and Europe by
Manchester University Press, Oxford Road,
Manchester M13 9PL, UK

Library of Congress Cataloging-in-Publication Data

The Ecosystem approach in anthropology : from concept to practice /
  Emilio F. Moran, editor.
      p. cm.
    Rev. ed. of: The Ecosystem concept in anthropology.
    Includes bibliographical references and index.
    ISBN 0-472-10178-1 (cloth : alk. paper). — ISBN 0-472-08102-0
  (pbk. : alk. paper)
      1. Anthropology—Methodology.   2. Human ecology—Research.
  3. Anthropology—Philosophy.   4. Environmental archaeology—
  Research.   5. Biotic communities—Research.   I. Moran, Emilio F.
  II. Ecosystem concept in anthropology.
  GN33.E27   1990
  301'.01—dc20   90-11306
                                                      CIP

British Library Cataloguing in Publication Data
The ecosystem approach in anthropology : from concept to
  practice.
    1. Anthropology. Research. Methodology
  I. Moran, Emilio F.
  301

ISBN 0-472-10178-1
ISBN 0-472-08102-0 pbk

The Ecosystem Approach in Anthropology

To James and Ann Vaughan

TABLE OF CONTENTS

ABOUT THE EDITOR AND CONTRIBUTORS

Emilio F. Moran is Professor of Anthropology and Professor in the School for Public and Environmental Affairs at Indiana University, Bloomington. He is a specialist in ecological anthropology, resource management and agricultural systems in the humid and dry tropics. He is the author of *Human Adaptability: An Introduction to Ecological Anthropology* (Duxbury 1979, reprinted by Westview 1982), *Developing the Amazon* (Indiana Univ. Press 1981), *The Dilemma of Amazonian Development* (editor, Westview Press 1983) and recently completed *The Human Ecology of Amazonian Populations* (Vozes 1990).

Daniel G. Bates is Professor of Anthropology at Hunter College, City University of New York. He has done field research in Africa and the Middle East and specialized in both ecological and economic anthropology. He is the author of *Nomads and Farmers: The Yoruk of Southeastern Turkey* (Univ. of Michigan Museum 1971), *Peoples and Cultures of the Middle East* (with A. Rassan, Prentice-Hall 1982) and is currently co-editor of the journal *Human Ecology* (with S. Lees).

John W. Bennett is Distinguished Scholar in Residence at Washington Univ., St. Louis. He has specialized in both economic and ecological anthropology. He has published on anthropological theory and methodology, agricultural decision-making, Third World development, cooperatives, Japanese society and modernization, and aspects of American culture. His books include *Northern Plainsmen* (Aldine 1969), *The Ecological Transition* (Pergamon 1976) and *Of Time and the Enterprise* (Univ. of Minnesota Press 1982).

Karl W. Butzer is Dickson Centennial Professor of Liberal Arts at the University of Texas, Austin. His work over the years has influenced Old World Archaeology and ensured the vigor of a human ecological approach within the field of Archaeology. He is the author of *Archaoelogy as Human Ecology* (Cambridge Univ. Press 1982), *Early Hydraulic Civilization in Egypt: A Study in Cultural Ecology* (Chicago 1979), *Environment and Archaeology: An*

Ecological Approach to Prehistory (Aldine 1971), and *Desert and River in Nubia* (Wisconsin 1968).

Francis Paine Conant is Professor of Anthropology at Hunter College, City Univ. of New York. His major field work has been in West and East Africa, where he studied the subsistence and settlement systems of farming and herding peoples. His specialization in ecological anthropology is the applications of remote sensing data to the study of resource utilization in the African Sahel. He has published on anthropological method and theory, as well as religion, kinship, community types, and the division of labor by gender in African societies.

Neville Dyson-Hudson, Professor of Anthropology at the State University of New York, Binghamton, has conducted research on pastoral populations in the Sudan, Uganda, and Kenya. He has long-standing interests in the ecology of East African pastoralists and has pursued this research for more than three decades. His present research deals with social networks and with violence. Among his publications are *Karimojong Politics* (Clarendon Press 1966) and *Perspectives on Nomadism* (edited with W. Irons, Leiden, 1972).

Rada Dyson-Hudson is on the faculty at Cornell University. A specialist in ecology, evolutionary biology and human ecology, she has done research on human territoriality and on East African pastoralism, migration and demography. Her current work on Turkana nomads focuses on nomadic movement and food production systems. She has edited *Rethinking Adaptation* (with M. Little, Westview 1983) and wrote *South Turkana Nomadism: Coping with an Unpredictably Varying Environment* (with J.T. McCabe, HRAFlex Books 1985).

Roy F. Ellen is Professor of Anthropology and Human Ecology at the University of Kent at Canterbury, United Kingdom. His current research interests span the ecology of subsistence behavior, ethnobiology, classification and the social organization of trade. His books include *Nuaulu Settlement and Ecology* (Nijhoff 1978), *Environment, Subsistence and System (Cambridge 1982), Social and Ecological Systems* (edited with P. Burnham, Academic

Press 1979), *Classifications in their Social Context* (edited with D. Reason, Academic Press, 1979), *Ethnographic Research: A Guide to General Conduct* (Academic Press 1984) and most recently *Malinowski between Worlds* (edited with E. Geller, G. Kubica and J. Mucha, Cambridge 1988).

James E. Ellis is Professor of Range Science and Acting Director of the Natural Resource Ecology Laboratory, Colorado State University. Recent research has emphasized dry tropical ecosystem dynamics, ecology of African pastoralists and ecology of grazing systems. He has conducted research in East Africa, northern China, and the Middle East, as well as semi-arid portions of N. America.

Paul R. Fish is Curator of Archaeology at the Arizona State Museum and Research Professor of Anthropology at the University of Arizona. His current research interests focus on the prehistory of the Greater Southwest and the analysis of regional settlement patterns.

Suzanne K. Fish is a research archaeologist with the the Arizona State Museum and is also associated with the Office of Arid Land Studies at the Univ. of Arizona, Tucson. She has conducted extensive fieldwork and published on the archaeology and ethnobotany of Mesoamerica and the Southwestern United States.

Kathleen A. Galvin is Assistant Professor in the Department of Anthropology and Research Scientist in the Natural Resource Ecology Laboratory at Colorado State University. A specialist in biological anthropology, she has conducted research on human adaptation to arid tropical ecosystems with emphasis on diet and nutrition of African pastoralists. She has published on nutrition, growth, health and energy requirements of pastoralists and on the nutrient composition of pastoral foods.

Daniel R. Gross was Professor of Anthropology at Hunter College, City University of New York and is currently a Consultant with the World Bank. His major fields of interest are nutrition, adaptation, rural development and energy policy. He is the editor of *Peoples and Cultures of Native South America* (Doubleday 1969) and of numerous articles in *Science*, *American Ethnologist*, *American Anthropologist* and *Human Ecology*.

Christine A. Hastorf is Assistant Professor of Anthropology at the Univ. of Minnesota, Minneapolis. Her major research interests have been the onset of political inequality, the human use of the landscape, agricultural change, and paleoethnobotany. She has focused most of her archeological field work in the central Andes of S. America. Among her publications on these subjects are *Current Paleoethnobotany* (edited with V. Popper, Chicago); "Prehistoric Agricultural Production in the Central Andes" in *Food and Farm* (edited by C. Gladwin and C. Truman, Univ. Press of America); "Corn and Culture in Central Andean Prehistory", *Science* (with S. Johannessen); and *Resources in Power: Agriculture and the Onset of Political Inequality before the Inka* (Cambridge, in press).

Michael Jochim is Professor of Anthropology at the Univ. of California, Santa Barbara. His primary research interests are anthropological archeology, European prehistory and New World Paleoindian and Archaic societies. Among his publications are *Hunter-Gatherer Subsistence and Settlement* (Academic Press 1976) and *Strategies for Survival: Cultural Behavior in an Ecological Context* (Academic Press 1981).

Susan H. Lees is a specialist in human ecology and Professor of Anthropology at Hunter College, City University of New York. She has written on various aspects of desertification, irrigation, environmental management, and rural development. She is co-editor of the journal *Human Ecology* (with D. Bates).

Paul W. Leslie is Associate Professor of Anthropology at the State Univ. of New York, Binghamton. He has pursued his interest in the interactions among the demographic, genetic, and cultural characteristics of human populations both through mathematical modeling and through studies of isolated Caribbean populations. More recently, his research has focused on the demography and reproductive ecology of East African pastoralists.

Michael A. Little, Professor of Anthropology at the State Univ. of New York, Binghamton, is a biological anthropologist with interests in adaptation to environmental stress. He has worked with high-altitude Quechua Indians in Peru and now directs his efforts to growth of children and reproductive ecology of Turkana nomads in northwest Kenya. Some of his publications include *Man in the Andes* (co-edited with P.T. Baker, Dowden, Hutchinson and Ross 1976); *Ecology, Energetics and Human Variability* (with G. Morren, W.C. Brown 1976), and *Human Population Biology: A Transdisciplinary Science* (edited with J. D. Haas, Oxford, 1989).

Robert McC. Netting is Professor of Anthropology at the University of Arizona, Tucson. He is a specialist in cultural ecology, social organization and historical demography. He is the author of *Hill Farmers of Nigeria* (Univ. of Washington Press 1968), *Cultural Ecology* (Waveland 1986), and *Balancing on an Alp: Ecological Change and Continuity in a Swiss Mountain Community* (Cambridge 1981).

Roy A. Rappaport is Leslie White Professor of Anthropology at the University of Michigan, Ann Arbor. He is a past President of the American Anthropological Association (1987-89) and one of the seminal figures in the development of an ecosystem approach to anthropology. He is the author of *Pigs for the Ancestors* (Yale University Press 1968, revised ed. 1984), the most influential and widely read book on ecological anthropology. *Ecology, Meaning and Religion* (North Atlantic Books 1979) is a collection of his original essays.

David M. Swift is Associate Professor of Range Science and Research Scientist at the Natural Resource Ecology Laboratory, Colorado State University. He is a specialist in nutrition and ecology of large herbivores, ecology of grazing systems, and in the development and application of simulation models. He has worked in Colorado, Kenya, Ethiopia, China and Pakistan.

Richard R. Wilk is Assistant Professor of Anthropology at Indiana University. He is an economic and ecological anthropologist with field experience in West Africa and Belize. He is the editor of *The Household Economy* (Westview 1989) and co-edited *Households: Comparative and Historical Studies of the Domestic Group* (California 1984).

PREFACE

This volume is a revised edition of *The Ecosystem Concept in Anthropology*– published in 1984 by the American Association for the Advancement of Science. Unlike the earlier edition, which purported to assess whether the ecosystem concept was still relevant and productive for anthropology, this new edition of the book takes a much broader view of its subject. It is no longer chiefly concerned with the relevance of the concept but, rather, moves quickly to demonstrate how it can be effectively used in designing ecological research and how it can be applied to ecological analysis.

The new edition has greater balance in its coverage with four papers in archeology, rather than just one, as in the earlier edition. The organization of the book makes clear that the new edition focuses more on the question of *when* and *how* this approach can be productive. It is certainly not always applicable or relevant. Several of the contributors make scant allusion to the ecosystem approach, preferring to apply other approaches to ecological analysis. Others use it chiefly in a conceptual fashion, to ensure the *systemness* of their research, while a few apply it in a very precise way, leading to simulation of real ecosystems in time and space.

This variety of approaches to ecological anthropology and human ecology does not reflect a lack of direction in the field but, rather, a healthy flexibility in research design and analysis. Today's ecological practitioners are much less likely to be dogmatic about the theoretical apparatus that they bring to bear on research problems than they might have been twenty years ago. They recognize that some problems are better studied by approaches other than the ecosystem approach. Thus, they take the ecosystem approach as a point of reference that reminds them of the importance of holism and contextuality. They recognize that to take account of local contingencies experienced by a specific population, other approaches may be better able to capture the variability of human responses of interest to them.

Part I of this edition begins with a critical introduction to the history of the ecosystemic approach in biology and ecology and its relation to other ecological approaches. This history is closely tied to the social and intellectual currents of their day-- most evident in the favorable view of ecosystemic approaches in the late 1960's, early 1970's, and its return to favor in the beginning of the 1990's. All these have been periods of concern with environmental crisis and publicly-shared concern with the future of our planet. By contrast, the critique of the ecosystemic approach saw its heyday between the mid-1970's and throughout the 1980's. This was a period that most observers have characterized as neoconservative: which saw a return to favor of neoclassical economics, that saw both biology and anthropology focus upon the individual as the object of importance and which saw the promotion of an evolutionary ecology in which some of its zealots did not seemed content to bring ecology and ecological anthropology back to a concern with individuals of the species, but seemed to want to deny a place to those concerned with group and ecosystemic processes. The importance of having conceptual constructs with different emphases is explored in chapter 2 by Roy Rappaport who, once again, responds to the critics of *Pigs for the Ancestors*. It is of more than trivial importance to be suspicious of theoretical approaches that reject the validity for certain questions of theoretical constructs focusing on group and ecosystem units of analysis. As Rappaport so aptly puts it in the conclusion to his chapter, "the banishment of any one of these levels--the individual, the social, the ecological-- from our analysis is a very serious mistake, conferring upon us no benefits and costing us dearly."

Part II of the new edition examines the place of ecosystemic approaches in the field of archeology and concretely presents, in three papers added for this edition, how archeologists use the ecosystem in designing research and in their analysis. Jochim and Butzer emphasize the conceptual and heuristic value of ecosystem, while Hastorf and Fish & Fish delineate the importance of ecosystem in regional approaches to study. The papers are rich in data and interpretation. Jochim illustrates how and when ecological approaches are appropriate by using three examples from European prehistory, while Fish & Fish demonstrate the operational requirements of ecosystem studies in archeology in the context of the multiple ecosystems in the Tucson Basin of Arizona exploited by the Hohokam. Fish & Fish, like Hastorf, point to the importance of cultural factors in explaining the changing relations between people and environment. For them, neither "environment" nor "culture" are privileged in explanation but, rather, must be seen as potential sources of the transformations observable in the archeological record in given settings. Hastorf advocates the enrichment of ecosystem with the recent work of interpretive anthropologists concerned with intentionality, contradiction and history-- and gives us a sweeping view of the social, economic and political transformations in the pre-Inka period in the central Andes. Karl Butzer, in what must be seen as a *tour de force*, examines archeological epistemology, and the dangers of inference, with and without the benefit of historical archival materials to correct errors likely to result from the fragmentary archeological record. His critique of certain kinds of analysis is "good to think" and relevant not only to archeology but to practitioners in the social and biological sciences.

Part III presents a series of cases from social and cultural anthropology that demonstrate the various ways in which contemporary ecological anthropologists practice their craft without falling prey to the presumed limitations presented by ecosystemic approaches (discussed in chapter one). Ellen and Netting demonstrate the importance of looking at ecosystems over time, and the necessary fluidity of boundaries over time and in space of these units (see also Fish & Fish, this volume). Lees and Bates give us an impressive review of the richness of approaches being used in contemporary human ecology. Their paper brilliantly highlights many of the important contributors to the journal *Human Ecology* in the past decade and the varied ways in which they have dealt with "the ecology of change." Moran demonstrates the ease with which analysts slide between levels of analysis, and the analytical consequences of doing so. He points to its relevance to understanding recent processes of human ecological change in the Amazon Basin. Gross introduces the reader to the rich variety of methods that can be used in ecological study, emphasizing cross-sectional, longitudinal, and comparative analysis.

Part IV of the volume brings together several papers that highlight the importance of inter- and multi-disciplinarity in ecological research. In fact, most papers in the volume could very well have been placed in this section. High quality ecological study almost necessarily requires a complexity of data collection and analysis that can be most effectively achieved through teams of

scientists complementing each other's expertise. This is true for the
archeology papers in Part II, as it is true of many of the papers in
parts III and IV. The long-term project of Little *et al.* is
an important one with multiple components and teams of systems
ecologists, social anthropologists, cultural anthropologists,
biological anthropologists, and range scientists. Rarely has the
complexity of human adaptive strategies in ecosystem management been
so completely studied as in this project among the Turkana. The paper
by Conant demonstrates the importance of anthropologists working with
geographers and other scientists expert in GIS and other types of
remote sensing technologies, if they are to be able to advance the
state of knowledge, and be a part of the current debates over the
human dimensions of global environmental change. The paper gives a
sound introduction to the technologies available, their limitations,
their relative cost, and their potential. The paper by Wilk brings
home the importance of looking not only at the ecology of nature but
also at the ecology of households. What such study contributes to
ecological analysis is attention to the structural forms of
households in different parts of the world, and their internal
functioning (reflecting gender, age, kinship and other sources of
variation). At least for living populations, it is necessary, but not
sufficient, to deal with households as undifferentiated residential
units. The internal dynamics of households are important to the
understanding of how labor is allocated into productive activities
and how goods are distributed within the household once produced. In
a final chapter, Bennett reminds all of us of our disciplinary
myopias and makes a call for abandonment of our comfortable ivory
towers as we seek to engage contemporary problems jointly with
colleagues of other disciplines. The role of advocacy in science has
rarely been made so well. This view takes on added importance as we
enter the 1990's, and the need for solutions to the global, and
local, environmental crisis become more deeply felt everywhere.

The book that is before you does not touch on every issue of
importance in ecological anthropology. It would be difficult for any
collection of authors to ever succeed in doing so. This collection
hopes to provide students and colleagues in anthropology, geography
and allied disciplines with examples of exciting recent ecological
research, with guidance to the rich literature on human ecology, and
with an understanding of the debates which have taken place over the
past few decades. If reading and discussing the contents of this book
encourages high quality ecological research, promotes less dogmatic

and more collegial cooperation in ecological studies, and leads to greater interdisciplinary research we will feel the effort more than amply rewarded.

I want to thank the contributors to the volume for the promptness with which they undertook the task of manuscript preparation and revision. I am particularly pleased that most of the authors of the original edition undertook substantial revisions of their papers to bring them up to date and in line with the new objectives– with very little prodding from me. I like to think that this alacrity reflects the vitality of ecological anthropology as it enters the 1990's.

I am grateful to the new authors joining us in this new edition: Karl Butzer, Suzanne and Paul Fish, Christine Hastorf, Rick Wilk, and Roy Rappaport for their superb contributions. They have helped make this volume richer and more complete than its predecessor. I am particularly delighted that Roy Rappaport, who was unable to join us in the earlier edition, found time to share his thoughts with us despite numerous conflicting obligations.

I wish to thank the anonymous reviewers of the new edition who provided us with many provocative suggestions that suggested the focus of the new edition, who called for strengthening the coverage of archeological studies, and particularly for the suggestion to focus more on the research design, analysis and practice of ecological anthropology . I am grateful to Colin Day, Director of the Univ. of Michigan Press, for his support of this project, and for making the book available in a paperback edition to facilitate its use in academic courses. Andrew Fisk and Joyce Harrison were helpful in the production and marketing phase of the book. Reggie Graham prepared the manuscripts into page-proofs with good humor and considerable skill despite a tight schedule.

The editor wishes to thank the College of Arts and Sciences at Indiana University for funds that facilitated the preparation of the final manuscript copy. I am grateful, too, to the Institute for Advanced Study at Indiana University for the opportunity to work undisturbed while I was a Fellow-in- Residence, and the John Simon Guggenheim Memorial Foundation for the Fellowship in 1989-90 that provided freedom to read and reflect on this and other subjects. None of these organizations should be held responsible for the views espoused herein.

May 1990

Emilio F. Moran
Institute for Advanced Study,
Indiana University, Bloomington.

PART I
ASSESSMENT OF THE
ECOSYSTEM APPROACH

CHAPTER 1

ECOSYSTEM ECOLOGY IN BIOLOGY
AND ANTHROPOLOGY:
A CRITICAL ASSESSMENT

Emilio F. Moran

Foundations of the Ecosystem Approach

The historical foundations of the ecological approach in anthropology are dual: on the one hand, the rejection of environmental deterministic explanations led by anthropologists in the first decades of the twentieth century and, on the other, the adoption of biological concepts in the 1960's to avoid over-dependence on the concept of culture. This polemic between culture and environment as the prime causes of observable social configurations forms the basis for the intellectual development of contemporary ecological anthropology.

Ecosystem Ecology in Biology

The term ecosystem generally refers to the structural and functional interrelationships among living organisms and the physical environment within which they exist. Indeed, the ecosystem is the total context within which human adaptation and biological evolution take place (Moran 1982). Its value derives from the fact that a system is an entity whose overall properties are different from the properties of its elements. Of all available ecological concepts, the ecosystem approach accords physical environmental factors the most attention. This attention to abiotic factors is, in itself, an important contribution to biology since evolutionary theory does not always recognize the importance of the physical environment in shaping species' behavior (Evans 1976). The physical environment tends to be viewed as a backdrop against which evolving species and adaptive responses are studied. "Ecosystems are different from plant and animal populations in having ion, carbon and energy cycles and fluxes and they can be separated from biomes by scale and uniformity of the abiotic environment" (Schulze and Zwölfer 1987:1).

The ecosystem concept was formally defined by Sir Arthur Tansley in 1935 (Golley 1984). Tansley drew attention to natural- physical interrelationships and suggested that it was in their "nature" to develop towards a dynamic equilibrium. As a philosophical stance, the concept can be found in many societies. Given the human capacity for language and culture-- and the consequent need to categorize and simplify the infinite variety of the external world-- all societies tend to develop systems of classification that represent associations of plants, animals and landscapes (see review in Major 1969). Terms such as biocoenosis (Möbius 1877), ecotope (Troll 1950), biogeocoenosis (Sukkachev 1960) and many others such as naturcomplex, holocoen, and biosystem approximate the ecosystem concept but emphasize different aspects of the physico-biological interactions of interest to each of its proponents. They have in common chiefly an emphasis upon homeostasis (Margalef 1981:281).

The ecosystem concept took time to develop in Tansley's own writings and grew from his earlier notion of "circles of affinity"-- defined as all those phenomena that are part of the total situation of an organism and that might influence it. The circle of affinity developed into the ecosystem concept. The concept was most useful as a didactic device in arguing for the unity of Nature and the importance of conservation-- rather than as a real research unit. This is important since Tansley was not only a gifted botanist but one of the founders of the Nature Conservancy and an activist in environmental issues (Golley 1984). In this he served as an early model for many ecosystem ecologists to follow who have combined scientific research with activism and advocacy in behalf of conservation of natural resources.

The ecosystem was conceived as applicable to the level of the plant community or biome, rather than lower levels of organization (Evans 1956). Tansley's ecosystem concept was considerably ahead of the state of ecological theory when it was first formulated and thus was slow to be adopted. Two publications were particularly significant in influencing the adoption of the concept in biology: the first edition of Eugene Odum's *Fundamentals of Ecology* (1953) and Evans' comment in *Science* (1956) in which the ecosystem was proposed as the basic unit of ecology. Evans argued that the ecosystem was a unit as important to ecology as species was to taxonomy and systematics. Odum's text offered the ecosystem as an organizing principle emphasizing obligatory and causal relationships. It influenced a whole generation of ecologists despite the difficulties faced by those trying to implement the concept in actual research and the assumption of steady progress towards equilibrium assumed for ecosystems. Long ago, specialists gave up the notion of *climax* as utopic and substituted this static notion with a focus on patch and vegetation dynamics. Increasingly, it has become more productive to conceive of ecosystems as "thermodynamically open systems that are out of equilibrium" (Margalef 1981:281).

The notion of ecosystem helped to bridge the then distinct fields of autecology and synecology[1]. Autecology, the study of the interrelations between individual organisms and two or three environmental variables, had been limited in the past by the difficulties of controlling experimental work due to the inconstancy of climate from one year to the next. Synecology, the study of groups of organisms, such as the community, had been characterized by two rather philosophical and deductive traditions: the American plant succession school led by F.E. Clements and the European plant sociology school of Braun- Blanquet, Cajander, and Sukkachev (Spurr 1969:5-7). Both of these traditions simplified the physical conditions of plant growth by postulating climatic stability and derived static taxonomies made up only of climax forms. Recognition of the role of climatic variation, of the in- and out-migration of organisms, and of ecological factors such as fire and animal browsing on biotic composition, however, led to abandonment of these approaches. Advances made since the 1940's, especially in cybernetics and engineering, for the first time allowed highly precise and controllable experimental research in growth chambers and the construction of large-scale deductive models through the use of computer simulations. The ecosystem approach provided an elegant basis for large-scale integrated modelling and a bridge between inductive and deductive modes of research in ecology.

Whereas the ecosystem concept, when defined by Tansley in 1935 and as used by Odum in 1953, was presented mainly as as didactic device to emphasize the interaction between living and non-living components of a system, the introduction of systems engineeering and cybernetics initiated conscious application of quantitative techniques to a purely theoretical statement of relationships (Wiener 1948; Shinners 1967, Odum 1957). In the process, the concept became frequently equated with biome units (see Bennett, this volume). It has proven difficult to quantify entire ecosystems. Ecosystem research requires a mode of research contrary to the conventions of both biological and anthropological research— dominated as they are by the single researcher with a few field assistants working at a single site (Golley 1984:40). Ecosystem studies require more often than not, teams of specialists and large budgets. As it grew in sophistication, the ecosystem approach has grown more concerned with modelling and with restoration biology and conservation (Allen 1988; Berger 1990; National Science Board 1989).

The use of the ecosystem approach is rooted in the use of the organic analogy (Spengler 1926; Odum 1971; Butzer 1980). To study ecosystems as real units, it became necessary to study their internal metabolism, since system metabolism is a measure of the collective processes which serve to maintain the integrity of the whole (Reichle and Auerbach 1979:91). Studies of trophic levels and energy flow, aided by the use of radioisotope techniques, were concerned with questions relevant to large scale ecosystems in units such as tundra, tropical forests, and grasslands. This interest is reflected in the development during the 1960's of the International Biological Program (IBP) and the choice of biome-level units for systems ecology research (see Little *et al.*, this volume). "The goal of ecosystem analysis is to develop quantitative ecosystem science which may provide new theoretical insights into the organization and function of natural systems at their most complex level" (Reichle and Auerbach 1979:92). Our concern with environmental problems brought the ecosystem approach to the fore in biological and anthropological ecology in the 1960's and again as we enter the 1990's. There is growing public feeling that we need to better understand the complex and integrating functions of ecosystems, such as wetlands and forests, their contribution as buffers to the pollutions emitted by our dense settlements and industrial activities, and of the need to bring about modifications in our role in the use and conservation of nature.

When reading the technical writings of systems ecologists such as Van Dyne (1969, 1979), it becomes clear that the ecosystem was conceived as a complex level of organization above the levels of cell, tissue, organ, organism, population and community. Ecosystems are tightly interwoven and coherent complexes of primary and secondary producers interacting in an abiotic medium. The goal of ecosystem research was to overcome the microscopic approach then dominant in biology and enrich it with a macroscopic view of general principles applicable to higher levels of organization. Unlike other approaches, it is impossible to reduce ecosystems to a common denominator such as energy flow or nitrogen flux, although some practitioners seemed to have tried to do so. Yet, even in its earliest usage, systems ecologists had to grapple with the problem of boundary definition. Odum (1953) clearly noted that ecosystems were defined by the needs of the investigator, and that the unit could include anything from a tiny pond to the entire biosphere. The only requirements to the application of the ecosystem approach would appear to be that at least several organisms be present and that the interaction of these organisms with the abiotic environment be taken into consideration.

Biologists developed what seemed clear criteria for defining an ecosystem: it must be a functional entity with internal homeostasis, have identifiable boundaries, and recognizable relationships between components (Reichle and Auerbach 1979:94- 5). "Ecosystem analysis is the integration of all knowledge available on all levels of organization to explain the matter balance of the system as a function of time" (Ulrich 1987:44). The goal was to focus on attributes of systems and to pose different questions than are normally undertaken by research on lower units or levels of analysis. Of concern are questions like: What homeostatic mechanisms ensure the self-perpetuation of ecosystems? What is the ecological significance of evergreenness? How does diversity relate to stability? One thing did emerge from the work of systems ecologists: the structural and functional characteristics of ecosystems may not be inferred from those of species, communities, or populations.

Ecosystem analysis usually begins by organizing available knowledge into models of the system in question. This includes all important system components and interactions but focuses on the more important "driving" function and aggregates the data so that it can be manipulated. Through modelling, systems ecologists seek to predict the behavior of a system from a basic understanding of its structural and functional relations (Odum 1971). The problem is that modelling, particularly at the community and ecosystem levels, has the inherent weakness that afflicts all deductive modelling-- it can only predict behavior under conditions for which the model was designed. Deductive models cannot predict the behavior of systems undergoing structural transformation, whether externally or internally induced. The tools of ecological modelling coming out of cybernetics and systems ecology proved excellent in describing an ecosystem under current and known conditions. However, difficulties arose in predicting future structural configurations and emergent functional relationships (cf. Golley 1983; Bourliére 1983; Jordan 1987; Lieth and Werger 1989).

To this day, ecosystem studies are characterized by a heterogeneity of approaches. Dominant are input/output analyses of fluxes in ecosystems and the interface of the biotic and abiotic environment (Schulze and Zwölfer 1987:1). These are excellent descriptive studies but they do not allow for biological adaptations and biological regulation of species within ecosystems (*Ibid.* p. 2). On the positive side, they permit the most far-reaching predictions of the future of ecosystems as units (in contrast to the future of its components). Among the goals of ecosystem analysis are prognosis and explanation. Prognosis is a major goal given the pervasive influence of the human species. Explanation generally tries to get at the temporal behavior of the system under varying conditions of environmental stress and human manipulations in order to derive principles for system change and/or stability (Ulrich 1987:46). Important, too, have been "attempts to deduce general principles for the functioning of the components within ecosystems" (Ibid.). Examples are Lindeman's work on energy flow (1942), Hutchinson's principle of niche dimensionality (1957), Margalef's principle of cybernetic regulation of ecosystems (1968), and Cowan and Farquhar's principle of optimization (1977). A notable result of ecosystem studies in ecology was the lack of long-term measurements of ecosystem parameters (McIntosh 1985:240). The National Science Foundation responded by creating a series of Long-Term Ecological Research (LTER) sites which will permit in time clearer understanding of time-dependent processes. Finally, ecosystem ecology has been particularly successful in dealing with applied problems. Practical objectives such as optimal harvesting of forest and fishery resources, integrated pest management, and conservation biology have contributed to an understanding of general principles of ecology and have depended on ecosystemic approaches (Altieri 1988; Gliessman 1989).

Ecological Approaches in Anthropology

Ecosystem ecology made its way into anthropology in the 1960's and seems to have been inspired by the writings of Eugene Odum (1953) and Marston Bates (1953). However, anthropological interest in ecology goes back to the very foundations of anthropology as a discipline. Actually, anthropology anticipated the now lively debates over the causality in systems during the acrimonious debates over the relative impact of environment in bringing about particular social or political forms (Thomas 1925).

Anthropology's origins are associated with an intellectual thrust to reject various forms of racial and environmental determinism and the over-generalizations of anthropogeography prevalent in the late nineteenth century (cf. review of this literature in Moran 1982; Ellen 1981; Glacken 1967; and Thomas 1925). The early work of American anthropologists was characterized by an emphasis on historical and cultural descriptions which focused on the uniqueness of human groups (Goldenweisser 1937). According to this view, which has come to be known as "historical possibilism", environment was seen as a passive force that limited human options but which played no active role in the emergence of observable human traits or institutions. In *The Mind of Primitive Man* (1911), Boas noted that the environment furnished the materials out of which people shaped and developed the artifacts of daily life but it was historical forces and diffusions which predominantly explained the particular forms that given artifacts took.

The possibilist position was sympathetic to the "cultural area approach" (Wissler 1926). In this approach, geographical regions were divided into culture areas based on shared cultural traits of residing cultures. This view was strongly espoused well into the 1930's. C.D. Forde (1934) and A. Kroeber (1939), for example, while acknowledging the need to collect ecological data because of its potential explanatory relevance, concluded that economic and social activities are products of the historical, but largely unpredictable, processes of cultural accumulation and integration.

Julian Steward was trained in this tradition in the 1930's but his views departed from that of his mentors. Whereas possibilism and cultural area approaches generated good descriptions of the historical past, they often failed in explaining the process that could be generalized beyond the particular case. Steward, on the other hand, was concerned with cross-cultural comparisons and with the causal connection between social structure and modes of subsistence. The crucial focus in Steward's approach was neither on environment nor culture. Rather, the process of resource utilization, in its fullest sense, was given research priority. The "cultural ecological" approach proposed by Steward involved both a problem and a method. The problem was to test whether the adjustment of human societies to their environments required specific types of behavior or whether there is considerable latitude in human responses (Steward 1955:36). The method involved three procedures: a) to analyze the relationship between subsistence system and environment; b) to analyze the behavior patterns associated with a given subsistence technology; and c) to ascertain the extent to which the behavior pattern entailed in a given subsistence system affected other aspects of culture (Ibid. p. 40-41).

This research strategy is all the more impressive if one considers its historical backdrop. From the broad generalities of the environmental determinists and the detailed inductive findings of the possibilists, Steward proposed a research method that paid careful attention to empirical details and that causally linked the *cognized environment*[2], social organization, and the behavioral expressions of human resource use. Steward delimited, more than anyone before him, the field of human/environment interactions. He viewed social institutions as having a functional unity that expressed solutions to recurrent subsistence problems. Steward's use of functionalism was concerned with the operation of a variable in relation to a limited set of variables, not in relation to the entire social system, and thus did not fall prey to the weaknesses of then current British functionalism. British functionalists emphasized the role of social institutions in the maintenance of structural equilibrium. Steward steered "cultural ecology" towards a concern with how single systems change through time and how the causal relationships within that system can actually lead to change.

Most attempts to operationalize the cultural ecological approach required modifications of the basic research strategy laid out by Steward (cf. Netting 1968; Sweet 1965; Sahlins 1961). His concept of the culture core proved to underestimate the scope, complexity, variability, and subtlety of environmental and social systems (Geertz 1963). The cultural ecological approach of comparing societies across time and space in search of causal explanations was judged to be flawed a decade later. Vayda and Rappaport (1968), among others, found the concept of the culture core, and the cultural ecological approach, to give undue weight to culture as the primary unit of analysis, and found the presumption that organization for subsistence had causal priority to other aspects of human society and culture to be both untested and premature (Geertz 1963).

Ecosystem Ecology in Anthropology

Critiques of Steward's cultural ecology paradigm led anthropologists towards a more explicitly biological paradigm. Geertz (1963) was the first to argue for the usefulness of the ecosystem as a unit of analysis. Its merits were eloquently stated: systems theory provided a broad framework, essentially qualitative and descriptive, that emphasized the internal dynamics of such systems and how they develop and change. The explicit adoption of biological concepts in anthropology led to provocative and sometimes productive results. As early as 1956, Barth applied the concept of the "niche" to explain the behavior of adjacent groups and the evolution of ethnic boundaries. Coe and Flannery (1964) noted the use of multiple ecological niches by prehistoric peoples of South Coastal Guatemala. Neither the niche nor other concepts from biology had as significant an impact on anthropological thinking, however, as did the ecosystem concept (with the possible exception of the concept of adaptation, see discussion in Little 1982).

The ecosystem approach was attractive to anthropologists for a number of reasons. It endorsed holistic studies of humans in their physical environment. It was elaborated in terms of structure, function and equilibrium that suggested the possibility of common principles in biology and anthropology (Winterhalder 1984). No less important was the connection between ecosystem ecology and advocacy of habitat and species preservation connected with concern for non-industrial populations at a time of deep environmental and social concern (i.e. the 1960's and 1990's).

Each subfield of anthropology was differentially affected by the ecosystem approach. Archeologists have always been conscious of the environmental context of society. However, in many cases the environment has been treated as a static background against which human dynamics occur (Butzer 1982:4). In part, the problem was the lack of "an adequate conceptual framework within which to analyze complex interrelationships among multivariate phenomena" (*Ibid.* p. 5). The seminal paper in archeology may have been Flannery's (1968) in which he postulated the useful applications of systems theory to archeological investigations. According to systems'-oriented archeologists, "culture is defined not as aggregates of shared norms (and artifacts) but as interacting behavioral systems" (Plog 1975:208). Emphasis was given to variability, multivariate causality and process (Clarke 1968).

In archeology, the ecosystem approach has proven to be a useful heuristic device leading archeologists to think in terms of systemic interrelationships. It was rarely used as a spatial unit of analysis. Thus, archeology did not fall into the trap of making ecosystems coterminous with biogeographical units or sites. Rather, the ecosystem approach encouraged the study of the landscape at large, the use of catchment analysis and a movement away from sites to larger regional surveys (see Jochim, Hastorf, Fish and Fish, this volume). Ecological archeology has benefitted from the breadth of the concept and appears not to have suffered from many of the problems that seem to have plagued ecosystem research in physical and social anthropology. Unlike energy flow studies (or decision-making studies), which emphasize present-time measurement, ecological archeology deals with spatio-temporal variability. As Butzer has put it: "the value of the human ecosystem as a framework for archeological research is explicitly conceptual. It is an interdisciplinary perspective that readily encompasses spatial variables, differences of scale, complexity and complex interactions, as well as equilibrium modes, including adaptive change and evolutionary transformation" (Butzer, this volume). The long time frames of the archeological record reflect aggregate changes in the physical environment and in the material manifestations of social and cultural change (Butzer 1982), thereby avoiding the pitfalls of synchronic equilibrium-oriented functionalism (E.A. Smith 1984).

Special note must be taken that archeology has found that ecosystems are particularly useful when they model regional-scale systems, rather than individual sites or communities (see Hastorf, Fish and Fish, Jochim, this volume). This is consistent with the higher level of organization which ecosystems represent in biological systems and may very well imply that social anthropologists and bioanthropologists may want to do likewise in the future. Processses like agricultural intensification may have multiple causes, not necessarily environmental ones (Hastorf, this volume). The ecosystem approach can accomodate such a view-- indeed, it always stood for modelling complex systems in which the forcing functions became clear only in the course of studying the whole gamut of interrelations.

In physical anthropology, Little (1982) has noted that in the 1950's interest developed in the study of adaptation to environment. This "new physical anthropology" focused on studies of body morphology and composition, physiological response to environmental stress, demographic and health parameters of adaptation and genetic attributes of populations (Harrison *et al.* 1964).

The research of the new physical anthropologists found support in the International Biological program (IBP) which began *circa* 1964. A "human adaptability" section was included in the program, intended to cover "the ecology of mankind" from the perspectives of health, environmental physiology, population genetics, developmental biology, and demography (Weiner 1965). Even though doubts were expressed at the 1964 symposium at Burg Wartenstein about the omission of social/cultural aspects of adaptability, the perceived gap between the methods of human biology and social science led to no solution to this problem (Weiner in Worthington 1975). Only a decade later did an IBP workshop begin to seek ways to bring together ecologists and social scientists so that humans could be incorporated into the IBP ecosystem approach (Little and Friedman 1973). This workshop was followed by one on the applicaton of energy flow to the study of human communities (Jamison and Friedman 1974).

The 1964-74 decade of IBP research led to more sophisticated methods and greater awareness of the limitations of original formulations. Practitioners now go beyond evaluating systems in terms of a single flow and, instead, consider multiple flows and constraints. Indeed, energy flow analysis[3] is seen as a method quite distinct from an adaptive framework or any other theoretical stance (Thomas 1973). The flaws of human energy flow studies carried out in the 1960's and early 1970's (cf. critique in Burnham 1982) resulted from preliminary efforts to test the utility of the new methods for anthropology. Indeed, energy flow analysis is a convenient starting point in understanding the complexity of human systems-- systems in which social relations and historical process play a primary role (Winterhalder 1984). To fully understand them, however, other methods are more appropriate to social and ideological analysis.

In social anthropology and human geography, ecological studies have become common since the 1970's. The majority of studies have not depended on the ecosystem approach, although some notable ones have (e.g. Rappaport 1967; Clarke 1971; Kemp 1971; Waddell 1972; Nietschmann 1973). For all intents and purposes, the use of ecosystems as units of analysis did not radically alter the scope of research: research still focused on small, non-urban communities.

A generation of anthropologists, trained in ecology and systems theory, went to the field to measure the flow of energy through the trophic levels of the ecosystems of which humans were but a part (Rappaport 1967). The choice of research site was still a local community, often treated as a closed system for the purposes of analysis. Emphasis on micro-level study in ecology was well argued by Brookfield (1970) who pointed out that an adaptive system can best be studied at this level because such a system model "acquires the closest orthomorphism with empirical fact" (1970:20). Micro-level studies using the ecosystem as a "unit of analysis" have provided valuable insights into flow of energy, health and nutritional status of populations, relative efficiency rates of various forms of labor organization and cropping practices, and into social organizational aspects of subsistence strategies (cf. discussion in Netting, 1977, Stone *et al.* 1990, Moran 1982, 1981; Ellen, Lees and Bates and others, this volume).

The impact of the ecosystem approach in social and cultural anthropology may have been most significant in increasing the degree of quantification thought desirable (see Gross, this volume). The earliest studies using the approach emphasized energy flow accounting, and more recent studies have emphasized the application of micro-economic approaches to adaptive strategies, time-allocation studies, and analysis of choice-making by individuals and groups (see Orlove 1980; Rappaport, Lees and Bates, Gross, Bennett, and Wilk, this volume).

Efforts to measure the flow of energy and the cycles of matter through human ecosystems served to detail more than before the environmental setting of specific populations. Energetics emphasizes the collection of data on a sample of components and flows so that the data may be aggregated and used in simulation models. The goal is to understand system dynamics by manipulating rates of flow given current conditions in the ecosystem. However, the value of these measures in studying small scale populations may have been overestimated in the 1960's. Flow of energy and cycles of matter are aggregate measures appropriate to macro-ecosystem description, but provide little insight into human variation in resource use in given localities-- a matter of great interest in anthropology (E.A. Smith 1984). These measures were never meant to address, or be relevant to, microecological or microeconomic processes or to evolutionary questions, as some critics of the ecosystem approach have suggested (Burnham 1982; E.A.Smith 1984).

Just as the ecosystem approach helped biology broaden its interests to include neglected physical environmental factors, so it affected anthropology. The ecosystem approach provided greater context and holism to the study of human society by its emphasis on the biological basis of productivity and served as a needed complement to the cultural ecology approach. By stressing complex links of mutual causality, the ecosystem approach contributed to the demise of environmental and cultural deterministic approaches in anthropology and took it towards a more relational and interactional approach to analysis even if practitioners preferred to dissociate themselves from the concept (cf. Johnson and Earle 1987; Grossman 1984; Richards 1985; Morren 1986; Little and Horowitz 1987; McCay and Acheson 1987; Sheridan 1988).

Limitations of the Ecosystem Approach

Perhaps no work has had a greater impact on the development of an ecosystem approach in anthropology than Roy Rappaport's study of a New Guinea population (1967), nor has any other study attracted as many critics of the ecological approach. Originally, many of the advocates of a more explicit ecological approach were satisfied with this study. It argued that a human population was a species within the ecosystem; that the system operated according to laws of nature that could be understood in the light of systems theory; and that major cultural processes, like ritual, could be understood to play cybernetic functions. The study was rich in ethnographic and quantitative data– despite the difficulties of collecting those data in the New Guinea Highlands of the early 1960's. Rappaport was able to relate the ritual cycle to the cycles of pig population growth, the fallow cycles of swiddens, and cyclical patterns of warfare and peace.

The merits of the ecosystem approach were evident in Rappaport's study: holism was stressed while, at the same time, specific relationships between human populations and the total environment served to give focus. The approach allowed Rapppaport both to eliminate detail at the macro-level and to examine in detail components at the micro-level. Empirically, the ecosystem approach, as employed by Rappaport, contributed to the accumulation of data on subsistence systems, the appreciation of the impact of those same systems upon the environment and the social system, and led to the acquisition of more sophisticated techniques for data analysis such as modelling and computer simulation. At least three articles have appeared which modelled and simulated the Maring ecosytem, based upon the data presented in *Pigs for the Ancestors* (Foin and Davis, 1987; Samuels 1983; Shantzis and Behrens 1973). This explicitness in data reporting and systematic explanation was a major advance that permitted others to test the validity of interpretations derived from the data.

A number of problems emerged in the process of applying the ecosystem approach to anthropology (see also the assessments by Vayda and McCay 1975; and Winterhalder 1984): a) a tendency to reify the ecosystem and to give it the properties of a biological organism; b) an overemphasis on predetermined measures of adaptation such as energetic "efficiency"; c) a tendency for models to ignore time and structural change, thereby overemphasizing stability in ecosystems; d) a tendency to neglect the role of individuals; e) lack of clear criteria for boundary definition; and f) level shifting between field study and analysis.

Reification of the Ecosystem

The tendency of some authors to reify the ecosystem and to transform the concept into an entity having organic characteristics appears to have been a product of the initial excitement generated by the notion of ecosystem. When the volume *The Ecosystem Concept in Natural Resource Management* (Van Dyne 1969) appeared, the editor and some of the contributors noted that they were at the threshold of a major development in the field of ecology. The concept was hailed as an answer to the divisions within bioecology and gained a large popular following during the "ecology movement" of the 1960's and early 1970's-- perhaps because of the very superorganic and equilibrium characteristics that were later to be faulted. It is evident that, for some, ecosystems became a shorthand for the biome or community and that this heuristically useful physical/biological construct was unwittingly endowed with purely biological attributes. As Golley has noted, it is generally understood that ecosystems are subject to the laws of biological evolution but they are also subject to laws not yet completely understood and that are not exclusively biological (1984).

When an ecosystem is viewed as an organic entity, it is assigned properties such as self-regulation, maximization of energy through-flow, and having "strategies for survival." This view is similar to earlier "superorganic" approaches in anthropology (Durkheim 1915; Kroeber 1917; White 1949). Few ecological anthropologists today would accept the notion that ecosystems "have strategies" and even fewer would suggest that energy maximization is always "adaptive" in human ecosystems. The notion of self-regulation is more problematic since it devolves around the question of whether ecosystems per se can be cybernetic, e.g. use information for self-regulation (Engelberg and Boyarsky 1979). Patten and Odum (1981) believe this to be a pseudoissue that distracts us from more fundamental concerns: how are we to think about ecosystems and how are we to place them within the scheme of known systems? Long-term selection favors as a matter of implicit design, structured relations or organizations that create order where there might be chaos (see Rappaport, this volume). However, this neither reduces nor enhances the potential for evolutionary change along orderly lines (Patten and Odum 1981:896).

Past anthropological ecosystem studies have tended to de-emphasize the capacity of *Homo sapiens* to transform the physical environment through organized social activity (Ellen 1981:91). Thus, studies have over-emphasized the self-regulatory features of ecosystems to the neglect of processes by which systems transform themselves in response to either external or internal dynamics. Inter-specific exchanges have been emphasized to the neglect of intra-specific exchanges and the role of labor exchange (Cook 1973:41-4). Cognitive dimensions of human behavior have been neglected despite the knowledge that cultural factors mediate such ecological dimensions as population size and resource use. Cognitive dimensions tended to be considered under the aegis of "ethnoecology" and were only rarely incorporated into systemic analyses. This oversight is currently undergoing correction as scholars point out significant alterations in the landscape brought about by populations, and the role of individual management strategies which show that individuals modify the environment and do not simply adapt to the constraints presented to them by nature (cf. Boster 1983, 1984; Richards 1985; Posey and Balee 1989; Moran 1990).

"The Calorific Obsession"

Perhaps no other problem has received more attention within anthropology in recent years than the charge that ecosystem studies were "obsessed with calories". Many young scientists took great pains to measure energy flow through ecosystems under the assumption that energy was the only measurable common denominator that structured ecosystems and that could serve to define their function. Energy flow studies conducted in the 1960's and 1970's demonstrated the descriptive usefulness of energetics before, during, and after field investigations. What they also proved was that the forcing functions of ecosystems varied from site to site and that it was naive to postulate energy as the organizing basis for all extant ecosystems (e.g. Kemp 1971; Rappaport 1971; Thomas 1973; Moran 1973; Baker and Little 1976; Vayda and McCay 1975; Ellen 1978).

The early energy flow studies delineated flows of energy and established magnitudes. They did not, however, give sufficient attention to the numerous decisions made which control those same flows (cf. Adams 1978). Winterhalder suggests that energy flow studies stand to benefit from joining hands with neo-Ricardian economics, given the latter's emphasis on the circular processes in which consumption feedsback into production. "Adapted to neo-Ricardian theory, energy flow methods could help to rigorously quantify and trace the partitioning of production" (1984:305). This has taken place in part in the study of optimal foraging strategies among hunter/gatherers (E.A. Smith 1984; Winterhalder and Smith 1981) and has been suggested as applicable to horticultural populations (Gudeman 1978; Keegan 1986).

The seminal work on the energetic basis of society was that of Leslie White (1949) who proposed that cultural evolution is determined by the control of progressively greater amounts of energy. In his scheme, social systems are determined by the use of technology to control energy. While White's scheme was purely theoretical, his successors began to measure food production in an effort to specify the relationship between harnessing of energy and social evolution. For macro-modelling, there is little doubt that the energetic approach to social evolution has considerable merit (cf. Harris 1977). However, such an approach submerges the anthropologically important concerns with human variability in time and space, and issues of causality, evident by use of other kinds of analysis (E.A. Smith 1984).

Today, few would suggest that measurement of energy flow ought to be *the* central concern of ecosystem studies. Concern has shifted, instead, to material cycling and to the impact of external factors upon given ecosystems (Shugart and O'Neill 1979; Barrett and Rosenberg 1981; Cooley and Golley 1984). Bioecologists are less concerned today with calories than with the loss of whole ecosystems, with loss of biotic diversity and with species extinction (Jordan 1987; National Science Board 1989).

Ignoring Historical Factors

Next to the "calorific obsession", ecosystem research has been faulted most often for ignoring time and historical change. Past construction of ahistorical models, in turn, led to an apparent overemphasis on stability and homeostasis rather than on cumulative change. The emphasis on self-maintenance and self- regulating characteristics of ecosystems (Jordan 1981) also contributed to a view that man's role was essentially disruptive of "natural processes." Ellen, Hastorf, Fish and Fish, Butzer, Bennett, Netting, Lees and Bates, Conant and others in this volume discuss the consequences of overlooking transformations over time and space. Their research also shows that attention to history is not incompatible with ecosystem research. Recent inclusion of a historical dimension in ecosystem studies provides an appreciation of the processes of stability and change in human ecosystems. Butzer (this volume) makes a particularly convincing case for linking historical studies to archeological studies, if errors in interpretation are to be avoided. At any given time, systems appear to be seeking, or be at, equilibrium, whereas over time they appear to be undergoing continuous and cumulative change leading to structural transformation (see Lees and Bates, this volume).

It is paradoxical that ecological anthropological studies have only rarely explored the population variable over time, given the importance of demographics in population ecology. In part, the reason must be sought in the very study of isolated small communities lacking historical records of births, deaths, and marriage. To see a human ecosystem in process, rather than as a synchronic snapshot, requires dependable, continuous, and relatively complete records for a population over a long period of time. Such ideal conditions are rarely found except in modern-period Western Europe and North America. Thus, it comes as no surprise that the few studies employing demographic data come from fieldwork in these areas, for example Netting's research in the Swiss Alps (1981, 1979, 1976, 1974, this volume).

Demographically-deep studies represent a relatively new direction in ecological anthropology (cf. N. Baker and Sanders 1971; Cook 1972; Polgar 1972; Zubrow 1976; Netting, 1981; Hammel 1988). Demographic studies lead us away from models emphasizing closure, constraints to energy flow and negative feeedback and toward questions emphasizing evolutionary change in systems (Zubrow 1976:21). Without such time depth, it is not possible to explain how systems come to be nor how they change. Additionally, population data have the advantage of being observable, replicable, quantifiable, and cross-culturally comparable (Zubrow 1976:4).

The change from a synchronic to a more diachronic ecological anthropology does not require an abandonment of the ecosystem approach. What it does imply is an extension of the tools of ecological analysis to include also the tools offered by historical demography. The seminal work on this topic is generally acknowledged to be Boserup's *The Conditions of Agricultural Growth* (1965). Cohen (1977), Basehart (1973), Bayliss-Smith (1974), Berreman (1978), Harner (1970), Netting (1973), and Vasey (1979), are but a few of the many who sought to test the validity of Boserup's thesis that population growth drove technological change and the move towards intensification. The tools of historical demography to date have required extensive records of property owned and controlled by households, records of household composition and labor supply, and both total production and marketable production. Whether what we learn about human population dynamics in these settings can be applied to the human/habitat interactions of preindustrial foragers and isolated horticulturalists remains to be seen. It can be argued, however, that the worldwide incorporation of scattered socio-political units within larger economic and political systems makes it impossible to treat local communities anymore as closed systems even for analytical purposes.

Vayda and McCay (1975) proposed that human ecology focus on life-threatening hazards, such as drought and flood, as a way of dealing with real problems faced by humans and to better understand adaptive strategies. Two problems result from this agenda for research: how to predict the occurrence of a hazard with sufficient anticipation to carry out a baseline study and how to justify such attention to basic research when the physical needs of the population demand attention to their basic needs for food and shelter. In addition, the question remains whether behavior under such severe stress is equivalent to behavior when conditions are not particularly stressful. More realistic appears to be the approach suggested by Lees and Bates (this volume). The process of rural development, they argue, and the interventions associated with it, provide an ideal context for testing hypotheses dealing with proximate causality (cf. also Grossman 1984; Little and Horowitz 1987). Such an approach does not free the investigator from the responsibility to seek, over time, additional evidence to gradually place the event observed within a historical matrix of physical, demographic and social change.

The Role of Individuals

Ecosystem approaches have tended to focus on the population and neglected the decision-making activities of individuals. In part, this resulted from the higher level of organization that ecosystems represent within the scheme of systems and from the cybernetic and equilibrium assumptions that usually accompanied it. Adoption of an individual, micro-economic and neo-Darwinian evolutionary approach, to the neglect of an ecosystem approach, is likely to create as many problems as it solves (see Rappaport, this volume). Evolutionary and ecosystem perspectives should be seen as complementary, rather than exclusionary-- e.g. energy flow studies would benefit from knowing how the actions of individuals choosing from among alternatives alters flow networks (Winterhalder 1984). On the other hand, some questions (e.g. desertification, global warming and tropical deforestation) demand that units larger than individuals be engaged in analysis (Schlesinger *et al.* 1990; Peck 1990).

Even the adoption of the household as a unit of analysis, as some have proposed, does not free one from trying to deal with the role of individuals. It is becoming increasingly clear that households do not act as undifferentiated collectives but, rather, embody individuals who engage in complex negotiations. These negotiations embody cultural expectations, social rank, gender hierarchies, age and other demographic considerations which shape the outcomes summarized as "household behavior" or "decisions" (see Wilk, this volume). Attention to the internal dynamics of households becomes necessary to understand the social relations of production, consumption and distribution-- although this may not be possible very often in archeological research, where "household" commonly refers to a "residential unit" (see Butzer, this volume, about the limits of archeological inference).

Problems of Boundary Definition

Just as the time dimension was long overlooked, so was attention to the criteria for boundary definition. The common wisdom was that the ecosystem was a flexible unit and that the boundaries were determined by the goals of the investigator. Any unit which provides the empirical conditions for defining a boundary may constitute an ecosystem for analytical purposes. However, most human ecosystems do not have the clear cut boundaries that a brook, a pond, or an island offers (see Ellen, this volume).

Rappaport (1967) defined the boundaries of the ecosystem he studied by using the concept of "territoriality". The Tsembaga Maring of New Guinea, as horticulturalists and as the ecologically dominant species, defined what the ecosystem, or territory, was through their regulatory operations (Rappaport 1967:148). This is a basically satisfying solution to the question of boundary definition except for two implicit problems: how do ecosystem boundaries change through time and how do shifts in boundary definition relate to internal and external structural or functional relations?

One of the most important steps in dealing with this problem is the identification of inputs and outputs and their measurement. Input/output analysis reveals the status of the system defined for investigation, indicates the system's storage capacity, its resilience to external variation in input, and helps identify structural changes likely to occur. The input/output fluxes of the whole system have specific properties which cannot be anticipated by investigating the system's component parts regardless of their importance (Schulze and Zwölfer 1987:8). Thus, the central problem of input/output analysis is the definition of the system's boundaries in space and time. The scale chosen will depend on the type of process under consideration. In some cases the system will be defined by the material cycles, in others by energy fluxes, in others by historical boundaries in terms of people-vegetation-abiotic interactions (see Ellen, Fish and Fish, and Netting, this volume). Contemporary conservation and restoration biologists define ecosystems as having integral and degraded patches and attempt to restore degraded patches in terms of the input/output relations that characterize the undegraded, or integral, parts of the ecosystem in question (Jordan 1987). This notion does not assume ecosystem equilibrium or a naive notion of reconstructing an "ideal climax" condition. Instead, it seeks to return the system to some degree of structural integrity and replication of functional inter-relations, although the actual species composition, and the "details" of the system may be quite different from any of its earlier states (Allen 1988; Berger 1990).

Bounding one's research is an ever present challenge to be faced by both biologists and anthropologists. By assuming that ecosystems are purely and subjectively definable, yet also somehow coterminous with biomes and other biogeographical units, creates real problems in defining clear sampling criteria. Environmental "patchiness" and heterogeneity, animal mobility, and massive ecosystem change due to natural and man-made disasters have received little attention as they affect one's sample population, for example. There has been progress in this regard, as several papers in this volume show. Ellen proposes the notion of "graded boundaries" that acknowledge temporal and spatial criteria for closure; Conant suggests the use of remote sensing over time to determine the boundaries of ecosystems; and Gross suggests the use of cross-sectional, comparative and longitudinal analysis to deal with the processes of stability and change in human societies. Clearly, time, space and hierarchical level all need to be accounted for in ecological analysis.

Level and Scale Shifting

Whereas it is normal and quite common to understand one level of analysis in terms of the other, such a tack may not be apppropriate. Indeed, this may be the most serious limitation of the ecosystem approach– although it has been rarely mentioned by the critics. All we have for most macro-ecosytems is data for a few sites, for a limited time period, and on only some aspects of the whole system of interactions. From an analytic perspective, one cannot confidently use site-specific studies as a basis for macro-ecosystem models. Geographers, of all scientists, have shown the most sensitivity to this constraint, particularly in reference to how one can understand a large region while only studying small areas within it (McCarthy *et al.* 1956; Dogan and Rokkam 1969).

Biologists and anthropologists deal with systems of very different scales in space and time. Commonly, biologists focus on particular components of ecosystems rather than on the whole system. The spatial scale can go from a few square kilometers to a whole watershed. Nevertheless, regardless of scale, the diversity and complexity of the system has to be reduced to a manageable model of the system, if analysis of the ecosystem is desired. On the other hand, if processes are to be understood, the reverse process is called for: isolating that process from the other system processes. The dilemma between the reductionist view of single processes and the deductivist view of systems is a persistent one -- although ultimately both approaches are necessary (Schulze and Zwölfer 1987:3). In addition, the stochasticity of many environmental parameters, such as rainfall and temperature, makes predictive models of uncertain accuracy.

Anthropologists and ecologists have shown less caution about the problems posed by scale and level shifting. Odum (1971) provides few cautionary words about the pitfalls of extrapolating evidence from single sites to macro-systems. Current trends in both ecology and anthropology suggest that the macro-ecosystem level may not be appropriate for dealing with questions of human impact and resource management except in very broad terms, like "seeking that industrial nations reduce CFC emissions by 20% by the year 2000." This global approach to environment is necessary, given that the problems posed by industrial emissions cut across national boundaries and require concerted, or global, agreement on what each nation will do to combat the problem (National Science Board 1989). On the other hand, it would be a mistake to think that resource management will be adequately addressed by these broad policies. Resource management is ultimately a site-specific task in which social, political, legal, and historical dimensions are at least as important as environmental ones. Local actions have global consequences when they converge in given directions, but corrective actions have to deal with the motives for the actions of individuals who act rationally, within the incentives and experience within which they live (Bennett, this volume, 1976). This is a very exciting arena to which ecological anthropologists could have much to contribute in the decades ahead, if they embrace multidisciplinarity (Dahlberg and Bennett 1986).

The Future

The above problems in past ecosystem studies do not justify that the ecological approach be discarded. There has been considerable progress in the assessment of systems of production due in no small part to the research of ecological anthropologists (cf. review in Ellen 1981). Research has shifted from an overemphasis on calories to a more multivariate approach to causality in human systems. The static functionalism and "vulgar materialism" that have attracted so many critics(Burnham 1982; Friedman 1974) is not evident in the papers found in this volume. Many ecological anthropologists, though by no means all, make use of the ecosystem approach without attributing teleological qualities to it. To them, the ecosystem is but one tool in the execution of research seeking to understand individual action and aggregate processes. The contributors to this volume highlight the conceptual usefulness of ecosystems, while rejecting earlier notions of equilibrium. Their attachment to ecosystem is utilitarian rather than dogmatic or "loyalist."

Jochim and Butzer make clear the heuristic advantages of ecosystem approaches, encouraging systemic thinking and inclusiveness-- before getting on with the task of understanding particular processes, requiring modes of research that are reductionist rather than systemic. While the ecosystem approach cannot claim credit for the development of regional analysis in archeology, its adoption by archeologists at the conceptual level certainly contributed to the design of studies which took archeology beyond the particular site to the understanding of catchment areas, watersheds, and larger regional units of study (cf. Hastorf, Fish and Fish, Butzer, this volume).

Little *et al.* show that biological anthropology was posed earlier than the rest of anthropology to interact with biologists and to participate in multidisciplinary studies using biomes as units of analysis. The results of the IBP studies while excellent in themselves, showed the importance of integrating social processes into the equation-- something that later Man and the Biosphere and more recently, the International Union of Biological Sciences have attempted to correct. The Turkana Project, discussed by Little *et al.*, is but one of many on-going studies that seek to model ecosystems through time. The difficulties of multidisciplinary research design and analysis cannot be overstated--but its potential in addressing problems caused by past myopias makes it necessary that it be given higher priority than has been customary. The U.S. Global Change Research Program makes "interdisciplinary science" one of its strategic priorities-- together with the study of human/environment interactions (Peck 1990:8)

Contemporary attention in policy circles with the human dimensions of global environmental change presents a rare opportunity to anthropologists. For the first time, policy makers acknowledge the central role of humans in environmental modification (Peck 1990) and thus *implicitly* accept what anthropology might have to say. On the other hand, this is an opportunity that seems to have largely bypassed anthropologists so far because of the small number of individuals who have experience with *resource management* as a research question and a policy issue. Anthropologists bring a rich perspective to these debates (Johnson and Earle 1987), and familiarity with many preindustrial populations who found ways to intensify production, in some cases without destruction. On the other hand, it is important that anthropologists recognize the difference in scale between the densities and technologies of contemporary societies and those which have been the subject of most of their studies if naive recommendations are to be avoided.

It has become increasingly clear that solutions to contemporary environmental problems will require the integration of experimental and theoretical approaches at various levels of organization. No single model or level will be adequate. We will need multiple approaches, at various levels of aggregation to deal with the task at hand. This will, in turn, lead to the development of theories that explain the patterns of response, as well as the magnitudes and characteristics of different ecosystems (Levin 1989:243). For example, approaches of the past emphasizing equilibrium and predictability were essential to the testing of null hypotheses. However, they do not serve us well as representations of real landscapes and hide the dynamic processes of patches within ecosystems (Levin 1989:248). For anthropological participation in the contemporary debates over human impact on global environmental change, ecosystem models are fundamental. An ecological anthropology lacking the ecosystem approach would be largely irrelevant to the debates over the processes of global environmental change--possibly the most important research agenda of the 1990's.

An area for likely advances is the cooperative research of anthropologists with agro-ecologists concerned with the growing vulnerability of contemporary agroecosystems (Peck 1990:13). Many of these systems have become so simplified and yield-oriented that they have lost stability and diversity. Future research on how to redesign agroecosystems to enhance their stability, reduce fossil fuel subsidies, and maintain yields at acceptable levels for current demand, challenges the application of ecosystem theory (cf. Odum 1984; Aldag 1987; Gliessman 1989; Berger 1990). One promising dimension has begun to be acknowledged in policy circles: the management of tropical rain forests by indigenous populations. The current rates of deforestation and species loss in these ecosystems have led to concern over predatory economic activities and to a rising concern with the "discovery" of sustainable approaches. The National Science Board (1989) has emphasized the importance of restoring damaged ecosystems and the importance of social and economic considerations in the emerging field of restoration ecology (Berger 1990; Allen 1988) -- perhaps the most exciting area for anthropologists to contribute to in the years ahead. This has been of interest to a small community of ecological anthropologists, including archeologists, who are now in a position to contribute to this concern with sound expertise and detailed knowledge of native systems of production and conservation (cf. Moran 1990; Posey and Balee 1989; Roosevelt 1989; Denevan and Padoch 1988). In short, the ecological approach in anthropology has gone through various phases, each of which has overemphasized environment, culture, techno-environmental features, energy flow, or natural selection. Each generation of anthropologists has found considerable limits to the causal models proposed by those before them. This is not uncommon in the sciences. However, the chief danger to practitioners of the ecosystem approach lies in allowing the concept to harden into fixed biotic boundaries, or to rely on any single index of adaptability as a measure. Attention to the context within which human societies exist, a biotic as well as abiotic context, is an important contribution of the ecosystem approach to anthropology. Although the ecosystem approach does not provide easy solutions on how to integrate the complex linkages between biotic and abiotic components of nature, it serves to remind us that real systems are far more complex than our models, whether evolutionary or ecosystemic, can ever fully conceive.

Acknowledgements

The revisions of this paper were carried out while the author was a Fellow at the Institute for Advanced Study at Indiana University, and a Fellow of the J.S. Guggenheim Memorial Foundation. These organizations should not be held responsible for the views espoused herein.

Notes

[1] Historically, autecology has been experimental and inductive in its approach, whereas synecology had been philosophical and deductive (Spurr 1969:5).

[2] Although the term "cognized environment" was introduced later, it is accurate in describing Steward's notion of "selected features of an environment of greatest relevance to a population's subsistence. "

[3] Energy flow analysis refers to methods that attempt to measure the chemical transformation of solar energy into biomass and its gradual diffusion and loss through a food web (cf. Odum 1971; Moran 1982).

References Cited

Adams, R.N.
   1978   Man, energy, and Anthropology: I can Feel the Heat but Where's the Light? *American Anthropologist* 80:297-309.
Aldag, R.
   1987   Simple and Diversified Crop Rotations: Approach and Insight into Agroecosystems. *Potentials and Limitations of Ecosystem Analysis.* Edited by E.D. Schulze and H. Zwölfer . pp. 100-114. Berlin: Springer-Verlag.
Allen, E. B. ed.
   1988   *The Reconstruction of Disturbed Arid Lands: An Ecological Approach.* Washington D.C.: AAAS.
   1976   *Man in the Andes.* Stroudsburg, PA: Dowden, Hutchinson and Ross.
Baker, P. and W. Sanders
   1971   Demographic Studies in Anthropology. *Annual Review of Anthropology* 1: 151-178.
Barrett, G.W. and R. Rosenberg eds.
   1981   *Stress Effects on Natural Ecosystems.* New York: Wiley
Barth, F.
   1956   Ecologic Relationships of Ethnic Groups in Swat, N. Pakistan.*American Anthropologist* 58: 1079-1089.
Basehart, H.
   1973   Cultivation Intensity, Settlement Patterns, and Homestead Forms among the Materngo of Tanzania. *Ethnology* 12: 57- 73.
Bates, M.
   1953   Human Ecology. *Anthropology Today.* Edited by A. Kroeber. Chicago: Univ. of Chicago Press.

Bayliss-Smith, T.
  1974   Constraints on Population Growth: The Case of the
          Polynesian Outer Atolls in the Pre-Contact Period.
          Human Ecology 2:259-295.
Bennett, J.
  1967   Microcosm-Macrocosm Relationships in N. American
          Agrarian Society. *American Anthropologist*
          69:441-454.
  1969   *Northern Plainsmen.* Chicago:Aldine
  1976   *The Ecological Transition.* London: Pergamon
Berger, J.J. ed.
  1990   *Environmental Restoration.* Washington D.C.:
          Island Press.
Berreman, G.
  1978   Ecology, Demography, and Domestic Strategies in the
          Western Himalayas. *Journal of Anthropological
          Research* 34:326-368.
Boas, F.
  1888   *The Central Eskimo.* Washington DC: Smithsonian
  1911   *The Mind of Primitive Man.* New York:Macmillan
Boserup, E.
  1965   *The Conditions of Agricultural Growth.* Chicago:
          Aldine
Boster, J.
  1983   A Comparison of the Diversity of Jivaroan Gardens with
          that of the Tropical Forest. *Human Ecology* 11
          (1):47-68.
  1984   Inferring Decision-making from Preferences and
          Behavior: An Analysis of Aguaruna Jivaro Manioc
          Selection. *Human Ecology* 12 (4):343-358.
Bourliére, F. ed.
  1983   *Tropical Savannas.* Amsterdam: Elsevier.
          Ecosystems of the World Series, No. 13.
Boyd, R. and P.J. Richerson
  1985   *Culture and the Evolutionary Process.* Chicago:
          Univ. of Chicago Press.
Brookfield, H.
  1970   Dualism and the Geography of Developing Countries.
          Presidential Address at the Australian and New Zealand
          Assoc. for the Advancement of Science.

Burnham, P.
1982   Energetics and Ecological Anthropology: Some Issues.
        Energy and Effort. Edited by G.A. Harrison. New
        York: Internat. Publication Service.
Butzer, K.
1980   Civilizations: Organisms or Systems? *American
        Scientist* 68:517-523.
1982   *Archeology as Human Ecology.* New York:
        Cambridge Univ. Press.
Clarke, K.
1968   *Analytical Archeology.* London: Methuen
Clarke, J.
1976   Population and Scale: Some General Considerations.
        Population at Microscale. Edited by L. Kosinski
        and J. Webb. New Zealand Commission on Population
        Geography.
Clarke, W.
1971   *Place and People.* Berkeley: Univ. of California
        Press.
Coe, M. and K. Flannery
1964   Microenvironments and Mesoamerican Prehistory.
        Science 143:650-654.
Cohen, M.
1977   *The Food Crisis in Prehistory: Overpopulation and
        the Origin of Agriculture.* New Haven: Yale Univ.
        Press.
Cook, S.
1973   Production, Ecology and Economic Anthropology: Notes
        towards an Integrated Frame of Reference. *Social
        Science Information* 12 (1):25-52.
Cooke, S.F.
1972   *Prehistoric Demography.* Reading, MA: Addison-Wesley
Cooley, J.H. and F. Golley eds.
1984   *Trends in Ecological Research for the 1980's.*
        New York: Plenum. Based on a NATO/INTECOL Workshop
        on the Future of Ecology after the Decade of the
        Environment.
Cowan, I.R. and G.D. Farquhar
1977   Stomatal Function in Relation to Leaf Metabolism and
        Environment. *Integration of Activity in the Higher
        Plant.* Edited by D.H. Jennings. pp. 471-505.
        Cambridge: Cambridge Univ. Press.

Dahlberg, K. and J.W. Bennett eds.
1986   *Natural Resources and People: Conceptual Issues in Interdisciplinary Research.* Boulder: Westview Press.

Denevan, W. and C. Padoch eds.
1988   *Swidden-Fallow Agroforestry in the Peruvian Amazon.* New York: New York Botanical Garden. Advances In Economic Botany Series, No. 5.

Dogan, M. and S. Rokkam eds.
1960   *Social Ecology.* Cambridge, MA: MIT Press.

Durkheim, E.
1915   *The Elementary Forms of the Religious Life.* London: Allen and Unwin.

Ellen, R.
1978   Problems and Progress in the Ethnographic Analysis of Small Scale Human Ecosystems. *Man* 13:290-303.
1981   *Environment, Subsistence and System.* Cambridge: Cambridge Univ. Press.

Engelberg, J. and L. Boyarski
1979   The Noncybernetic Nature of Ecosystems. *American Naturalist* 114:317-324.

Evans, F.C.
1956   Ecosystems as the Basic Unit in Ecology. *Science* 123:1127-8.

Evans, G.
1976   A Sack of Uncut Diamonds: The Study of Ecosystems and the Future Resources of Mankind. *Journal of Ecology* 64:1-39.

Flannery, K.
1968   Archeological Systems Theory and Early Mesoamerica. *Anthropological Archeology in the Americas.* Edited by B. Meggers. Washington DC: Anthropological Society of Washington.

Foin, T.C. and W.G. Davis
1987   Equilibrium and Nonequilibrium Models in Ecological Anthropology: An Evaluation of "Stability" in Maring Ecosystems in New Guinea. *American Anthropologist.* 89 (1): 9-31.

Forde, C.D.
1934   *Habitat, Economy and Society.* New York: Dutton.

Friedman, J.
   1974   Marxism, Structuralism and Vulgar Materialism.
           Man 9: 444-469.
Geertz, C.
   1963   *Agricultural Involution.* Berkeley: Univ. of
           California Press.
Glacken, C.
   1967   *Traces on a Rhodian Shore.* Berkeley: Univ. of
           California Press.
Gliessman, S. ed.
   1989   *Agroecology.* Berlin: Springer-Verlag.
Goldscheider, C.
   1971   *Population, Modernization, and Social Structure.*
           Boston: Little, Brown.
Goldenweisser, A.
   1937   *Anthropology.* New York: F.S. Croft
Golley, F.
   1984   Historical Origins of the Ecosystem Concept in Biology.
           The Ecosystem Concept in Anthropology. Edited
           by E.F. Moran. pp. 33-49. Washington DC: AAAS
Golley, F. ed.
   1983   *Tropical Rain Forest Ecosystems: Structure and
           Function.* Amsterdam: Elsevier. Ecosystems of the
           World Series, No. 14a.
Grossman, L.S.
   1984   *Peasants, Subsistence Ecology and Development in
           the Highlands of Papua New Guinea.* Princeton:
           Princeton Univ. Press.
Gudeman, S.
   1978   *The Demise of a Rural Economy.* London:
           Routledge, and Kegan Paul.
Hammel, E.
   1988   A Glimpse into the Demography of the Ainu. *American
           Anthropologist* 90(1): 25-41.
Harner, M.
   1970   Population Pressure and the Social Evolution of
           Agriculturalists.   *Southwest   Journal   of
           Anthropology* 26:67-86.
Harris, D. ed.
   1980   *Human Ecology in Savanna Environments.* London:
           Academic Press.

Harris, M.
1977   *Cannibals and Kings.* New York: Vintage books
Harrison, G. *et al.*
1964   *Human Biology.* London: Oxford Univ. Press.
Hutchinson, G.E.
1957   Concluding Remarks. *Cold Spring Harbor Symposium in Quantitative Biology* 22:415-427.
Jamison, P. and S. Friedman
1974   *Energy Flow in Human Communities.* University Park, PA: US/IBP Human Adaptability Coordinating Office.
Jochim, M.
1981   *Strategies for Survival.* New York: Wiley
Johnson, A.W. and T. Earle
1987   *The Evolution of Human Societies.* Stanford: Stanford Univ. Press.
Jordan, C.F.
1981   Do Ecosystems Exist? *American Naturalist* 118:284-287.
Jordan, C.F. ed.
1987   *Amazon Rain Forests: Ecosystem Disturbance and Recovery.* Berlin: Springer-Verlag. Ecological Studies, No. 60.
Keegan, W.
1986   The Optimal Foraging Analysis of Horticultural Production. *American Anthropologist* 88:92-107.
Kemp, W.
1971   The Flow of Energy in a Hunting Society. *Scientific American* 224:104-115.
Kroeber, A.
1939   *Cultural and Natural Areas of Native North America.* Berkeley: Univ. of California Press.
1917   The Superorganic. *American Anthropologist* 19:163-213.
Lee, R.B.
1979   *The !Kung San.* New York: Cambridge Univ. Press.
Levin, S.A.
1989   Challenges in the Development of A Theory of Community and Ecosystem Structure and Function. *Perspectives in Ecological Theory.* Edited by J. Roughgarden, R. May and S.A. Levin. pp. 242- 255. Princeton: Princeton Univ. Press.

Lieth, H. and M.J.A. Werger eds.
  1989   *Tropical Rain Forest Ecosystems: Biogeographical
            and Ecological Studies*. Amsterdam: Elsevier.
            Ecosystems of the World Series, No. 14b
Lindeman, R.
  1942   The Trophic Dynamic Aspect of Ecology. *Ecology*
            23: 399-418.
Little, M.
  1982   The Development of Ideas on Human Ecology and
            Adaptation. *A History of American Physical
            Anthropology*, 1930-1980. Edited by F. Spencer. New
            York: Academic Press.
Little, M. and S. Friedman
  1973   *Man in the Ecosystem*. University Park, PA:
            US/IBP Human Adaptability Coordinating Office.
Little, M. and G. Morren
  1976   *Ecology, Energetics and Human Variability*.
            Dubuque, Iowa: W.C.Brown
Little, P. and M. Horowitz (with A.E. Nyerges)
  1987   *Lands at Risk in the Third World: Local Level
            Perspectives*. Boulder: Westview Press.
Major, J.
  1969   Historical Development of the Ecosystem Concept.
            *The Ecosystem Concept in Natural Resource
            Management*. Edited by G. M. Van Dyne . New York:
            Academic Press.
McCarthy, H., J.C. Hook, and d. S. Knos
  1956   The Measurement of Association in Industrial Geography.
            Dept. of Geography, Univ. of Iowa.
McCay, B. and J.M. Acheson eds.
  1987   *The Question of the Commons: The Culture and
            Ecology of Communal Resources*. Tucson: Univ. of
            Arizona Press.
McIntosh, R.P.
  1985   *The Background of Ecology: Concept and Theory*.
            Cambridge: Cambridge University Press.
Margalef, R.
  1968   *Perspectives in Ecological Theory*. Chicago:
            Univ. of Chicago Press.

1981   Stress in Ecosystems: A Future Approach. *Stress Effects on Natural Ecosystems.* Edited by G.W. Barrett and R. Rosenberg . pp. 281-289. New York: Wiley.

Milan, F.
1979   *Human Biology of Circumpolar Populations.* Oxford: Cambridge Univ. Press. Vol. 21 IBP Synthesis Series.

Mobius, K.
1877   *Die Auster und die Austernwirtschaft.* Berlin: Wiegundt, Hempel and Payey.

Moran, E. F.
1973   Energy Flow Analysis and Manihot esculenta Crantz. *Acta Amazonica* 3 (3):28-39.

1981   *Developing the Amazon.* Bloomington: Indiana Univ. Press.

1982   *Human Adaptability: An Introduction to Ecological Anthropology.* Boulder: Westview Press. Originally publ. 1979 by Duxbury Press.

1990   *A Ecología Humana das Populações da Amazônia.* Petrópolis, Rio de Janeiro (Brasil): Editôra Vozes.

Moran, E.F. ed.
1984   *The Ecosystem Concept in Anthropology.* Washington DC: AAAS.

Morren, G.E.B.
1986   *The Miyanmin: Human Ecology of a Papua New Guinea Society.* Ann Arbor: UMI Research Press.

National Science Board
1989   *Loss of Biological Diversity: A Global Crisis Requiring International Solutions.* Washington, D.C.: National Science Board, Committee on International Science (Task Force on Global Biodiversity).

Netting, R.
1968   *Hill Farmers of Nigeria.* Seattle: Univ. of Washington Press.

1973   Fighting, Forest and the Fly: Some Demographic Regulators among the Kofyar. *Journal of Anthropological Research* 29: 164-179.

1974   The System Nobody Knows: Village Irrigation in the Swiss Alps. *Irrigation's Impact on Society.* Edited by T. Downing and M. Gibson . Tucson: Univ . of Arizona Press.

1976   What Alpine Peasants have in Common: Observations on Communal Tenure in a Swiss Village. *Human Ecology* 4:135-146.

1977   *Cultural Ecology.* Menlo Park: Cummings.

1979   Household Dynamics in a 19th century Swiss Village. *Journal of Family History* 4:39-58.

1981   *Balancing on an Alp.* New York: Cambridge Univ. Press.

Nietschman, B.
1973   *Between Land and Water.* New York: Seminar Press.

Odum, E.
1953   *Fundamentals   of   Ecology.*   Philadelphia: Saunders.

1977   The Emergence of Ecology as a New Integrative Discipline. *Science* 195:1289-93.

1984   Properties of Agroecosystems. *Agricultural Ecosystems.* Edited by R. Lowrance, B.R. Stinner and G.J. House. pp. 5-11. New York: Wiley.

Odum, H.T.
1957   Trophic Structure and Productivity of Silver Springs, Florida. *Ecol. Monog.* 27:55-112.

1971   *Environment, Power, and Society.* New York: Wiley.

1983   *Systems Ecology.* New York: Wiley.

Orlove, B.
1980   Ecological Anthropology. *Annual Review of Anthropology* 9:235-273.

Patten, B. and E. Odum
1981   The Cybernetic Nature of Ecosystems. *American Naturalist* 118:886-895.

Peck, D.L.
1990   *Our Changing Planet: The FY 1991 U.S. Global Change Research Program.* Washington DC: Office of Science and Technology Policy, Committee on Earth Sciences.

Plog, F.
1975   Systems Theory in Archeological Research. *Annual Review of Anthropology* 4: 207-224
Polgar, S.
1972   Population History and Population Policies from an Anthropological   Perspective.   *Current Anthropology* 13:203-211.
Posey, D. and W. Balee eds.
1989   *Resource Management in Amazonia.* New York: New York Botanical Garden, Advances in Economic Botany Series, No. 7.
Rappaport, R.
1967   *Pigs for the Ancestors.* New Haven: Yale Univ. Press.
1971   The Flow of Energy in an Agricultural Society. *Scientific American* 224:116-132.
1977   Ecology, Adaptation, and the Ills of Functionalism. *Michigan Discussions in Anthropology* 2:138-90
Reichle, D. and S. Auerbach
1979   Analysis of Ecosystems. *Systems Ecology.* Edited by H. Shugart and R. O'Neill. Stroudsburg, PA: Dowden, Hutchinson and Ross. Orig. publ. in 1972 by American Inst. of Bio. Sciences.
Richards, P.
1985   *Indigenous Agricultural Revolution.* Boulder: Westview Press. Published jointly with Hutchinson (London).
Ricklefs, R.
1973   *Ecology.* Portland, Oregon: Chiron Press.
Roosevelt, A.C.
1989   Resource Management in Amazonia Before the Conquest: Beyond Ethnographic Description. *Advances in Economic Botany* 7:30-62.
Sahlins, M.
1961   The segmentary Lineage: An Organization for Predatory Expansion. *American Anthropologist* 63:322-345.
Salt, G.W. ed.
1984   *Ecology and Evolutionary Biology: A Roundtable on Research.* Chicago: Univ. of Chicago Press.
Samuels, M.
1983   A Simulation of Population Regulation among the Maring of New Guinea. *Human Ecology* 10:1-45.

Schlesinger, W.H. et al.
1990   Biological Feedbacks in Global Desertification.
         Science 247:1043-1048.
Schneider, J. and R. Schneider
1976   *Culture and Political Economy in W. Sicily.* New
         York: Academic Press.
Schulze, E.D. and H. Zwölfer eds.
1987   *Potentials   and   Limitations   of   Ecosystem
         Analysis.* Berlin: Springer-Verlag. Ecological
         Studies, No. 61.
Shantzis, S.B. and W.W. Behrens
1973   Population Control Mechanisms in a Primitive Society.
         Toward Global Equilibrium. Edited by D. Meadows
         and M. Meadows. Cambridge: Wright-Allen.
Sheridan, T.
1988   *Where the Dove Calls: The Political Ecology of a
         Peasant Corporate Community in Northwestern Mexico.*
         Tucson: Univ. of Arizona Press
Shinners, S.M.
1967   *Techniques   of   Systems   Engineering.* New
         York:McGraw-Hill
Shugart, H.H. and R.V. O'Neill eds.
1979   *Systems   Ecology.* Stroudsburg,   PA:Dowden,
         Hutchinson and Ross. Benchmark Papers in Ecology. Vol.
         9.
Smith, E.A.
1984   Anthropology, Evolutionary Ecology, and the Explanatory
         Limitations of the Ecosystem Concept. *The Ecosystem
         Concept in Anthropology.* Edited by E.F. Moran . pp.
         51-86. Washington DC: AAAS
Spengler, O.
1926   *Decline of the West.* New York: Knopf.
Spurr, S.H.
1969   The Natural Resource System. *The Ecosystem Concept
         in Natural Resource Management.* Edited by G.M. Van
         Dyne. New York: Academic Press.
Steward, J.
1955   The Concept and Method of Cultural Ecology. *Theory
         of Culture Change.* Urbana: Univ. of Illinois Press.
Sukkachev, V.N.
1960   Relationship of biogeocoenosis, ecosystem and facies.
         Soviet Soil Science 6:579-584.

Sweet, L.
  1965   Camel Pastoralism in N. Arabia and the Minimal Camping Unit. *Man, Culture and Animals*. Edited by A. Leeds and A.P.Vayda. Washington DC: AAAS

Tansley, A.G.
  1935   The Use and Abuse of Vegetational Concepts and Terms. *Ecology* 16:284-307.

Thomas, F.
  1925   *The Environmental Basis of Society*. New York: Century Press.

Thomas, R.B.
  1973   *Human Adaptation to a High Andean Energy Flow System*. University Park, PA: Pennsylvania State Univ., Dept. of Anthropology, Occasional Paper Series.

Troll, C.
  1950   Die Geographische Landschaft und ihre Erforschung. *Studium Gen.* 3 :163-181.

Ulrich, B.
  1987   Stability, Elasticity and Resilience of Terrestrial Ecosystems with Respect to Matter Balance. *Potentials and Limitations of Ecosystem Analysis*. Edited by E.D. Schulze and H. Zwolfer. pp. 11-49. Berlin: Springer-Verlag.

Van Dyne, G.M.
  1979   Ecosystems, Systems Ecology and Systems Ecologists. *Systems Ecology*. Edited by H.H. Shugart and R.V. O'Neill. Stroudsburg, PA: Dowden, Hutchinson and Ross. Article reprinted from the original publ. in 1966 by Oak Ridge National Labor.

Van Dyne, G.M. ed.
  1969   *The Ecosystem Concept in Natural Resource Management*. New York: Academic Press.

Vasey, D.E.
  1979   Population and Agricultural Intensity in the Humid Tropics. *Human Ecology* 7:269-283.

Vayda, A.P. and R. Rappaport
  1968   Ecology, Cultural and Non-Cultural. *Introduction to Cultural Anthropology*. Edited by J. Clifton. Boston: Houghton and Mifflin.

Vayda, A.P. and B. McCay
  1975   New Directions in Ecology and Ecological Anthropology. *Annual Review of Anthropology* 4: 293-306.

Waddell, E.
   1972   *The   Mound-Builders.*   Seattle: Univ.   of
           Washington Press.

Watson, W.
   1964   Social Mobility and Social Class in Industrial
           Communities. *Closed Systems and Open Minds.*
           Edited by M. Gluckman. Chicago: Aldine.

Weiner, J.S.
   1965   *IBP Guide to the Human Adaptability Proposals.*
           London: ICSU, Special Committee for the IBP.

White, B.
   1973   Demand for Labor and Population Growth in Colonial
           Java. *Human Ecology* 7:217-236.

White, L.
   1949   *The Science of Culture.* New York: Farrar,
           Strauss and Giroux.

Wiener, N.
   1948   *Cybernetics.* New York: Wiley.

Winterhalder, B.
   1984   Reconsidering the Ecosystem Concept. *Reviews in
           Anthropology* 11(4):301-330.

Winterhalder, B. and E.A. Smith eds.
   1981   *Hunter-Gatherer Foraging Strategies: Ethnographic
           and Archeological Analyses.* Chicago: Univ. of
           Chicago Press.

Wissler, C.
   1926   *The Relation of Nature to Man in Aboriginal
           America.* New York: Oxford.

Worthington, E.
   1975   *The Evolution of the IBP.* Cambridge: Cambridge
           Univ. Press

Zubrow, E. ed.
   1976   *Demographic Anthropology.* Albuquerque: Univ. of
           New Mexico Press.

CHAPTER 2

ECOSYSTEMS, POPULATIONS AND PEOPLE[1]

Roy A. Rappaport

One of the fundamental problems for any ecological anthropology worthy of the name is to develop a truly synthetic conceptual structure, one which does not merely squeeze together the untempered terms of distinct intellectual traditions but is commensurable with the discourse of anthropology on the one hand, and with general ecology on the other. This general consideration has informed the line of thought expressed most recently in the "epilogue" to the second edition of *Pigs for the Ancestors* (1984) in which I argued, as I had earlier, that ecosystems, local populations, regional populations, and individuals all have legitimate places in ecological analyses done by anthropologists. Other essays in this volume make it clear that agreement on these candidates for inclusion in ecological formulations is by no means universal (see Lees and Bates). They must therefore be discussed yet again.

Ecosystems

In the conclusion of *Pigs for the Ancestors* (1968, 1984) I defined "ecosystem" as "a demarcated portion of the biosphere that includes living organisms and non-living substances interacting to produce a systematic exchange of materials among the living components and with the non-living substances" (p. 225). This definition, as it stands, seems almost unexceptionable, but questions have arisen concerning the nature or even the reality of whatever it is that it claims to distinguish. There are questions, first of all, about the systematicity of such associations. That is, do they have properties other than the sum of the properties of their

41

constituents? Are they, to put this in slightly different terms, *organized* as such?

The "classical view", as expressed by figures like Eugene Odum (1969) and Ramon Margalef (1968), held that what are called "ecosystems" do have emergent, holistic properties. First, it is claimed, they possess well-known structural characteristics. Regardless of what their constituent species may be, ecological systems are roughly cyclical with respect to material flow and pyramidal with respect to the productivity, trophic structure, and regulation of constituent populations. They have also been said to possess "self-organizing" properties. That is, they transform themselves as wholes in response to changes in external conditions, sometimes replacing their constituent species with populations of other species and sometimes through mutual adaptation of their constituent species. Although ecosystems include species that may have come together accidentally, forming relationships that are, at the outset, only crudely articulated, the constitution of such systems are likely to become increasingly complex and coercive through time as mutual dependencies among the species are elaborated. The classical conception further holds that ecosystemic successions, unless they are arrested or deflected, exhibit holistic tendencies and that these tendencies are similar in systems differing radically in species composition. Under conditions which are both enduring and stable, it has been proposed they move toward a "climax" state in which the numbers of species present increases, perhaps to some rough maximum; these species are likely to become increasingly specialized and an increasing proportion of them larger, longer-lived and slower-breeding. The system as a whole requires less and less energy flux per unit of standing biomass to sustain itself, but productivity per unit area increases. Material and energy pathways proliferate, as do regulatory mechanisms. Systemic redundancy thus offsets, at least partially, the loss of stability that might otherwise be a concomitant of the increasing specialization of the species present.

The classical view has been seriously challenged over the past decade and a half. Within anthropology, for instance, Jonathan Friedman has written that "an eco-system is not organized as such. It is the result of the mutual and usually partial adaptation of populations each of which has laws of functioning that are internally determined" (1974:466). Vayda and McCay (1975:229-ff) are in rather close agreement. Citing Colinvaux's remark that "nowhere can we find discrete ecosystems, let alone ecosystems with the self-organizing properties implied by the concept of climax" (1973:549), they assert that "the ecosystem is an analytic, not a biologic entity." They propose that "interactions observed in complex ecosystems need not be regarded as expressing self-organizing properties of the systems themselves; instead, they can be understood as the consequences of the various and variable adaptive strategies of individual organisms living together in restricted spaces." While Friedman favors the "social formation" as an analytic unit, Vayda and McCay assign priority to the individual organism because natural selection operates on individuals.

There are two general points to be made before proceeding. First, the distinction that Vayda and McCay make between the *analytic* and *biological* seems to me to be a mistaken one. It is surely the case that there are few if any ecosytems less inclusive than the solar system that are hermetically sealed to flows of matter, energy, and information across their borders. Their boundaries must, therefore, be specified analytically, as must those of social formations or, for that matter, social units of any sort. The question of the criteria used to discriminate ecosytems from the continuity of natural phenomena is, of course, strategic and I have discussed it elsewhere (1969; 1971a), proposing that in some instances it is useful to see a single population participating in more than one ecosystem, while in others two or more human populations may be taken to be participating in a single ecosystem. In *Pigs for the Ancestors*, the criterion for establishing the boundaries of local ecosystems, in what was a continuous biotic association, was human territoriality.

The second point concerns the degree to which the systems under consideration have self-regulating properties. It is important to keep in mind in this regard that, in the example just mentioned, the Maring are horticulturalists and, as such, ecological dominants. They set the conditions encouraging or discouraging the presence of other species, and they attempt to construct anthropocentric ecosystems within areas in which the engagement of humans in interspecies exchanges is conventionally regulated. Maring local groups are regulating the ecosystems within which they participate, or to put it in the converse, the domain of the regulatory operations of a local group in this instance defines an ecosystem. Because a Maring local group is a component of the ecosytem which it regulates (and upon the perpetuation of which its own persistence is contingent) the ecosystem is by definition self-regulating.

It might be argued that self-regulatory properties are peculiar to anthropocentric ecosystems. I think this is not the case. Self-regulating mechanisms, I believe, inhere in ecosystems *qua* ecosystems *as well as* in their constituent populations. That there are self-regulating mechanisms at one level of organization does not mean that there are none at others. Every population in every ecosystem must have "internally determined laws of functioning" of its own, but this does not mean that there are not self-regulating mechanisms emerging out of relations between and among populations-- as is the case of the mutual regulation of predator and prey populations. Ecosystemic self-regulation may be a product of dynamic interaction among a number of species none of whom exercises central control or is even dominant in a less active way. Such diffuse regulation is not unfamiliar to us; it animates, at least conceptually, the "perfect market" of economic mythology.

But the critique of the ecosystem concept in general ecology did not begin and end with Paul Colinvaux. In a recent review of theoretical developments in that field, the environmental historian Donald Worster (1989) tells us that this" 'individualistic' view [which] was reborn in the mid-1970's [its earlier incarnation dates back to the 1920's] by the present decade had become the core idea of what some scientists hailed as a new revolutionary paradigm in ecology." In brief, whereas the older ecology was concerned with revealing nature's order and regularity, the newer ecology is concerned not only to discover disorder, disturbance and randomness but to replace conceptions of order with them. Thus, Drury and Nisbet (1973) attacked general assumptions of successional direction, and they were joined in this attack several years later by Connell and Slatyer (1977). The word "disturbance" appears more and more often in the literature, and the disturbing agent is less often human and more often natural: fire, wind, invading microorganisms, gophers. The general message of the papers composing the volume *The Ecology of Natural Disturbance and Patch Dynamics* (1973), Worster observers, is that "the climax notion is dead, the ecosystem has receded in usefulness and in their place we have the idea of the lowly 'patch.' Nature should be regarded as a landscape of patches, big and little... a patchwork quilt of living things, changing continually through time and space, responding to an unceasing barrage of perturbations. The stitches in that quilt never hold for long."

Anthropologists need to be very careful in entering into domestic arguments in other people's houses but it is reasonable to observe that disturbance and disorder may well have been given insufficient attention in the ecology of Clements and his followers (see Moran, preceding chapter, this volume), but that attention to disorder and disturbance does not preclude attention to order and regularity. It can even be argued that the terms "order" and "disorder" are each meaningless in the absence of the other. That ecological order may resemble the hidden *Logos* of Heraclitus, within which flux takes place and by which flux is bounded does not make it disorderly. It may be that the conceptions of order with which ecologists work need to be expanded to include perturbations, disturbance and randomness, not only as environmental conditions amidst which all living systems maintain orderliness but as themselves constituents of order-maintaining processes. This is for the ecologists to decide. What anthropologists, I think, can decide is whether or not the ecologists have presented them with sufficient grounds to abandon the ecosystem concept. The answer to that question, I believe, is "No." Whether we should abandon it for our own reasons is another matter. Again, I think the answer is "No." For one thing, as I have already suggested, whatever the case may be in systems which are not dominated by humans, anthropocentric ecosystems, to the extent that they are managed, do show holistic tendencies. More importantly, the ecosystem concept is, at all its levels of inclusiveness up to the level of the biosphere, and in all its degrees of specificity and vagueness, "*good to think.*" On the one hand it provides a framework within which humans can think well about their general relation to the world, while on the other it provides a framework within which specific problems can be rigorously formulated and approached. Given the systemic nature of contemporary environmental problems--ozone depletion, green house warming, deforestation, desertification, pollution, acid rain, nuclear waste contamination and so on-- the ecosystem concept, or something very much like it, is indispensable and, to be as frank as possible, it would seem to me to be politically and socially absurd to abandon it.

I say "politically and socially absurd." In my view anthropology is not now, nor has it ever been, nor can it ever be, disengaged from its subject matter, a species that lives and can only live in terms of meaning it itself must construct in a world without intrinsic meaning but subject to natural law. Its most profound problems flow from disconformities between law and meaning. On the one hand, it is

unlikely that the laws of nature will ever be fully known, but even
if they are the outcomes of their operations will remain, in large
degree, as unpredictable as the weather. On the other hand, there is
nothing in the nature of human thought to prevent it from
constructing self-destructive or even world-destroying errors. The
ecosystem concept mediates between law and meaning. It is a concept
that can at once serve to organize *symbolic constructions* of
the nature of the world and humanity's place in it *and* to
formulate procedures for *discovering* aspects of the world
that are tectonically, genetically and environmentally constituted
rather than symbolically constructed.

I think, then, that it is as proper as it is unavoidable to give
weight to ideological as well as scientific factors in deciding
whether or not to retain an ecosystem approach in anthropology.
Worster, I might note, suggests that ideological factors may also
underlie contemporary developments in general ecology (1989). In
searching for reasons for the shift in emphasis from order to
disturbance, after noting that most of the "disturbance boosters are
not and have never been ecosystem scientists" but have been trained
as population biologists, he says:

> For some scientists, a nature characterized by highly
> individualistic associations, constant disturbance, and
> incessant change, may be more ideologically satisfying than
> Odum's ecosystem, with its stress on cooperation, social
> organization, and environmentalism.

He takes Paul Colinvaux's chapter on succession in his popular
book *Why Big Fierce Animals Are so Rare* (1978) as a case in
point. It begins

> If the planners really get hold of us so that they can stamp out
> all individual liberty, and do what they like with our land,
> they might decide that whole counties full of inferior farms
> should be put back into forest.

The political point of view is patent. The same chapter ends as
follows:

We can now... explain all the intriguing predictable events of plant succession in simple, matter of fact, Darwinian ways. Everything that happens in successions comes about because all the different species go about earning their livings as best they may, each in its own individual manner. What look like community properties are in fact the summed results of all those bits of private enterprise

It is of interest that the rise of disorder in general ecology coincides with the emergence of sociobiology in biology and an increased emphasis upon the individual actor in anthropology. It is probably not coincidental that the late seventies and the decade of the eighties were also a period during which neoconservatism emerged. We will return to this later.

Ecosystems and Interlevel Contradiction

Whatever the outcome of the debate in general ecology may be, there are sufficient grounds for anthropologists to retain the ecosystem concept not only for ideological, political, and practical reasons but because there are sufficient grounds to take ecosystems to be organized as such and, at least in the case of anthropocentric systems, to be self-regulating and self- organizing. They differ, of course, from other classes of such systems in important respects including not only the sharpness of their boundaries but relative coherence and the relative autonomy of their subsystems, matters to which we shall return.

A statement of Robert Murphy's could, nevertheless, be taken to be an argument against using the ecosystem as an object of functional or systemic statements even if the organized status of such systems is granted:

Higher order phenomena arrange lower order phenomena to their purposes, though they may not change their properties. Correspondingly, human social systems reach out and embrace ecosystems rather than the reverse proposition, and culture reorders nature and makes appendages of the parts of it that are relevant to the human situation (1970:169).

The only qualification I would voice with respect to this statement is that it may not apply to hunting and gathering populations. With food production-- or at least with plant cultivation-- men become ecological dominants, setting the conditions encouraging or discouraging the presence of populations of other species. The burden of regulating anthropocentric ecosystems rests largely upon the humans dominating them. That humans attempt to put nature to their own purposes (i.e. to regulate ecosystems in accordance with what they take to be their self-interest) is true, but this is hardly the end of the matter. We want to know how this is done; what the purposes and understanding of the actors may be; to what degree these purposes are themselves constituted, coerced, or constrained by environmental characteristics; the degree to which they conform to, or are even explicitly concerned with or informed by, an awareness of the requirements of ecosystemic perpetuation and, of course, whether actions guided by such purposes are compatible with ecosystemic requisites or are ecologically degrading or destructive. The compatibility or incompatibility of human purposes and ecosystemic imperatives should, in fact, be central to the "problematic" of any ecologically-concerned anthropology.

Cultivation demands that complex climax communities of plants and animals be replaced by simpler communities composed of smaller numbers of species selected by humans according to criteria of apparent usefulness, and arranged by them in limited numbers of short food chains of which the humans themselves strive to be the *terminus*. Such communities are likely to be less stable than the climax communities they replace. The relatively degraded nature of anthropocentric ecosystems is in part a function of their simplicity, in part a function of the nature of their constituent species: often poorly adapted to local conditions, often helpless, frequently unable to reproduce without some assistance. And humans are poor dominants. It is interesting to note that dominants in nonanthropocentric ecosystems are almost always plants, like oak trees in temperate forests or animals resembling plants like corals in reef-lagoon communities. They are well-suited to their role, for their mere nonpurposive existence fulfills well the needs of their associated species. Humans, on the other hand, must maintain their dominance through behavior. Because their behavior is less reliable than the simple existence of oak trees, because they are capable of acting selfishly and maliciously, and because their purposes may not coincide with the needs of the systems which they dominate, humans set conditions that tend toward instability (Rappaport 1970a, 1970b).

To insist upon the self-regulating, self-organizing characteristics of ecosystems is not to claim that only ecosystems are organized. Well-defined social units and individual organisms are also constituted systems with "laws of functioning that are internally determined." Observing that ecosystems are constituted does not not deny to the organisms and populations of organisms participating in them their relative autonomy. Conversely, recognizing the relative autonomy of a social unit does not deny the organization of the ecosystems of which it is a part, any more than recognizing the relative autonomy of individuals denies the organization of the social units of which individuals are always members. A refusal to recognize the organization of more inclusive systems is tantamount to arguing for the absolute autonomy of systems-- at one or another level of inclusiveness-- operating in larger fields which are not systematic or systematic only in a derivative sense. The complexity of the world does not warrant such a view even for analytic or heuristic purposes. We must recognize that more inclusive entities are indeed organized systems, made up of components that are themselves relatively autonomous. Ecosystems, for instance, include populations which are, in turn, composed of individuals. While it may be, as systems theory proposes, that *generally* similar principles organize systems at various levels of inclusiveness, important differences in their more specific "laws of functioning" should be recognized.

A difference to which little attention has been given is that of the differential degrees of coherence that systems of different classes both require and can tolerate. By "relative coherence" I refer to the degree to which changes in one component of a system effect changes in other components of the system. A fully coherent system is one in which any change in any component results in an immediate and proportional change in all other components. As no living system could be totally incoherent, neither could any living system be fully coherent, for disruptions anywhere would immediately spread everywhere. Perhaps because their functioning depends upon fine, quick, and continual coordination of parts, organisms are, and must be, more coherent than social systems. Conversely, the degree of coherence continually required by organisms would probably be intolerable for social systems (which may attain levels of coherence comparable to that of organisms only in extraordinary circumstances, such as rituals, for relatively brief periods). Ecosystems are probably less coherent than social systems, at least human social systems (perhaps because orderly relations within ecosystems depend more on increasing redundancy than upon coordination). Their low degree of coherence may well be in large part responsible for the frequent failure to recognize their systemic characteristics.

It is obvious that the maintenance of systems and the purposes of their relatively autonomous components do not always coincide. Humans can surely do violence to the structure and function of the ecosystems that they come to dominate, just as certain subsystems of societies, such as industries, can do violence to the social entities that they come to dominate. Although such violation is not unexampled among tribal peoples, I have suggested elsewhere (1978b) that its likelihood is increased and its effects extended and intensified by increased differentiation of society, by alienating or at least separating the economic rationality of individuals from direct ecological imperatives, by production for gain rather than use, and by industrialization.

The imperatives of individual existence often bring individuals into conflict with the social systems of which they are members; the cultural imperatives of social systems may lead to actions at variance with ecological principles. Contradictions between constituted systems on various levels of inclusiveness -- between individuals and societies, between societies and ecosystems-- are inevitable. Sahlins (1969) and Friedman (1974) err, I think, in taking ecological or adaptive formulations to be "innocent of a concern for contradiction." Elsewhere (1970b, 1978b, 1979a) I have discussed maladaptation in structural terms, proposing that it is to be understood as, or as resulting from, among other things, interlevel contradiction. Indeed, it seems to me that it is Friedman's argument, not adaptive and ecological formulations, that is analytically innocent of a concern for interlevel contradiction. He is right in insisting upon the relative autonomy of certain social entities, yet he denies organization or even reality to the systems that include the systems he recognizes. But if reality or organization is denied to ecological systems how is it possible to discover contradictions between them and the entities participating in them? We may not only miss much of explanatory importance but also much of what is problematic or even poignant or tragic in the human condition. The impoverishment of the perspective is increased if relevance is also denied to the relative autonomy of the individuals composing the social entitites with which Friedman is concerned. It should not be forgotten that one of the perennial concerns of human thought about the human condition is with the problems congregating in the relationship between individuals and the societies of which they are members, problems which are conceptually summarized in such phrases as "the problem of freedom"-- or of happiness, duty, honor, authenticity, ambition, responsibility, obligation, or ethics. Even granting the systemic nature of ecosystems, the question remains, however, whether or not ecosystems are properly thought of as units of analysis. There are difficulties in establishing their boundaries, and, more generally, their ultracomplexity leads to problems in comprehending interactions among their constituents. I do not believe that any natural ecosystem has ever been anything like fully modeled (see Golley 1984). The accounts or descriptions of ecosystems which ecological anthropologists may produce, even in cooperation with specialists, are the merest sketches. We can do little more than outline the general structure of such systems, try to measure a few important interactions, and attempt to locate the species with which we are concerned in the web. In practice, it seems, we may not be far from Steward's (1955) "significant environment."

That ecosystems are complex beyond realistic possibilities of comprehensive modeling does not distinguish them radically from other entities with which we deal and which we take to be units of analysis: organisms, social groups, social formations, populations, cultures. Moran (1984) proposes that the ecosystem is conceptually important even if it presents difficulties as a unit of analysis. (He adds the matter of variable scale to the problems of boundary and complexity). I agree. If the ecosystemic concept is abandoned we are faced with a choice. We either find some other way of representing the systemic character of surrounding nature or we confine ourselves to the study of decontextualized interactions between humans as individuals or groups on the one hand, and one or a very few nonhuman elements in their surroundings on the other. To the extent that ecology is defined by a commitment to a holistic view of nature, the latter choice is not ecological but *anti-ecological.*

Ecological Populations

The analytic use of populations as environed units distinguishes what others have called "the new ecology" most clearly from the "cultural ecology" of Julian Steward (1955), in which cultures are taken to be the environed units. The choice has attracted considerable criticism. Sahlins, for instance, has declared that in

the "translation of a 'social order' into a 'population of organisms'
... everything that is distinctively cultural about the object has
been allowed to escape" (1976b:298). This is, in my view, a
misunderstanding and because this misunderstanding persists it is
well to make clear some of the considerations that entered into the
analytic choice of the *population* two decades ago.

First, difficulties are entailed by the use of cultures or their
parts (e.g. Steward's "culture core") as environed units in
ecological analyses, and when cultures are combined with an
ecosystemic concept of the environment these difficulties are
increased. Such a choice is the result of a rather subtle
methodological or even logical confusion, a confusion between an
explicandum on the one hand and, on the other, the primary
units of analysis, the major components of analytic or descriptive
models, the elements seen to stand in relation to each other in the
system under study.

This confusion is easy to fall into, and it is understandable
that anthropologists interested in the formative effects of
environments on cultures would take cultures, or parts of cultures,
to be their primary units of analysis. The conception of culture as
an order of phenomena distinct from the psychological, biological,
and inorganic has been one of anthropology's important contributions
to Western thought. Inasmuch as cultural phenomena may be
distinguished from other phenomena and inasmuch as *culture*,
however it is understood, is what most cultural anthropologists wish
to elucidate, cultures or their constituents have seemed obvious
choices for referent units in ecological as well as other
anthropological formulations. Indeed, such a choice seems almost
inevitable, considering that culture is conceived by many
anthropologists not only to be ontologically distinct from
biological, psychological, and inorganic phenomena, but processually
independent of them as well. Culture, it is said, "obeys laws of its
own," laws distinct from those governing organic and inorganic
processes.

At the same time that cultural ecology took cultures, distinct
from the organisms bearing them, to be interacting with environments,
it borrowed the concept of ecosystem from general ecology. In the
resulting formulation, cultures simply interact with ecosytems. But
cultures and ecosystems are not directly commensurable. An
ecosystem is a system of matter and energy transactions among unlike
populations or organisms and between them and the nonliving
substances by which they are surrounded. "Culture " is the label for

the category of phenomena distinguished from others by its contingency upon symbols. The incommensurability of ecosystems and cultures becomes clear when one considers the analogy implicit in the notion of cultures independent of, or at least conceptually separated from, culture- bearing organisms interacting with other components of ecological systems:

CULTURE: ECOSYSTEM: ANIMAL POPULATION: ECOSYSTEM

It is ironic that the choice of cultures as primary units of analysis in cultural-ecological formulations aimed to protect the uniqueness of culture against the degrading power of ecological principles, but it has the opposite effect, for the processual equivalence of cultures and animal populations is logically entailed by it. The analogy not only implies that they have similar requirements which must be fulfilled in similar ways and that they are similarly limited by environmental constraints, but also that cultures, far from obeying laws of their own, are directly subject to the same laws as those governing animal populations. No cultural ecologist (in the strict sense) has ever taken such a position, of course, and, I am sure, none ever intended to. It is nevertheless intrinsic to the eclectic conjunction of incommensurable and incongruent terms that characterizes the cultural-ecological conception. Eclectic formulations bring together the disparate concepts they subsume only by violating some or all of them. In the case at hand, the violence wreaked is upon the notion of culture, for it, or some of its aspects, such as the social order, may be conflated with biological phenomena.

This problem is illustrated by difficulties experienced by Marshall Sahlins in *Social Stratification in Polynesia*. It was his thesis that certain differences in social stratification were to be accounted for by differences in productivity. In societies such as those of pre-contact Polynesia, the term "productivity" must refer to the yields of horticulture and fishing. Lacking direct production data, Sahlins attempted to estimate the comparative productivity of the societies in question by comparing the size of the largest networks through which garden, grove, and fishing yields were distributed in each society, making the assumption that the larger the network and the more frequently it operated, the greater the surplus and therefore the higher the productivity. The relationship between surplus and productivity is problematic, however, and there is no reason to believe, *prima facie*, that a society of a certain size organized into a single distributive network produces more per capita than a smaller society, or more than one of the same size organized into a number of smaller networks. Moreover, if the scope of the distributive prerogatives vested in chiefs is an aspect of stratification, and if stratification is to be accounted for by productivity, to read productivity from the size of redistributive networks brings the argument perilously close to circularity.

The dubiety of a long inferential chain and the dangers of circularity can be avoided, however. Productivity in an ecological sense is understood as the amount of biomass or energy produced per unit of area per unit of time. The horticultural productivity of societies can thus be compared in such terms as tons per acre per year. In the absence of harvest and landing data, indices may or must be used, but the size of redistributive networks is not an appropriate one. In relatively undifferentiated nonindustrial societies like those of Polynesia, it is reasonable to infer comparative productivity, or production, from a comparison of the densities of the populations supported by that production. Such a comparison indicates that there was probably no correlation between social stratification and productivity in aboriginal Polynesia. But the point I wish to make here is that I believe Sahlins's difficulties have followed from a failure to distinguish an aspect of *social* orders --the size of redistributive networks-- from a *biological* characteristic of populations, namely their densities.

We may recall here Bennett's charge (1976) that the analysis developed in *Pigs for the Ancestors* was, in its use of a general ecological paradigm, "merely an analogic operation." Bennett is mistaken. The analysis represented a deliberate, self-conscious attempt to replace the inappropriate and deforming analogy underlying "cultural ecology" (an analogy which takes human cultures and animal populations to be equivalents) with a homology. As such it represented an attempt to replace a mode of analysis which is merely eclectic with one which is appropriately synthetic. Synthetic formulations, in constrast to those which are merely eclectic, subsume the subject matter of what had previously been separate realms of discourse under terms of sufficient generality to accommodate both without distortion. They do not ignore or deny distinctions but employ terms of sufficiently high logical type to encompass seemingly disparate phenomena as more or less distinct members of subclasses of a common class.

The first step in moving towards any synthesis is to find terms expressing commonalities. In the case of the place of humans in ecosystemic processes, rather than attending only to that which distinguishes the human species from other species, synthesis begins with what is common to all of them and then proceeds to whatever may distinguish them. We thus begin with the simple observation that the human species is, after all, a species among species and that, as such, humanity's relations with its physical and biotic environments are, like those of other animals, continuous, indissoluble, and necessary. It follows, then, that it is not only possible but proper to take populations of the human species to be environed units in ecological formulations.

In *Pigs for the Ancestors* and elsewhere I have defined *ecological population* as an aggregate of organisms sharing distinctive means for maintaining a common set of material relations with the other components of the ecosystem in which they together participate. An ecosystem, as we have already noted, is the total of ecological populations and nonliving substances bound together in material exchanges in a demarcated portion of the biosphere. The term *ecosystem* and *human population*, taken in the ecological sense are fully commensurable and congruent.

But what of culture? Is its uniqueness ignored, violated, degraded or reduced by the synthesis represented in this general form of ecological anthropology? I think not. For purposes of ecological accounts cultures or their constituents can be included among the properties of populations. In this view, cultures are not analogous to animal populations, but constitute the major and most distinctive means by which human populations maintain common sets of material relations with other components of the ecosystems in which they participate. Culture, in other words, is central to the adaptive characteristics of the species; cultures, in turn, are central to the adaptations of the many more or less distinctive populations into which the species is organized.

If the concept of culture-as-adaptation is the truth it is not the whole truth. While it would be hard to imagine the emergence of culture out of a precultural substrate as other than a strongly-selected-for innovation in the adaptive process itself, it would be equally difficult to argue that, having emerged, culture has remained nothing more than an adaptive apparatus in the service of the organic processes of a particular species. Once emerged, cultures both developed needs of their own and established the goals, values, and purposes of humans. It becomes no longer clear whether culture is a symbolic means to organic ends, or whether organisms are living means to culture's ends, for humans come to serve and preserve their cultures as much as, or even more than, their cultures serve and preserve them.

This account proposes that a culture's own requisites, and the values, purposes, and goals which it ordains, may come to be at odds with the organic needs of the humans striving to live by and fulfill them, and may also violate the ecosystems in which they are fulfilled. That the cultural properties of human populations may be inimical to their organic characteristics is as inherent to the view expressed here as is the recognition that aspects of cultures are properly regarded as central to the adaptations of their bearers. The contradiction, perhaps inevitable, between the cultural and the biological is, in my view, among the most fundamental problems to be addressed by an ecologically-aware anthropology.

That the cultural and organic properties of humanity are mutually dependent needs no discussion. It may however, be worth noting that their relationship is not symmetrical. Although no group or assemblage of human organisms could survive physically in the absence of culture, it is not the case that any *particular* convention or set of conventions is indispensable to that survival. In contrast, the persistence of all cultural properties is contingent upon their realization by living humans. Cultures do not, as is sometimes loosely said, "develop lives of their own." They can "live" only in the lives of those whose thought and action they form. Organic processes and cultural order are bound together in human life and its evolution.

That the realization of culture is contingent upon the lives of organisms does not mean that all of culture is to be accounted for in ecological or adaptive terms. Such theories cannot, by themselves, constitute a general theory or explanation of culture, although it may be that some anthropologists, including me (Rappaport 1971b; Vayda and Rappaport 1967), have come close to trying to make them do so from time to time. The perspectives of ecology, adaptation, and evolution do, however, have a contribution to make to a general understanding of culture and to understandings of particular cultures.

It does not seem to me, in sum, that any violation of culture's unique characteristics is entailed, nor its significance diminished, when cultures are taken to be properties of populations. The social order is not thereby reduced to the status of the organic, as Sahlins (1976a:298) has charged. Indeed, the organic characteristics of human populations are more clearly distinguished from the cultural conventions ordering those populations than they are in the cultural ecology of Julian Steward or the early Sahlins. To distinguish is not to separate, of course. In this view the cultural and organic properties of populations are neither inappropriately conflated nor radically separated. The attempt to find terms commensurable with all of nature while giving full consideration to the uniqueness of human life. It is synthetic rather than eclectic.

There is a further conceptual effect which requires comment. In our formal attempts to understand our relation to the rest of the world, we are encouraged to seek special explanations, explanations which, founded upon assumptions of uniqueness, can cover only one, or at best a narrow range, of cases, and we are dissuaded from attempts to understand cultural phenomena in terms of principles that apply to other species as well. It may be suggested that formulations that illuminate similarities rather than emphasize differences have a proper place in our epistemologies, not always as alternatives to more special explanations but as complements to them. It may even be proposed that the exposure of similarities among phenomena such as organisms, populations, or even living systems generally precedes adequate understanding of the differences among them. Unless similarities are recognized and understood, the magnitude and significance of differences cannot be comprehended. What to a more general perspective appear to be variations on a common theme, may seem to a narrower perspective to be enormous differences of kind.

It seems to me, in sum, that the use of the human population in ecological analyses preserves a view of humanity as part of nature at the same time that it recognizes the uniqueness conferred upon the species by culture. As such it preserves the terms defining the condition of a creature that can live only in terms of meanings, largely culturally constituted, in a world to which law is intrinsic but meaning is not.

Local and Regional Populations

One of Roy Ellen's criticisms of *Pigs for the Ancestors* is useful in approaching the distinction between local and regional populations:

Rappaport's analysis is based upon an assumption that the
Tsembaga represent a closed spatial and temporal system, at least
as far as the variables which he discusses are concerned. He had
to assume that the Tsembaga and Maring are relatively isolated
and constitute a population in dynamic balance with its
environment. The data clearly suggest that his assumptions are
not valid. (1982:185-186).

I find this criticism mystifying. Although I argued that the
identification of local groups as ecological populations is
appropriate to the analysis of localized interspecific trophic
exchanges among the Maring, *I did not* argue that such
transactions exhaust the exchange relationships of any Maring local
group. All local groups of tribal people, with the possible exception
of a few isolated on remote islands or in the deep reaches of the
Arctic, engage in exchanges with other human groups living in other
localities. The materials exchanged may include foodstuffs, but trade
goods, valuables, personnel, and such services as ritual support,
military reinforcements, and hospitality in times of emergency are
often more important. Moreover, territory may be reallocated from
time to time among the local groups occupying a region through such
processes as warfare, migration, and alliance formation. Such
extralocal relations were not slighted in *Pigs for the
Ancestors*. They were, in fact, crucial to the account presented,
occupying a chapter of their own, figuring prominently in the chapter
on the ritual cycle, and receiving considerable attention in the
final chapter.

Although I adduced evidence to show that the Tsembaga alone derive
subsistence from their territory and (almost) only from their
territory, and although I claimed that a similar degree of local
trophic self-suffificiency is general among the Maring, my account
did not indicate and *certainly* did not assume that
Maring local groups are "relatively isolated" from each other nor
that the Maring as a whole are isolated from other peoples in the
general area.

I did argue in the conclusion of *Pigs for the Ancestors*
and elsewhere (1969, 1971b) that "the concept of the ecosystem (and
of the ecological population) which provides a convenient frame or
model for the analysis of trophic exchanges between ecologically
dissimilar (i.e. occupying different niches) populations occupying

single localities can accommodate only by the introduction of analogy nontrophic ... exchanges between ecologically similar populations occupying separate localities" (1984:225). To deal with transactions of the latter sort I suggested that we recognize that local groups (and by implication the separate subgroups and individuals composing them) not only participate as ecological populations in local ecosystems but also constitute elements or components of regional systems. Among tribal peoples, ecosystems are likely to be more or less localized systems of interspecies exchanges with trophic exchanges being the most important class of transactions taking place in them. Regional systems are more dispersed systems of intraspecies exchanges of personnel, information, and services of various sorts, as well as trade goods and valuables.

Ecosystems and regional systems can *usually* be distinguished among tribal peoples by differences in the degrees to which they are spatially limited. Local territorial groups and the territories they claim can often be taken to constitute ecosystems. The aggregate of local groups interacting in an area may constitute a regional system. Such clarity does not always prevail, however, because territoriality is not always strongly developed among tribal peoples and is often weakly developed or nonexistent among hunters and gatherers. Even in instances in which particular local groups are associated with particular areas as, for instance, among bushmen bands (Lee 1976; Yellen 1976) individuals are able to move from one group to another with great ease. In such cases I have suggested (1971b:250) that the ecological population may have to be taken to include all the bands living in the region, and the region as a whole viewed as the ecosystem. Observations and measurements performed on one or several local groups can then be used as statistical samples of the environmental relations of the entire population. The local group, this is to say, is in such cases *a sample and not an analytic object in its own right* (see Arensberg 1961).

There are grounds for distinguishing ecosystems and regional systems, grounds which are, in fact, more fundamental than simple local concentration versus regional dispersion. First, there is the matter of the relationships among constituent units. Ecosystems are composed of more or less continuous interactions among ecologically dissimilar populations, that its, populations, usually of different species, occupying distinctive positions in networks of trophic exchanges. Regional systems are composed of interactions among ecologically similar populations, that is, distinct populations of the same species occupying similar or equivalent ecological niches. Maring local groups, for instance, constitute units in both (local) ecosystems and a wider regional system.

Ecosystems and regional systems are "about" different things. Ecosystems are, esssentially, systems of matter and energy exchanges among unlike species. These exchanges are, in the main, trophic and the system is "concerned with" sustenance. The transactions typical of regional systems are, in contrast, mainly nontrophic. Food may, of course, figure in regional exchanges, but exchanges of personnel, valuables, trade goods, services, and information are frequently of greater material importance. The system is "about" population dispersion, defense, aggression, non-food necessities, luxuries and valuables, reproduction, social relations and ritual. It is economic, social, political and religious in nature.

Third, the formal characteristics of the exchanges are different in the two sorts of systems. The trophic exchanges defining ecosystems are two-party exchanges in which the material exchanged is one of the parties --one eats the other. The exchanges defining regional systems are also exchanges between two parties, but the objects exchanged, be they material goods, information, services, or men or women given away in marriage, are separate from the exchanging parties. Exchanges in the regional system are intrinsically *social* in nature-- they are between similar parties that, as a rule, survive the exchange and usually benefit from it in one way or another. (Warfare is an exception). Ecosystemic exchanges, in contrast, do not seem intrinsically social. They are between dissimilar parties, one of which usually does not survive the exchange. It is, however, no less striking because it is well known that ecosystemic exchanges are, despite their intrinsically antisocial nature, often viewed in social terms. Marsupials, for instance, are taken by the Maring to be "the pigs of the Red Spirits" and we may recall here Wagner's (1977) discussion of Melanesian conceptions of humanity.

It is useful to discriminate between ecosystems and regional systems even in instances in which the two are geographically and demographically coextensive because such a discrimination focuses attention on the manner in which the discriminated systems are articulated. Articulation between unlike processes is not to be taken for granted. Questions concerning the matter include the extent to which events in one system affect events in the other. Regional and ecosystemic relations are articulated through local organization and local ritual cycles which, in *Pigs for the Ancestors*, were taken to be outcomes of participating in both ecosystemic processes and regional relations.

In sum, local groups participate in, and are the locus of conjunction between, local ecological processes and regional social, economic, and political processes. These two sets of processes are incommensurable and it therefore *reduces* confusion to distinguish them. I must emphasize, however, that to distinguish is not to separate, and to take them to be incommensurable is not to take them to be unrelated. Events in regional systems can and do affect local ecological processes and vice versa, and both are affected by relations with the even more inclusive national and world systems. Distinguishing these systems in terms of what they are "about", differences in their "languages" and "metrics", so to speak, forces us to focus upon the precise ways in which they are articulated, and such foci may enlarge our understandings of particular institutions as well as of systemic processes generally. In *Pigs for the Ancestors*, for example, I proposed that Maring ritual operates as a binary transducer, translating, as it were, changes in the states of local ecological systems into information meaningful in the regional social and political systems.

I have labeled the aggregate of local groups that together comprise regional systems "regional populations." As I noted in the conclusion of *Pigs for the Ancestors*, regional populations may resemble or even be identical with other associations identified by anthropologists and other scientists-- the populations of demographers, the breeding populations of geneticists, the societies of sociologists and anthropologists. Like these units the boundaries of regional systems are typically vague rather than sharp. I have also suggested that whereas regional populations persist indefinitely, the *local groups* which are their constituents *are more or less ephemeral* (see Lowman- Vayda 1980).

Individuals: Their Decisions and Actions

Since the publication of *Pigs for the Ancestors* increased emphasis has been given in ecological formulations to individuals and their decisions and actions (Ellen 1982; McCay 1978; Orlove 1980; Vayda and McCay 1975). This trend is not confined to ecological anthropology, but is a more or less special manifestation of a movement in the field generally. Ortner (1984) notes that "For the

past few years there has been growing interest in analysis focused
through one or another of a bundle of interrelated terms: practice,
praxis, action, interaction, activity, experience, performance. A
second and closely related bundle of terms focuses on the doer of all
that doing: agent, actor, person, self, individual, subject." She
further suggested that this trend promised to be the most salient
characteristic of the anthropology of the eighties. The development
deserves a qualified welcome. Earlier analyses in ecological
anthropology, reflecting a general deficiency in cultural and social
anthropology, did not pay sufficient attention to the purposes
motivating individual actions, actions which, when aggregated,
constitute group events; nor did they pay sufficient attention to
behavioral variations among individuals, to differences in the
understandings of the world entertained by individuals, to
individuals as adaptive units, or to conflicts between individual
actors or between individuals and the groups to which they belonged.

Pigs for the Ancestors has been faulted on these grounds,
and, like most analyses of its time, it was open to these criticisms.
It is worth noting, however, that the book was centrally concerned
with ritual, secondarily with warfare, two of the most highly
integrated forms of social activity undertaken by humans. In ritual
and war, the purposes of individuals are more comprehensively
aggregated, their concerns more rigorously subordinated, their
actions more tightly coordinated than in other sorts of human
enterprise. In no other situations do groups act more as groups and
less as aggregates of autonomous individuals than they do in the
performance of rituals, and warfare is characterized by the
enunciation of common purposes, high degrees of coordination, and
frequent replacement of self- interest by self-sacrifice. Preparation
for rituals commemorating warfare occupied most of the time and
thought of most people during the fourteen months that I lived among
the Tsembaga. Ritual is a mode of action conforming more closely than
any other to "ideal structure" and that is what I saw. I did give a
great deal of attention in *Pigs for the Ancestors* and in
subsequent publications (especially 1971b, 1979b) to the "cognized
model" -- the shared understandings underlying individual
understandings, choices and actions-- but not to those individual
understandings, choices, and actions themselves. The analysis would
have been richer had I done so.

Although the development of what is being called "practice theory" may ameliorate deficiencies now apparent in earlier anthropological accounts, some questions concerning it must be raised. Many of these questions are asked and discussed more fully in Ortner's essay (1984) and in Hastorf's paper in this volume.

First, it may be asked what the choices and actions of individuals are meant to account for in anthropological analyses. Human choices and actions are informed by understandings which are culturally constituted. Most of these understandings have an assumed or explicit public core, a core of understandings upon or around which there is general agreement. Individual variations are expressed in details or, more importantly, in differences among the conclusions that different individuals may derive from what they take to be the same sets of facts or differences in the strategies employed by individuals standing in varying relationships to the same facts. Accounts of events should, of course, take into consideration the individual variations that are implicated in precipitating them. An adequate account of the occurrence of the *kaiko* would, for instance, have attended to the details of consensus formation. Consideration of such matters may show significant processual differences between events of different sorts. For instance, the decision of a local group to begin its *kaiko* comes out of a more or less protacted process of consensus formation that must, eventually, enlist the support of a preponderance of the membership. In contrast, the decision to go to war, a decision binding upon a corporate group, can be preempted by any individual or small group of individuals who take it upon themselves to avenge the death of a kinsman by killing a member of another group.

Of more interest and importance than the extent to which individual actions and decisions account for events is the extent to which they can account for conventions and institutions and for changes in them. To put this differently, to what extent may practice account for structure and changes in it? An older anthropology which took action to be largely the enactment of culturally-prescribed rules would have replied "very little." More recently, however, interest in the ways in which practice may modify or transform social and cultural forms has grown.

The increased concern with *practice* which seems widespread in contemporary anthropology generally recognizes the existence of sociocultural structures, systems, or formations with properties distinct from the individuals realizing them. The novelty, such as it is, is that the relationship between social process and individual action is now taken to be a "problematic" in its own right. It is now being seen in more reciprocal or mutually formative terms than it had been, but outside of ecological anthropology and sociobiology there has been little attempt to reduce social phenomena to aggregated individual behavior. Herein lies what may be rather special about developments in sociobiology and the current in ecological anthropology influenced by sociobiology. Whereas the roots of the "practice movement" in anthropology generally lie squarely within the traditions which take group phenomena to be central, the theoretical ground of the new concern in ecological anthropology lies, at least in large part in evolutionary biology and in formal economics. Orlove (1980:248) has written that the recent emphasis on decision-making models in ecological anthropology "corresponds to recent developments in biological ecology, with its stress on natural selection at the level of individual organisms as a principle which organizes populations and communities." In more particular terms, it represents an attempt to make anthropological formulations compatible with the general rejection in biology of group selection as a process figuring significantly in organic evolution.

Earlier anthropologies erred in undervaluing action, individual or otherwise, which they were inclined to dismiss as little more than the enactment of culturally encoded patterns for behavior— the mere realization of cultural and social forms in practice. In contrast, implicit in what Orlove calls "processual ecology" is a tendency to take social and cultural forms to be little more than the sums of decisions made by individuals acting in terms of their own self-interest. To put this in slightly different terms, in accounting for conventions-- publicly accepted understandings, rules and practices-- explanatory priority is given to the advantages they confer upon those *individuals* accepting them. Calculi of advantage are formalized in "microeconomic models (which) resemble economic models of choice-making" (Orlove 1980:247) although "satisfizing", "minimaxing", "optimizing" and "hierarchies of strategies" may in some instances replace simple maximizing.

Although it may be that neglect of the individual in particular and practice in general required rectification, it seems to me that the correction entailed in such a program is in danger of becoming excessive, not simply replacing one error with another, but replacing neglect of the individual with something even more serious-- the dissolution of the social as such. There are several more particular points to be made in this regard.

First, on the face of it at least, conventions do not necessarily aid and abet individuals in their competition against each other or against their environments. More typically, in fact, they constrain the actions of individuals in their dealings with others. It can, of course, be argued that some individuals or groups are always more favored by prevailing conventions than are others, that such inequities are inevitable, and that competition for control of the apparatus for reproducing and transforming convention animates many societies. Nevertheless, to account for convention (or if you prefer, culture) in general in such terms leads to a notion of explanation as nothing more than expose, a concept of culture as nothing more than a set of species-specific modes of social domination. Such accounts, to put it as kindly as possible, are naive oversimplifications that will neither illuminate the world's undeniable injustices nor contribute to their rectification.

The second point is related to the first. It may be that, given any set of conventions, people will behave in accordance with what they take to be their own interests. This hypothesis concerning the behavioral propensities of individuals does not, however, account for the particular forms or conventions prevailing in any society. It proposes what people will do given the existence of any such forms. The hypotheses of humans as maximizers of individual advantage or as maximizers of inclusive fitness are behavioral hypotheses devoid of any specific cultural content, and cannot account for that content.

A third related point is that the hypothesis that individuals will behave in terms of what they perceive to be their self- interest does not account for what they perceive those interests to be. Even if individuals' perceptions of self-interest are more or less idiosyncratic formulations of universal, general, and natural "maximizing" or "optimizing" proclivities, it can hardly be doubted that culturally particular definitions of that self- interest intervene.

Fourth, it follows that if concepts of self-interest are culturally formulated, then self-interest can be defined in terms which are at variance from, or even contradictory to, the immediate material advantage of the living individual or the long-term advantage of his or her genotype. The use of language even makes it possible to define individual advantages in ways which separate it from personal survival. Individuals can easily be led to sacrifice themselves in terms of "their own best interests." History is generous in bestowing examples upon us. Attempts to account for all these cases in terms of the increased reproductive success of the hero's brothers and sisters seem to me pathetic. We may conclude, I think, that whatever may be the case among other species the use of language, which makes possible such conceptions as honor, morality, altruism, valor, righteousness, prestige, gods, heaven and hell, makes group selection important among humans. By group selection I mean selection for and perpetuation of *conventions* enhancing the persistence of groups, even though these conventions can be disadvantageous to those individuals whose actions accord most closely with them.

Fifth, the subordination of the properties of groups to the interest of individuals in adaptive formulations is unwarranted by observation. If adaptation is a process of response to perturbations, adaptive units may be defined by unitary response to such disturbances. *Response to perturbation is not a property exclusive to individual organisms, but inheres in groups* at virtually all levels of inclusiveness: families, sub-clans, clans, local populations and even states. In dismissing the adaptive properties that may inhere in any particular level there is not only a risk of misconceiving the structure of adaptive processes as wholes, but also of missing both deformations of that structure and interlevel conflicts in adaptive processes.

Sixth, Orlove himself notes that "actor-based models have tended to treat environmental variables as part of a relatively static set of external constraints to which individuals respond and adapt. . . . They have thus omitted some of the concerns of ecological anthropology" (1980:248). I would go further. There is a danger here, particularly when the concept of ecosystem is abandoned, that environmental variables will be construed as little more than sets of natural resources and natural hazards, raw commodities or goods on the one hand, and dangers, potential liabilities, and "bads" on the other. *Ecological anthropology is in danger of being replaced by environmental economics.* Economic activity entails valuation: decisions are made, presumably to apply scarce resources to differentially graded ends. Microeconomic models, which "resemble economic models of choice making" (Orlove 1980:247), models that have been developed in societies with money and markets, may be inappropriate for the representation of such values. Cost-benefit analysis that transforms distinctions between unlike things into mere quantitative differences by reducing all things to a common metric is, I think, especially inappropriate for understanding choices in societies without money.

Reliance on actor-based models alone does not merely omit some of ecological anthropology's concerns. It ignores what I take to be its most fundamental problem. We are concerned with a species that lives in terms of meanings in a universe devoid of meaning but subject to physical law. The relationship of actions formulated in terms of meaning to the systems constituted by natural law within which they occur is, in my view, *the essential problematic of ecological anthropology*. It is missed by approaches which focus upon actors and ignore or deny the systemic character of an encompassing nature.

I earlier suggested that this problematic not only describes the human condition but provides a set of supracultural criteria in terms of which human actions can be assessed. It could be argued that the social sciences have no right to make assessments and that the general approach is ideological. If we are going to address contemporary problems we are, of course, going to make assessments and we may as well make them upon explicit grounds. Assessment is, I suppose, in some degree ideological, even when the attempt is to assess the adaptiveness of understandings, conventions, and actions in terms which are as objective as possible.

We have already noted that theory in anthropology, and perhaps more generally in the sciences, seems to be rather sensitive to events in the larger society. As the ecosystemically-oriented ecological anthropology of the early 1970's may have been responsive to the ecology movement, so may the more recent actor-oriented perspective in anthropology be an expression of the "new individualism" of the late 1970's and 1980's, an individualism encouraged by the exhaustion of liberalism and the rise of neoconservatism in politics-- expressed more generally in what Christopher Lasch (1978) calls "The New Narcissism". Recent trends in ecological anthropology are no less subject to the ideological fashions prevailing in the society at large than were earlier ones.

Conclusion

I have noted a series of units ranging in inclusiveness from the individual to the ecosystem that figure in ecological anthropology. Moran (1984), following prevailing thought in general ecology, has reminded us that ecosystems themselves can be discriminated at many levels, from small systems occupying only portons of the territories of local groups to the biosphere. If this is the case, human aggregations of various magnitudes of inclusiveness, up to the human species as a whole, may, under certain circumstances, be regarded as ecological populations. The important point here, however, is that attention to units at one level does not preclude the existence or significance of those at others. This would not need to be asserted were it not for the fact that some modes of analysis have ignored individual choices and actions, others would dissolve the social into no more than the aggregation of individual behavior, and some have denied the existence and significance of ecosystems. The banishment of any one of these levels -- the individual, the social, the ecological-- from our analyses is a very serious mistake, conferring upon us no benefits and costing us dearly.

At the end I want to emphasize that the world humans inhabit is in part symbolically constructed, in part constituted by cosmic, geologic, ecologic and genetic processes. But although it is common-sensical and perhaps even correct in more rigorous senses, the extent to which the distinction between the symbolically constructed and the "naturally" constituted can be imposed upon the things and processes of the world is very limited and its application would demand high tolerance for ambiguity, arbitrariness, or both. How would one classify a rose? Or radioactive waste? We generally conclude, commonsensically if not tacitly, that although the two classes may be conceptually distinct, they are, in the actual world, both inseparable and mutually formative. Dominance has shifted through time, however. With technological evolution the capacities of symbolic constructions to shape the "natural" consituents of the world have increased by magnitudes over what they were when all humans lived only by hunting and gathering.

The concept of ecosystem, like the concept of the "free market," is itself a symbolic construction. As such, it reflects what ecologists have discovered in their environments, but it does more than that. It also integrates the "discovered" with the "symbolic" in ways that are not inevitable, for it grasps the naturally lawful through *particular* human meanings. In a world in which the lawful and the meaningful, the discovered and the constructed, are inseparable the concept of the ecosystem *is not simply a theoretical framework* within which the world can be analyzed. It is itself an element of that world, one that is crucial in maintaining that world's integrity in the face of mounting insults to it. To put this a little differently, the concept of the ecosystem is not simply descriptive or, in Austin's (1962) sense, "constative." It is also "performative"; the ecosystem concept and actions informed by it are *part of the world's means for maintaining, if not* indeed *constructing, ecosystems.* Note its central role in the emerging field of restoration ecology (see Moran, previous chapter).

It is obvious that the ecosystem concept is not an ineluctable extrapolation from observations of nature. Other meanings or interpretations are always possible (e.g., the ecosystem concept could be abandoned in favor of environmental economics), and it is never possible to avoid conceptual choices. It is a matter of history that the conceptual choices favored by societies, those that become dominant or hegemonic, also shift and oscillate. We privilege certain of our symbolic constructions and not others during particular times and in particular places. This is hardly less the case in science than it is in social life generally, and it is obvious that the two are not clearly separable. I have already noted the more or less simultaneous rise, during the 1970s, of neo-conservatism and the "new narcissism" on the one hand and, on the other, the assault on the ecosystem concept in the names of various forms of individualism and disorder.

I both predict and encourage another swing of the pendulum. I predict revitalization of the ecosystem concept because it seems to accord with a general public's commonsense experience of a world beset by multiplying and interrelated environmental disorders, most of which it can attribute to humanity itself. I encourage this revitalization, with appropriate modifications, because the ecosystem concept itself is a vital element in the construction, maintenance and reconstruction of the webs of life upon which, by whatever name we call them, we are absolutely dependent.

Notes

[1] This essay is a substantial revision of Section 10, *Units and Their Implications, Epilogue 1984*, in *Pigs for the Ancestors. A new enlarged edition.* Yale University Press: New Haven 1984.

References Cited

Arensberg, C.
   1961   The Community as Object and Sample. *American Anthropologist* 63:241-264.

Austin, J.L.
   1962   *How to Do Things with Words.* New York: Oxford
          University Press.
Bennett, J.W.
   1976   *The Ecological Transition.* London: Pergamon
          Press.
Colinvaux, Paul
   1973   *Introduction to Ecology.* New York, NY: Wiley.
   1978   *Why Big Fierce Animals are so Rare: An Ecologist's
          Perspective.* Princeton, NJ. Princeton University
          Press.
Connell, Joseph and Ralph Slatyer
   1977   "Mechanisms of Succession in Natural Communities and
          Their Role in Community Stability and Organization."
          The American Naturalist 111:1119-1144.
Drury, William and Ian Nisbet
   1973   "Succession." *The Ecology of Natural Disturbance
          and Patch Dynamics.* Orlando: Academic Press.
Ellen, R.
   1982   *Environment, Subsistence and System.* Cambridge:
          Cambridge Univ. Press.
Friedman, J.
   1974   Marxism, Structuralism, and Vulgar Materialism.
          Man 9:444-469.
Lasch, C.
   1978   *The Culture of Narcisism.* New York: Norton.
Lee, R.
   1976   !Kung Spatial Organization. *Kalahari
          Hunter-Gatherers.* Edited by R. Lee and I. DeVore .
          Cambridge:Harvard Univ. Press
Lowman-Vayda
   1980   Environment, Society and Health. Ph.D. Dissertation,
          Columbia University. Dept. of Anthropology.
McCay, B.
   1978   Systems Ecology, People Ecology and the Anthropology of
          fishing Communities. *Human Ecology* 6:397-422.
Margalef, R.
   1968   *Perspectives on Ecological Theory.* Chicago:
          Univ. of Chicago Press.
Moran, E.F.
   1984   Limitations and Advances in Ecosystems Research.
          The Ecosystem Concept in Anthropology. Edited

by Emilio Moran. Washington DC: American Assoc. for the Advanc. of Science.

Murphy, R.
1970   Basin Ethnography and Ecological Theory. *Languages and Cultures of W. North America*. Pocatello: Idaho State Univ.

Odum, E.
1969   The Strategy of Ecosystem Development. *Science* 164:262- 270.

Orlove, B.
1980   Ecological Anthropology. *Annual Review of Anthropology* 9:235-273.

Ortner, S.
1984   Theory in Anthropology since the sixties. *Comparative Studies in Society and History* 26:126-166

Picket S.T.A. and P.S. White eds.
1985   *The Ecology of Natural Disturbance and Patch Dynamics*. Orlando, FL: Academic Press.

Rappaport, R.A.
1968   *Pigs for the Ancestors*. New Haven: Yale Univ. Press.
1969   Some Suggestions concerning concept and method in Ecological Anthropology. *Contributions to Anthropology: Ecological Essays*. Edited by D. Damas. Ottowa: Nat. Museum of Canada. Bulletin #230.
1970a   Purpose, Porperty and Environmental Disaster. *Science Looks at Itself*. New York: Scribners
1970b   Sanctity and Adaptation. Paper presented at Wenner-Gren Conference on the Moral and Aesthetic Structure of Human Adaptation.
1971a   The Flow of Energy in an Agricultural Society. *Scientific American* 225:116-132.
1971b   Nature, Culture and Ecological Anthropology. *Man, Culture and Society*. Revised Edition. Edited by H. Shapiro. New York: Oxford
1978a   Adaptation and the Structure of Ritual. *Human Behavior and Adaptation*. Edited by N. Blurton-Jones and V. Reynolds. London: Taylor and Francis.
1978b   Maladaptation in Social Systems. *Evolution in Social Systems*. Edited by J. Friedman and M. Rowlands. London: Duckworth.

1979a   Adaptive Structure and Its Disorders. *Ecology, Meaning and Religion.* Richmond, CA: N. Atlantic Books.

1979b   On Cognized Models. *Ecology Meaning and Religion.* Richmond, CA: North Atlantic Books.

1984   *Pigs for the Ancestors.* Revised edition. New Haven: Yale University Press

Sahlins, M.
1958   *Social Stratification in Polynesia.* Seattle: Univ. of Washington.

1969   Economic Anthropology and Anthropological Economics. *Social Science Information* 8:13-33.

1976   *Culture and Practical Reason.* Chicago: Univ. of Chicago Press.

1976a   Comment on A.H. Berger. *Structural and Eclectic Revisions of Marxist Strategy: A Cultural Materialist Critique. Current Anthropology* 17:298-300.

Steward, J.
1955   *The Theory of Cultural Change.* Urbana: Univ. of Illinois Press.

Vayda, A.P. and B. McCay
1975   New Directions in Ecology and Ecological Anthropology. *Annual Review of Anthropology* 4:293-306.

Vayda, A.P. and R.A. Rappaport
1967   Ecology, Cultural and Non-Cultural. *Introduction to Cultural Anthropology.* Edited by J. Clifton. Boston: Houghton- Mifflin.

Wagner, R.
1977   Scientific and Indigenous Papuan Conceptualizations of the Innate: A Semiotic Critique of the Ecological Perspective. *Subsistence and Survival:Rural Ecology in the Pacific.* Edited by T. Bayliss-Smith and R.G. Feachem. London: Academic Press.

Worster, Donald
1989   "The Ecology of Order and Chaos." Plenary address, Conference on Solving Environmental Problems, Evergreen State College, Olympia, WA, April 25.

Yellen, J.
1976   Settlement Patterns of the 1:21 !Kung. *Kalahari Hunter- Gatherers: Studies of the !Kung San and their Neighbors.* Edited by R. Lee and I. DeVore. Cambridge: Harvard Univ. Press.

PART II
ECOSYSTEMIC APPROACHES
IN ARCHAEOLOGICAL RESEARCH
DESIGN AND PRACTICE

CHAPTER 3

THE ECOSYSTEM CONCEPT IN ARCHAEOLOGY

Michael Jochim

Current Ecological Research in Archaeology

Although archaeology has a long history of interest in environmental and cultural relationships, only the most recent project designs and field techniques have begun to allow for the complexity involved in an ecological approach to past behavior. Ecology is the study of dynamic relationships between organisms and their total environments. An ecological approach to cultural behavior requires, consequently, that any particular aspect of behavior be examined within its cultural and natural context, keeping in mind that this context may be varying in space and time. With the realization that past behavior must be explained, in part, in relation to its context, the concept of the ecosystem, as the structure of dynamic interrelationships of behavior and its context, has emerged as important in archaeological method and theory.

The ecosystem concept has been useful to archaeologists primarily as an heuristic device, encouraging us to think in terms of the systemic interrelationships among cultural and natural factors. A knowledge of the characteristic structural elements and processes of natural ecosystems has directed our attention to such features as the flow of energy and the diversity of interacting species in prehistoric human ecosystems. As a concrete unit of analysis, however, the ecosystem has had little role in archaeological research. In dealing with the past we can hope to reconstruct only small portions of past ecosystems. Moreover, much of ecological anthropology is moving toward Darwinian theory more explicitly, with a consequent shift of perspective from populations to individuals, from large regions to local areas. Along with a shift in analytical and geographic scale is an expansion of temporal scale, with the short- and long-term, and their interrelationships, assuming importance. Archaeology, however, is limited by differential preservation and by our present inability to resolve fine-scale environmental variability in time and space.

Nevertheless, the use of the ecosystem concept has had many positive implications for archaeology. First of all, new data have become relevant to prehistoric behavior. Archaeological research designs now often are truly interdisciplinary and seek to include the gathering of information pertinent to the study of ecosystem traits and processes, such as the spatial distribution of environmental characteristics and the stability of their distribution through time. From this standpoint, new interactions among variables emerge and new determinants of behavioral variation may be examined.

Secondly, archaeology has moved away from the emphasis on prime movers to a systems orientation in explaining prehistoric behavior, so that relationships among multiple variables must be examined simultaneously. Monumental construction projects, for example, require the procurement of building materials and the organization of labor. The former, in turn, depends on the structure of raw material distributions and trade routes and networks, while the latter is affected by population distribution and the seasonal scheduling of subsistence activities. Neither the Pyramids of Egypt nor Stonehenge are comprehensible without consideration of such factors. Similarly the development of complex political bureaucracies can be examined in light of governmental roles in controlling the distribution of materials and personnel and in monitoring and buffering environmental fluctuations (Isbell 1978; Jorde 1977). Increasingly, an array of environmental variables has become an important set of factors in explaining cultural behavior.

A third implication of the use of the ecosystem concept in archaeological research is that many projects regularly include regional surveys and the analysis of site surroundings in order to investigate more fully the nature of human interaction with the environment. These approaches represent an improvement over the use of simple site locations as the object of study since they attempt to deal with the distribution of activities throughout the environment, rather than at a few points in space. In fact, some archaeologists now deemphasize the value of sites as analytical units, preferring to examine the distribution of materials across a larger landscape (Foley 1981; Thomas 1975). The use of techniques such as Geographic

Information Systems facilitates this broader perspective and is growing in importance, together with various spatial modeling procedures (Kvamme and Kohler 1988). Furthermore, physical techniques have been developed to help determine more precise chronological and seasonal relationships among sites. Archaeological methods are increasingly attempting to gather information about spatial and temporal variation in prehistoric behavior.

Examples: Increasingly Complex Approaches to Simple Questions

A commitment to ecological archaeology, therefore, with its unifying concept of the ecosystem, introduces spatio-temporal variability and complex interrelationships as important elements of the problems investigated. These requirements of an ecological approach have significant implications for both techniques of archaeological research and the types of explanation acceptable. In order to illustrate the research demands and the potential of such approaches, three problems of European prehistory will be discussed. In each case, alternative interpretations will be presented along with their support in the current data and their implications for future research.

1. *Development of Microlithic Technology.* The first problem concerns a change in lithic technology. Beginning in the late Pleistocene and accompanying a general miniaturization of stone implements, there begin to appear small microblades and microliths throughout much of Europe. These implements were probably inserted into shafts to form composite tools, and they increase in frequency until they form the dominant characteristic of the early postglacial Mesolithic industries. A variety of hypotheses have been advanced to explain this change in technology, but these may be simplified into three general views according to emphasis.

The first view sees this change as related to a modification of a traditionally important activity---hunting. That is, microliths are interpreted largely as points, barbs, and cutting edges forming part of the hunting and butchering equipment and thus represent functional replacements for earlier points and knives (Chard 1969:173; Clark 1967:104). Suggestions about the reasons for this replacement have been varied. Most emphasize the development of the bow and arrow as a more efficient hunting tool and its requirements for projectiles of lighter weight. Additional factors of possible importance include a need for new and more complex projectile shapes made possible by composite tools, a need for more efficient use of smaller sources of stone raw material, and the more efficient repair and replacement of composite tools.

A second view is that this technological change represents not just a modification of an existing activity but, more importantly, the development of an essentially new activity altogether---the collection and processing of plant foods (Clarke 1976). In this view, the development of microliths met the need for longer and more complex cutting edges required for harvesting and shredding activities.

A third suggestion, drawing upon recent work with hunter-gatherer ecology (e.g. Binford 1980), is that this technological change reflects largely an organizational change. Myers (1989), for example, interprets British Mesolithic technological changes in light of organizational responses to environmental shifts affecting prey distribution and regularity of movement.

In all three views, the technological change can be related to changes in the natural environmental context. With the late and postglacial warming and reforestation of central and northern Europe, mobile and gregarious animals of open country became rarer and were gradually replaced by more sedentary and solitary species of closed forests. Hunting methods and equipment may have changed accordingly, such that stalking became more efficient than driving or ambushing. If so, an advantage would have been presented by the bow and arrow with its longer striking distance and by projectiles with more complex barbs to hold fast despite snagging in the denser vegetation. Moreover, the use of arrows in forests would probably have resulted in more frequent breakage of points, so that the ease of replacing only a portion of a composite tools may have been an additionally significant factor. Simultaneously, the economic adjustments of this period, including the shift to less mobile game and a broadening of the resource base, may have led to a smaller effective range of the local groups, perhaps causing a shift to local but smaller and less abundant sources of stone.

At the same time, the postglacial warming led to an increase in the variety and abundance of plants and thus to an increase in the potential plant foods available. If this potential was utilized, then some technological accommodations may have been necessary. Since game was becoming both less productive and riskier, a shift to the more secure plant foods may have occurred. An alternative or additional response to this situation, however, could have been an adjustment in the technological organization of hunting. Portable, multi-purpose tools may have become more important as greater mobility and flexibility became more significant (Torrence 1989).

All three views have some *a priori* probability, therefore, and the existing archaeological evidence offers meager support to any. Scattered finds show microliths clearly mounted as points and barbs of arrows or embedded in the bones of animals but also hafted to form knives of unknown function. The three hypotheses, however, should have some contrasting implications that should be considered in further research. The overall role of plant foods in Mesolithic diets is still elusive, despite the development of chemical analyses of human skeletal material (Brown 1973; Price 1989). Techniques developed in Germany may allow the identification of concentrations of animal and plant fats in site sediments and could be used where possible (Rottlaender and Schlichtherle 1979). Surveys could be directed more specifically to bogs and other regions favoring preservation of plant foods. Microscopic use-wear analysis of microliths should certainly be pursued but in the future supported by experimentation with the gathering and processing of roots and other wild plant foods as well as other activities. In addition, spatial and temporal patterns of association and covariation of tools and activities should be given greater attention. Environmental reconstructions should include both animal and plant food potentials, perhaps separately. Regional surveys could then stratify according to each criterion and predictions about the relative importance of microliths in each stratum be generated, as was done for southern England (Mellars and Reinhardt 1978). Situations with low probability of plant use, such as high elevations or winter camps, could then be examined with specific implications in mind. If microliths represent largely hunting equipment, then their varying frequency in relation to other hunting equipment, such as stone points and bone harpoons, could be examined for temporal patterns. Moreover, the varying size and spatial distribution of stone raw material should be examined for support of the assumptions for greater localization. Patterns of faunal butchery and transport, as well as of settlement, need to be examined for evidence of changes in the organization of hunting.

2. *Early Neolithic Settlement Changes.* A second problem concerns the change in residential and settlement patterns as an agricultural economy and population moved from southeast Europe toward the north and west. In simplified terms, Neolithic villages of small, single-family houses were transformed into settlements of fewer but larger houses as the Neolithic spread into central Europe. Interpretation of this change is hampered by the fact that the occupation floors of all known Early Neolithic Linear Pottery houses in Central Europe have disappeared through erosion, so that the reconstruction of the houses rests solely upon evidence of postholes.

Again, simplified hypotheses about the function of these larger houses may be suggested (see Milisauskas 1978:99-105). The first is that nuclear families were still the major residential unit but that larger houses were necessary in the harsher central European climate to provide shelter for animals and storage for grain. This view places the change within the context solely of climatic differences between southeast and central Europe.

A second view is that the housing shift represents a change in the organization of the economy and family units. That is, the larger houses are interpreted as reflecting extended family organization accompanying an economic transformation. This argument considers not just the climatic differences between the two regions but rather the greater environmental context. A number of cross-cultural ethnographic surveys have suggested some contexts favoring extended family organization, including the practice of pioneering slash-and-burn farming with labor-intensive forest clearance, the integration of diverse activities or multiple garden locations within economic units, and a frontier situation with general labor shortages (Netting 1969; Pasternak et al. 1976; Thompson 1973). However poor our understanding of the Neolithic economies of southeast and central Europe, there are some indications that less sedentary agriculture and settlement characterized central Europe, although not as much so as was once thought.

Unlike the first hypothesis, this second view examines the change in housing in relation to the complex interaction of environment, economy, and settlement. Changes in population density, land availability, and labor demand become important in understanding these interactions. Moreover, this second hypothesis, unlike the first, can address other changes in the archaeological record. The major domestic animals show a shift from sheep and goats in the southeast to cattle in central Europe. The latter require more land

and perhaps more work and thus would increase the overall demand for labor, perhaps in conflict with the practice of agriculture, and thus favoring a larger economic unit. Furthermore, by comparison with the southeast, central Europe shows more evidence for joint activities in the form of communal drying pits and ovens, suggesting a change in the organization of economic units. The few true cemeteries outside of villages known for central Europe (and absent in the southeast) may represent one component of a communal system of land tenure in the context of shifting active farm plots. Recent work in Belgium reveals fortifications around Linear Pottery sites, suggesting that demands on labor for construction and for defense may have been greater than was once supposed (Keeley and Cahen 1989).

Future archaeological research into this question should take a number of courses. The analysis of soil nitrogen within houses in an attempt to locate concentrations of livestock has been used sporadically with differing results (Milisauskas 1978:105); this approach could be pursued in an attempt to find patterns of concentration. The excavation of complete Linear Pottery sites as opposed to single houses and the survey of large regions should be given high priority. This has been done in research connected with brown coal strip mining in northwestern Germany (see, for example, Kuper *et al.* 1975). Building on such research, variation among Linear Pottery sites could be examined for systematic relationships between house size and variability in climatic factors or population density, land availability, and in relation to the presence and distribution of storage pits and cemeteries.

3.   *The Localization of Palaeolithic Cave Art.*   A third problem that can profitably be examined in an ecological context is the restricted geographical distribution of cave art in the Palaeolithic. With a few exceptions, painted caves of west and central Europe are confined to the region of northern Spain and southwestern France. Neither the suitable caves nor the practice of decorating smaller, portable objects shows such a limited distribution. Ignoring earlier interpretations based on assumptions of ethnic differences in artistic sophistication and appreciation, two suggestions may be offered that attempt to place this art in an ecological context. Both view the painted caves as fixed ritual spots, and both address the question of the roles of such ritual locations rather than the specific contents of the rituals themselves.

The first hypothesis stresses one well-documented aspect of the economy: hunting. In this view, ritual served to coordinate and integrate the normally dispersed and potentially competitive groups whose periodic cooperation was necessary for communal hunting (Hammond 1974). The painted caves represent the locations of these rituals. Support is provided by the economic importance of reindeer and the small site sizes suggesting small labor pools in individual camps.

One major problem with this suggestion is that the importance of reindeer hunting extends beyond the region of painted caves to areas such as southern Germany with many cave sites but no decorated walls (Hahn 1979). Because periodic aggregation and cooperation of dispersed groups for reindeer hunting has been proposed for this area as well (Weniger 1989), a supplementary hypothesis must be proposed: that similar integrating rituals took a different form in Germany. For example, sites with impressive concentrations of portable art, such as have been excavated in the Rhineland (Bosinski et al. 1978), may represent functional alternatives to the French and Spanish painted caves as the locus of communal rituals, and the distribution of the two types of sites may be examined for similar patterning.

An alternative hypothesis focuses on the interrelationships between other economic activities, climatic and demographic changes, and patterns of land use (Jochim 1983; 1987). This hypothesis views the painted caves as fixed ritual locations serving to assert land tenure and coordinate increasingly dense and sedentary populations associated with population influx and the growing importance of salmon as a back-up resource.

Apparently coincident with the onset of cave painting is the beginning of the last glacial maximum. The worsening climate led to progressive abandonment of Britain, northern France, the Low Countries, and much of Germany. Southwestern France is the most northerly part of central and west Europe to show substantial, continuous occupation throughout the entire Upper Palaeolithic, and may have served as a refuge for northern populations. Such a population influx may have led to increased population densities, greater land scarcity, decreased human mobility, intensification of hunting and its technology, and an increased use of more reliable resources such as salmon. Studies of human skeletal material suggest that salmon may have played a small absolute role in the diet (Hayden et al. 1987), but their presence and occasional use as a back-up resource may have added considerable stability to the subsistence economy.

Future research should be directed to the determination of both reindeer hunting techniques and salmon utilization, as these vary among regions. Excavations must regularly include screening and flotation techniques to recover fish remains. Dating of the cave paintings must continue to be pursued. In addition, the spatial distribution of the painted caves might be examined in the context of environmental reconstructions based alternatively on reindeer habitat and behavior and on probable salmon abundance. The spatial relationships of the caves to one another and to other contemporary sites should also be studied since both hypotheses imply certain regularities in these patterns. Evidence for population influx into the southwest must also be sought, both in stylistic patterns of artifacts and perhaps in raw material distributions. Finally, the complex array of changes during the late glacial when the cave art ceases must be given more attention. Not only do reindeer eventually disappear, but also there is a rapid out-migration and reoccupation of central and northern Europe. Attention should also be given to southeastern Europe since it, too, may represent a potential refuge area during the glacial maximum and might show functionally similar responses to a population influx.

Obviously, the problems, the alternative interpretations, the available evidence, and the possibilities for future research have, in each example, been presented only superficially and simplistically. These topics were chosen, however, for a number of reasons:

1) They all demonstrate the necessity for dealing with *complex ecosystemic interrelationships* and patterns of spatial and temporal variation;

2) They all demonstrate the value of recent *methodological developments*, such as techniques of chemical analysis and programs of extensive regional survey;

3) They demonstrate the *wide scope* of ecological approaches in archaeology, whereby technology, settlement behavior, and even art can be examined in systemic environmental contexts;

4) They all demonstrate the utility of having *contrasting working hypotheses*, even though oversimplified, in order to isolate critical factors for study;

5) They all suggest the need for *multiple environmental reconstructions* in ecological research, depending on the problem and factors of interest;

6) They all indicate that more attention must be given to

economic and social *organization*; humans adapt by
organizing their behavior in different ways, and
archaeologists must continue to develop ways of detecting
subtle organizational differences.

Problems Facing Ecological Research in Archaeology

Despite the promise of new research approaches, archaeology faces
a number of problems in trying to examine prehistoric ecosystemic
relationships. First of all, reconstructions of prehistoric
environments depend largely upon two methods: 1) the gathering of
information about plant and animal resources from particular sites
and 2) the analysis of pollen diagrams. Neither of these is
especially suitable for reconstructing the spatial patterns of
environmental features. Materials derived from sites represent a
biased sample of only those resources of economic importance from a
surrounding region of unknown size. The selectivity of human
interaction with the environment cannot be evaluated. (This problem
is similar to that facing studies of optimal foraging when the costs
of only those resources actually gathered can be be measured).
Pollen diagrams give a representation of vegetational composition
from an area of uncertain relationship to the region of economic
importance and, moreover, provide no information about the spatial
distribution of the vegetational components. The relative patchiness
of prehistoric environments is difficult to assess. Thus, we can
identify only coarse-grained, major biotic zones and must depend upon
analogy with modern habitats to estimate the spatial distribution of
the plant and animal components.

Secondly, our control of fine-scale temporal variations is poor.
We rely largely on stratigraphic and pollen evidence to reconstruct
temporal changes in the environment. As a result, we can reconstruct
only major cycles of variation and largely miss briefer episodes and
shorter cycles. It is unfortunate that we miss much environmental
variability and uncertainty, given the growing emphasis on these
factors in explaining some aspects of behavior. People do not adapt
to average conditions, yet our environmental reconstructions tend to
produce just such an average.

Furthermore, given our poor control of fine time differences, our
reconstructions of behavior tend to be aggregates of behavior over
uncertain lengths of time. In some cases these reconstructions may
resemble a type of average: if a region had nine good years for

every one drought year, the evidence for behavior during good years might be nine times more abundant than that of drought years. In other cases, however, the extraordinary behavior, such as population aggregation or a major ceremonial event, might have high archaeological visibility, and we would wrongly consider it as either the most important behavior or part of each year's behavior. Such problems of spatial and temporal resolution present major obstacles to the reconstruction of dynamic ecological relationships in the past.

New Directions in Ecological Research in Archaeology

Despite such obstacles, there are a number of promising new directions in ecological research in archaeology. We have rightly broadened our focus from sites to larger regions; we must now rethink our techniques of environmental classification in regional surveys. Currently regional survey techniques often stratify the region according the present environmental variation, but past and present vegetational zones may not have coincided, and small differences in environmental patterning may be of enormous significance. The shifts of vegetational zones during glacial periods, for instance, may not simply have resulted in unique low-latitude tundras, but also have created much steeper latitudinal gradients in vegetation than is seen at present, providing access to a great variety of resources within a small region (see, for example, Stein and Wright 1978). In warm, arid regions slight variations in temperature or humidity may have had profound effects on vegetational distributions. For example, by converting patchy, mosaic forests and grasslands into more homogeneous woodland-savannas, such climatic changes might have led to great changes in animal and human distributions. In assessing climatic changes, therefore, archaeologists must recognize both quantitative and qualitative distributional changes and attempt to gather information at a much finer spatial scale. The use of remote sensing and geographic information systems promises to refine our abilities of spatial resolution.

A second problem with current techniques of regional stratification and survey is that the reasons for selecting criteria for stratification are frequently not clear. Often a few obvious environmental characteristics are used, such as elevation, vegetation, or proximity to water, without consideration of their relevance to the research topic (see Moran, in this volume, for a discussion of how this has affected systems ecology as well). It is doubtful that a single environmental reconstruction and classification will be suitable for all interests. Attention to large-scale elevational variation may be sufficient for discussions of transportation routes, geographic isolation, or altitudinal effects of climatic conditions on different crops, whereas small-scale topographic investigations may be necessary for examinations of drainage problems and field distributions. We must specify the necessary environmental information to be collected, with the realization that different research topics may demand different criteria, so that the environmental variables used in one valley survey may be of little value to another study.

Finally, perhaps the most important focus of future research should be the differences in scale of temporal resolution in archaeological and modern studies. As mentioned earlier, archaeological reconstructions of both natural environments and human behavior tend to produce only gross patterns that ignore fine-scale temporal variation. The study of tree-rings, varves, and microstratigraphy promise to increase our powers of resolution but may be applicable only in restricted contexts. Similarly, temporary hunter-gatherer sites may offer greater potential than permanent sites for the isolation of brief episodes of behavior, but the determination of the length of occupation and number of episodes is notoriously difficult for such sites. Our control of chronology is crude and our reconstructions of past ecosystemic relationships limited to the identification of relatively gross correlation of variables---a *macro* palaeoecology.

This situation is indeed unfortunate because our interpretations of such correlations derive largely from modern ethnographic studies. These studies monitor short-term behavior on a small scale; they are fine-grained studies of *micro* ecological relationships. Our vocabulary and entire mode of reasoning derive from such studies, and yet the archaeological data are not appropriate to such reasoning. If we derive hypotheses from such fine-grained studies, their implications may be archaeologically undetectable.

Clearly, we need to bridge this gap in scale of analysis. One approach is to turn to modern aggregate data that may be more comparable to the sort of archaeological data we can obtain. That is, we could turn to the historical past as well as to the ethnographic present (see Ellen, Butzer, Hastorf, and Netting, in this volume). We might sample the historical record at intervals and attempt to reconstruct an interrupted "archaeological" picture of changing ecological relationships. The underlying dynamic causes of observed changes could then be suggested on the basis of additional, more detailed historical and ethnographic evidence. Another approach is to use simulations and other modeling techniques to create aggregates of small-scale behaviors. Bettinger (1977), Mithen (1988), Thomas (1973), and Wobst (1976) provide examples of such simulations, but the potential of this approach for allowing the creation of different sorts of archaeological record remains great. Such approaches might help us to bridge the gap between fine-scale processes and large-scale patterns, between a micro- and a macro-ecology of past human behavior.

References Cited

Bettinger, R.
   1977   Aboriginal Human Ecology in Owens Valley: Prehistoric Change in the Great Basin. *American Antiquity* 42:3-17.
Binford, L.
   1980   Willow Smoke and Dogs' Tails: Hunter-Gatherer Settlement Systems and Archaeological Site Formation. *American Antiquity* 43:4-20.
Bosinski, K. (ed.)
   1978   Geowissenschaftliche Untersuchungen. *Goennersdorf*. Wiesbaden: Franz Steiner Verlag.
Brown, A.
   1973   *Bone Strontium Content as a Dietary Indicator in Human Skeletal Populations*. Ph.D. dissertation, University of Michigan, Ann Arbor.
Chard, C.
   1969   *Man in Prehistory*. New York: McGraw-Hill.
Clark, G.
   1967   *The Stone Age Hunters*. New York: McGraw-Hill.
Clarke, D.
   1976   Mesolithic Europe: the Economic Basis. *Problems in Economic and Social Archaeology*. Edited by G. Sieveking, I. Longworth, and K. Wilson, pp. 449-481. London: Duckworth.

Foley, R.
  1981   *A Model of Regional Archaeological Structure.*
          Proceedings of the Prehistoric Society 47:1-27.
Hahn, J.
  1979   Essai sur l'ecologie du Magdalenien dans le jura
          souabe. *La Fin des Temps Glaciaires en Europe.*
          Edited by D. de Sonneville-Bordes, pp. 203-213. Paris:
          CNRS.
Hammond, N.
  1974   Palaeolithic Mammalian Faunas and Parietal Art in
          Cantabria: A Comment on Freeman. *American*
          Antiquity 39:618-619.
Hayden, B., B. Chisholm, and H. Schwarcz
  1987   Fishing and Foraging: Marine Resources in the Upper
          Paleolithic of France. *The Pleistocene Old World:*
          Regional Perspectives. Edited by O. Soffer, pp.
          279-292. New York: Academic Press.
Isbell, W.
  1978   Environmental Perturbations and the Origin of the
          Andean State. *Social Archaeology: Beyond*
          Subsistence and Dating. Edited by C. Redman *et*
          al., pp. 303-313. New York: Academic Press.
Jochim, M.
  1983   Palaeolithic Art in Ecological Perspective.
          Hunter-Gatherer Economy in Prehistory. Edited
          by G. Bailey, pp. 212-219. Cambridge: Cambridge
          University Press.
  1987   Late Pleistocene Refugia in Europe. *The*
          Pleistocene Old World: Regional Perspectives.
          Edited by O. Soffer, pp. 317-331. New York: Academic
          Press.
Jorde, L.
  1977   Precipitation Cycles and Cultural Buffering in the
          Prehistoric Southwest. *For Theory Building in*
          Archaeology. Edited by L. Binford, pp. 385-396.
          New York: Academic Press.
Keeley, L. and D. Cahen
  1989   Early Neolithic Forts and Villages in NE Belgium: A
          Preliminary Report. *Journal of Field Archaeology*
          16:157-176.
Kuper, R. *et al.*
  1975 Untersuchungen zur neolithischen Besiedlung der
        Aldenhovener Platte. *Bonner Jahrbucher* 175:191-229.

Kvamme, K. and T. Kohler
1988 Geographic Information Systems: Technical Aids for Data Collection, Analysis, and Display. *Quantifying the Present and Predicting the Past.* Edited by J. Judge and L. Sebastian, pp. 493-548. Denver: U.S. Department of the Interior, Bureau of Land Management.

Mellars, P. and S. Reinhardt
1978 Patterns of Mesolithic Land-Use in Southern England: A Geological Perspective. *The Early Postglacial Settlement of Northern Europe.* Edited by P. Mellars, pp. 243-294. London: Duckworth.

Milisauskas, S.
1978 *European Prehistory.* New York: Academic Press.

Mithen, S.
1988 Simulation as a Methodological Tool: Inferring Hunting Goals From Faunal Assemblages. *Computer and Quantitative Methods in Archaeology 1987.* Edited by C. Ruggles and S. Rahtz, pp. 119-137. Oxford: BAR International Series 393.

Myers, A.
1989 Reliable and Maintainable Technological Strategies in the Mesolithic of Mainland Britain. *Time, Energy and Stone Tools.* Edited by R. Torrence, pp. 78-91. Cambridge: Cambridge University Press.

Netting, R.
1969 Ecosystems in Process: A Comparative Study of Change in Two West African Societies. *Ecological Essays.* Edited by D. Damas, pp. 102-112. Ottawa: National Museums of Canada Bulletin No. 230.

Pasternak, B., C. Ember, and M. Ember
1976 On the Conditions Favoring Extended Family Households. *Journal of Anthropological Research* 32:109-123.

Price, T.
1989 The Reconstruction of Mesolithic Diets. *The Mesolithic in Europe.* Edited by C. Bonsall, pp. 48-59. Edinburgh: John Donald Publishers Ltd.

Rottlaender, R. and H. Schlichtherle
1979 Food Identification of Samples from Archaeological Sites. *Archaeophysika* 10:260-267.

Stein, J. and H. Wright, Jr.
    1978   Review of Amerinds and Their Paleoenvironments in Northeastern North America. *Science* 200: 306-307.

Thomas, D.
    1973   An Empirical Test for Steward's Model of Great Basin Settlement Patterns. *American Antiquity* 38: 155-176.
    1975   Nonsite Sampling in Archaeology: Up the Creek Without a Site? *Sampling in Archaeology*. Edited by J. Mueller, pp. 61-81. Tucson: University of Arizona Press.

Thompson, S.
    1973   *Pioneer Colonization: A Cross-Cultural View.* Addison-Wesley Modular Publications 33. Menlo Park: Cummings Publishing Co.

Torrence, R.
    1989   Re-Tooling: Towards a Behavioral Theory of Stone Tools. *Time, Energy and Stone Tools*. Edited by R. Torrence, pp. 57-66. Cambridge: Cambridge University Press.

Weniger, G.
    1989   The Magdalenian in Western Central Europe: Settlement Pattern and Regionality. *Journal of World Prehistory* 3:323-372.

Wobst, H.
    1976   Locational Relationships in Paleolithic Society. *Journal of Human Evolution* 5:49-58.

CHAPTER 4

A HUMAN ECOSYSTEM FRAMEWORK
FOR ARCHAEOLOGY

Karl W. Butzer

Anthropological or Interdisciplinary Archaeology?

Archaeology is, by its nature, ultimately empirical. But archaeologists strive to go beyond the elucidation of material culture by attempting to identify social aggregates and to generate evidence of, and explanations for, change. Considering the difficulty of understanding group ecological behavior in a contemporary setting, or of identifying cause and effect relationships in the historical record, this seems an ambitious and challenging goal. But like many other social scientists, archaeologists have proposed to achieve such ends by borrowing theoretical concepts from the natural sciences, so as to develop and ostensibly to test hypotheses, based on a particular set of assumptions. Such assumptions are justified by disciplinary "mapping" of familiar, empirical configurations which, in the case of archaeology, are generally grounded in cross-cultural, ethnographic experience.

During the last 30 years, archaeology has moved from an essentially empiricist stance to an explicitly theoretical paradigm, with a strong, positivist and quantitative bias, only to be followed in recent years by the post-modernists' anti-positivistic reaction. These philosophical shifts parallel those in other social sciences. Apart from their direct significance for the formulation and execution of archaeological research, they affect the interdisciplinary relationships of archaeology. In the case of American archaeology, its academic context has been within the field of anthropology, with archaeology representing a subdiscipline, explicitly or implicitly charged with the temporal elaboration of anthropology. This has the advantage of exposing archaeology to the intellectual cross-currents of anthropology. But it has also been disadvantageous, exposing archaeology to disciplinary fads and limiting effective contacts with other sciences.

An example of this is the long preoccupation of archaeologists with site-specific excavations, reflected in a closed-system approach, similar to that adopted by decades of ethnographic case studies and even early examples of energetics research (see Rappaport 1968). Archaeological survey of larger areas was considered a revolutionary innovation, and still is seen by some as a deviation from a "purer" form of research. The open-system perspective can dramatically change macro-scale interpretation, as Netting (this volume) and Adams and Kasakoff (1984) show. The traditional, disciplinary penchant for closed systems not only contrasts the perspectives of anthropology and human geography, but also helps explain why these disciplines communicate today so poorly.

In regard to ecology, the approaches of anthropologists and geographers complement each other. Anthropologists tend to be interested in the processes and structures whereby human groups match resources with their needs, and incorporate them into cultural behavior (Steward 1955; Barth 1956; Geertz 1963; Rappaport 1968; Orlove 1980; Ellen 1982; Adams 1988; Moran, 1982 and in this volume). Geographers, on the other hand, focus on a broad sphere of interaction with respect to resources, emphasizing the spatial matrix of the socioeconomic and biophysical environment (Brookfield 1964; Stoddart 1965; Clarkson 1970; Chorley 1973; Butzer 1976, 1982, 1989, n.d.; Denevan 1983). Evidently, anthropology and geography share many basic interests, at least at the macro-scale, and they have much to learn from each other. This is particularly so for archaeology, where macro-scale implies resource interdigitation, spatial dynamics, and long-term transformation of cultural systems.

Archaeologists generally accept that geography provides useful data and techniques, but rarely learn how to incorporate them effectively. One reason for this is that their intellectual priorities leave them insufficient time to acquire the requisite experience and technical skills. These factors may account for the frequently simplistic treatment of environmental variables in archaeology. The empiricists of the 1950s used environmental reconstruction or biological identifications with genuine interest but little sophistication. Their positivistic counterparts of the

1960s and 1970s actively explored concepts from biology and human geography (see Clarke 1968), but most eventually failed to assimilate them as more than mechanical constructs or peripheral variables, ultimately to be held constant (e.g., Gould 1978; Renfrew et al. 1982; Watson 1986; Schiffer 1987). The post-processualists simplified their search by abdicating interdisciplinary perspectives and downplaying the significance of socioeconomic change, in favor of social context and the symbolic meaning of data (e.g., Hall 1977; Hodder 1982; Miller and Tilley 1984; Leone 1986).

The argument of this paper is that archaeology cannot hope to achieve the degree of cultural resolution demanded by sociocultural anthropology. Firstly, a more productive eco-systemic focus is proposed for archaeology. Secondly, the limitations of archaeological inference are illustrated by historical cross-checking at the general and the detailed level. This is followed by a critique of how different levels of archaeological inference are constructed. Finally, a selection of fundamental questions is identified, and formulated in systemic terms, that demonstrate the vital importance of archaeological research for the social sciences.

The Human Ecosystem as a Focus for Archaeological Research

Both the advantages and difficulties of the ecosystem concept have been discussed by Moran (see introduction to this volume). As he defines it, the ecosystem refers to the structural and functional interrelationships among living organisms and their physical environment. Transfer of the concept to the social sciences requires several explicit restrictions. It is not a concrete unit of analysis but a dynamic perspective that facilitates the articulation of complex, interdependent relationships, characterized by positive and negative feedbacks, and variable equilibrium properties. It also provides insight into long-term changes, thresholds, "simplification," or "catastrophic" readjustment. But simulation or even quantification of entire ecosystems may be impossible. More fundamental is that human ecosystems differ from biological ecosystems in kind as well as degree (Butzer 1982:32). Information, technology, and social organization play inordinately greater roles in human ecosystems, and a primary "regulator" function must be introduced to accommodate the "steering" role of human cognition, value systems, and goal orientation (Bennett 1976: ch. 3). In doing so we are not creating a dichotomy between human beings and nature, but rather singling out one human component--the mentalistic process--from the energetic processes of the system (Adams 1988:89).

Within these specific conditions, it is possible to accept the definition that *human ecosystems* represent the interlocking of social systems with ecosystems (Chorley and Kennedy 1971:4). For archaeology, a practicable general goal is to study and elucidate the archaeological data base, as derived from particular sites or site networks, as part of a human ecosystem. Such a focus serves to draw attention to the systemic interactions among cultural, biological, and physical factors or processes (Butzer 1982:7; see also Jochim, this volume).

The human ecosystem concept can be explicitly extended to include the roles of individual and aggregate human behavior, decision-making with respect to alternative possibilities, screening by the experience and deeper values encoded in culture, as well as the fundamental tension between individual goal-conflicts and human unpredictibility (see also Rappaport, this volume). These powerful variables for change vie with the community and institutional structures that favor stability, to maintain a fundamental tension that is incompatible with a homeostatic, ahistorical view, even in the absence of external perturbations (Butzer n.d. 1).

The human ecosystem, as more broadly defined above, has great heuristic value to examine socioeconomic behavior, but its applicability for strictly systemic goals is limited. Complex systems are almost impossible to simulate effectively, as exemplified by the failure of almost all economic prognoses. Prediction, whether of long-term evolutionary change or of rapid modification, is difficult, even in probabilistic terms. Retrodiction is almost as difficult, with historical processes remarkably intractable to generally accepted modes of explanation.

The value of the human ecosystem as a framework for archaeological research is explicitly conceptual. It is an interdisciplinary perspective that readily encompasses spatial variables, differences of scale, complexity and complex interactions, as well as equilibrium modes, including adaptive change and evolutionary transformation (Butzer 1982:7-11; 286-313). It is also flexible. In situations where the detail of sociocultural resolution is poor, it can be used in its narrow sense to deal primarily with socioeconomic inferences. And in cases where such resolution is good, it can be applied in its expanded form to incorporate a range of cognitive concerns.

Lessons from Historical Archaeology

Without contemporaneous, written records, and working with configurations well removed from the ethnographic "present," archaeologists can only postulate how economic structures functioned, and they have little prospect of grasping the values and goals of a society by whatever leap of ethnographically-based optimism. These assertions can be justified by two examples from historical archaeology.

Historical archaeology is a relatively little understood subfield, that receives only grudging National Science Foundation support. To North Americanist archaeologists it represents research in very late time ranges, usually Colonial or Industrial era sites. Many such historic sites are excavated in the course of salvage projects, and some involve only a single high-rise foundation in an inner-city area. Historical documentation here represents supplementary information, that rarely elucidates fundamental questions about cultural activities, since time and place are already well understood. There are no big surprises, as a rule, from either the historical or archaeological side, and relatively few, basic contradictions arise. Excavation discoveries mainly serve to enhance an already richly-textured historical appreciation. A different twist is provided by excavation of historic Indian sites, to ask new questions and discover new insights, independent of Euroamerican written records (Trigger 1980).

In the Old World, the perspective is very different, because a potential time range of 5000 years is involved in some areas, and because the cultural record prior to the 16th century is incompletely and imperfectly understood. The history of study of ancient Near Eastern civilizations, or of the Greek and Roman worlds, reveals a genuine complementarity between excavation results, on the one hand, and inscriptions, historical texts, or archival documents on the other. "Classical" archaeologists and philologist-historians, long ago converged their attention on problems of common interest. Since each field worked with a different set of assumptions, neither of them foolproof, the net effect was to mutually correct flaws in interpretation.

A question that has never been raised is whether "anthropological" archaeology by itself could have done (1) a credible job in reconstructing the broad outlines of cultural configurations in the Mediterranean Basin and Europe prior to A.D. 1500, and (2) offered reasonably correct interpretations for observed changes? My reply to the first part of the question would be a qualified yes, to the second, a definite no. This opinion is based on a long-term, macro-study of ancient Egypt and an equally protracted micro-study of a small ecosystem across a thousand years in a mountain cluster of eastern Spain.

Limitations of the Archaeological Record for Ancient Egypt

The historical record of Egypt has, since classical times, been divided into three eras, eventually known as the Old, Middle, and New Kingdoms. They were followed by 2500 years of foreign domination, interrupted by only one episode of independence. This cyclicity of the Egyptian historical record poses particularly interesting questions in regard to what I would call system growth and periodic discontinuity, superimposed on a trajectory otherwise characterized by socioeconomic continuity. Over some 30 years I have been periodically engaged in archaeological survey or excavation in Egypt, complementing this experience by in-depth examination of the textual evidence, in translation. From this I have developed a series of theoretical, regional, or thematic publications addressing subsistence-settlement change, socioeconomic developments, recurrent ecological crises, and systemic response (see Butzer 1960, 1976, 1980b, 1984a, 1984b). The following conclusions are drawn from this background.

In regard to basic configurations, even without the ability to decipher inscriptions, the shifts of royal residence from the Old to Middle Kingdom would have alerted archaeologists to a major dynastic change, and the progressive impoverishment of elite tombs during the late Old Kingdom would have signalled economic problems. The Middle Kingdom breakdown, resulting from the Hyksos invasion, would not have been recognized, although Asiatic settlement sites in the eastern Delta might have raised postulates about a temporary loss of that region. The Ethiopian, Assyrian, and Persian conquests or occupations would have remained unknown, since there is no evidence other than texts or inscriptions in Egyptian. The New Kingdom control of Syria-Palestine would have been interpreted as nothing more than trade contacts, with Egyptian commercial outlets at two Lebanese ports.

At a more cognitive level, the appreciation that the same elite families controlled and deliberately held Egypt together as a nation from the Old to the New Kingdoms, despite dynastic replacement and changes of the official deities, would have been lost, leading to gross misconstructions. The multitude of factors responsible for cyclic discontinuity—dynastic fragility, bureaucratic inadequacy or corruption, administrative dysfunction, social change, foreign invasion, Nile failure--would not have been grasped. Instead, simplistic interpretations would have been touted every few years, without an appreciation for how the system really worked or why it periodically faltered or recovered.

On the other hand, historical study alone would have been sterile indeed without the unusually informative legacy of architecture and artistic expression, or the record for progressive social and technological change evident in the archaelogical record. Equally fundamental is that the Old Kingdom capital had at most 15,000 inhabitants, with less than 50,000 people nationwide living in towns of 1,000 or more; during the New Kingdom there were two capitals, with 150,000 people each, and a half-million urban residents overall. The historical evidence by itself did not suggest substantial differences in urbanization or socioeconomic complexity.

These examples from Egypt, deliberately chosen at a macro-scale, illustrate the severe limitations of both archaeology and history if they were theoretically isolated.

Limitations of the Archaeological Record in a Spanish Micro-Study

These problems carry right down to the detailed level of standard archaeological interpretation. How fundamental the problems are became evident during the course of a project in the Sierra de Espadán, 50 km north of Valencia, Spain. Settled by Muslims around A.D. 1100, this area was reconquered by Christian Aragón A.D. 1238, but the Muslims were allowed to remain until their expulsion and replacement by Christian settlers A.D. 1609-10.

The Espadán Project was focused on one municipality, Aín, but was extended to exploration of a dozen others, with discovery and recording of 15 "lost" but historically-known Medieval villages. One such satellite village, called Beniali, and belonging to Aín, was occupied 1342-1526. It was partially excavated during two seasons. Another excavation was devoted to the adjacent castle of Aín, occupied during the 12th and 13th centuries, as was a second castle, used as a Bronze Age and then as a Muslim settlement site, and converted into a Christian fortress during the 14th century. The project included detailed archival documentation of local Medieval life, ethnographic research, and delineation of social and land-use history from the 12th century to the present (Butzer et al. 1985, 1986, 1989; Butzer and Butzer 1989; Butzer n.d.1; Butzer and Ferrer 1987).

The objective of the Espadán Project was not simply to "understand" continuity and change in a unique setting, but to consider the micro-region as a laboratory to explain the interplay of culture-ecological variables involved in the social maintenance of resilient minority communities and their succeeding Christian counterparts, in terms of socio-cultural dynamics and adaptive strategies. The discussion that follows relates to a by-product of this project--the unexpectedly valuable complementarity of archaeological and archival research.

Given two good, parallel records, from excavation and from the archives, it was for the first time possible to compare such records to examine whether and how archaeological, hypothethico-deductive procedures of the 1970s would have succeeded or failed. Several examples can be singled out. Some are relatively trivial but instructive; others are fundamental.

The castle of Aín was abandoned in the 13th century, according to the pottery and geoarchaeological evidence, only to be soon but briefly reoccupied by common villagers, then besieged and destroyed. Given the general historical context, <u>one would have "predicted" that the nearby people of Aín sought refuge in and defended the castle during the great Muslim revolt against Aragón 1276</u>. *Instead, documents demonstrate that Aín was one of the villages that did not participate in this revolt: the rebels besieged here came from other places.*

<u>The founding of the satellite village of B8eniali,</u> deep within Aín's municipal lands and directly above its major water source, logically <u>suggests a daughter colony,</u> set up in greater proximity to relatively distant fields. *But instead, a document indicates that another town, across the mountains, had received a royal charter to colonize the site.* Further, the surnames of the 15th century inhabitants verify that the settlers did not come from Aín, but from several other Sierra villages, and considerable rivalry between Aín and Beniali is witnessed by litigation records.

Excavation showed that Beniali consisted of about two dozen thatched, but mortared, stone houses with average interior dimensions of only 11.2m^2. So small, and located in the mountainous hinterland of a rural village with less than 100 families, <u>one would "expect" a high degree of isolation from the outside world and a homogeneous farming economy.</u> *Yet 13 tax-exemptions (franquicias), essentially licenses for commerce, were granted to residents of Beniali in the early 1400s, and one of them was a Jew.* In the case of Ain, we know that some of its residents actually did craftwork on contract in the distant city of Valencia. That Beniali was neither isolated in an economic or social sense, nor socioeconomically homogeneous, is supported by other lines of evidence.

<u>The abundance of luxury pottery, imported from ceramic centers 80-120 km away, would normally be explained by indirect trade, via middlemen and local market centers.</u> *However, in one case a resident of Ain posted bail for a potter in an 80 km-distant pottery center, indicating much more intimate contacts, once again arguing against social isolation.* The distribution of luxury wares among the structures of Beniali shows that the number of decorated sherds is proportional to pottery abundance, and does not allow the inference of a concentration of wealth in any particular house or zone. *Yet the archives show that eight families received commercial licenses while two or three others, at various dates, were indigent.* <u>The material record failed to demonstrate this differentiation of status and wealth.</u>

<u>Parish registers after 1624 show a mean radius of 2.7 km for exogamous marriages</u> in the case of Christian Ain, concentrated in two other villages, suggesting a logical ethnographic premise. *However, comprehensive study of hundreds of patronyms in 14 local villages between 1380 and 1563 suggest that the Muslims of Ain shared mates with five villages, at an average distance of 8.9 km.* In effect, Muslim family networks were much larger and involved a radius three times as large. By extension, *the Muslim villages maintained wider and tighter interconnections than did Christian ones.* <u>This could not have been anticipated by normal archaeological procedures.</u>

Abundant ethnobotanical materials from the Beniali excavations include "hard" and bread wheat, sorghum, oil seeds, green beans, olives, almonds, and various orchard fruits. <u>Against the "predictable" inference, that the residents of Beniali were</u>

self-sufficient, Mediterranean-style farmers, stand *numerous documents from the 15 km-distant market center of Onda, verifying that residents of Aín and other nearby villages frequently purchased large quantities of grain there on credit, and sometimes even oil or beans.* This startling fact reveals that the Sierra was not self-sufficient in staples, possibly selling only orchard products and flax (not present at Beniali, but verified by its commensal, weedy oil plant, Camelina). Further light is provided by private economic records of the early 1800s, showing that grain was purchased at the market centers in some years, sold there in others. Hypothetico-deductive procedures from an exclusively archaeological data base would have offered an erroneous, or at least inadequate, interpretation.

A rich archaeozoological inventory indicates that goats represented the main livestock, but that female goats were slaughtered at a somewhat earlier age than would be ideal for optimal herd reproduction, a trait symptomatic of people living near the hunger mark. *Archival resources shed a great deal more light on this phenomenon. The annual tithe income from flour mills and from the butchery in a larger, adjacent village show that more animals were slaughtered in those years when less grain was ground: goats were mainly eaten during poor harvest years.* This is indeed consonant with ethnographic experience, but the dietary implications could not have been specifically inferred from the excavation data. The convergent lines of evidence underscore the marginal character of the resource base deduced from site catchment analysis: only 24 ha of cultivable land (10% of that potentially irrigated) for 20 to 25 families.

Finally, the archaeological record provided minimal evidence for differences between Muslim villagers and Christian castellans of the 14th century. Both used the same pottery, with luxury classes produced in a Valencian center where Muslim potters crafted wares steeped in Islamic tradition, but symbolically modified to incorporate popular, Gothic themes. Such pottery was used in the region by both ethnic groups, but also exported to both Christian France and Muslim Egypt. The only hint of a difference in preference was the presence of a Gothic "rose" in a Christian castle level--a specifically Christian symbol--and a highly stylized mano de Fatima, (a hand to ward off the evil eye), in a Muslim village level. *This suggests that stylistically-informative pottery is not necessarily as specific to identity conscious groups as it is sometimes claimed to be.*

These discordances between, or lacunae of the archaeological and archival records were only resolved when the two sources of information were combined. In isolation, both the economic and social inferences that would normally be drawn from the archaeology would, in several significant ways, be either wrong or misleading. These caveats go well beyond the so-called cautionary tales about interpreting micro-activities from artifactual distributions (see Gould 1978). They raise serious questions about the reliability of seemingly logical conclusions drawn from deductive socioeconomic principles that have been taken for granted. Post-processual archaeologists have begun to appreciate the difficulties of generalization from universal categories, but primarily for theoretical reasons. These case studies also show that matters of cognition are almost hopelessly elusive to archaeological investigation alone.

This is not to suggest that the archaeology of the Espadán project was in any way inconsequential. It was essential, in that it elucidated living arrangements and subsistence patterns only vaguely hinted at in the written documents, thus illuminating the material culture, the way of life, the basic social structure, relative values placed on material goods, and the overall poverty of the Benialí community. The archival documents, however, revealed the people as individual actors in the dynamics of local and regional processes or events. The archival sources also demonstrated direct contacts with the world outside of the Sierra, implying a vision of and an active participation in a larger, regional arena.

An Interim Assessment: The Emperor Has No Clothes

The lesson to be drawn from these Egyptian and Spanish examples is that archaeology can never be a direct analog to ethnography, to be achieved by excavation, regional survey, and sophisticated deductive models. Archaeology simply cannot provide the range of sociocultural information possible in an anthropology of contemporary populations. Socioeconomic theories, as well as ethnographic inferences from living peoples, must be treated with much greater caution than has usually been the case.

In further suggesting that it is counterproductive to place social context and, especially, meaning and symbolism at the center of archaeological aspirations, I risk being compared with Hawkes (1954). But having recently been labelled as both a logical positivist and a hyperparticularist, by different archaeologists, I have little to lose to reductionism. While the "real past" is not inaccessible (to apply the terms of Watson 1986), substantial and consequential parts of that past are. The primary value of "structural archaeology" is that it serves to draw attention to prehistoric ideological or cognitive systems. These are legitimate concerns for archaeology, in as far as the possibility of at least sketching their outlines as potential variables in broader interpretation must be explored. Individual contributions to this genre of research also demonstrate that attention to culturally-sensitive phenomena allows valuable insights (e.g., Hodder 1984; Pearson 1984; Young 1988; Hastorf, this volume). But the limitations of prehistoric archaeology are such that cognitive dimensions can never represent primary concerns, and it is improbable that they could be "reconstructed" as more than component parts of larger, mainly hypothetical structures.

A Critique of Positivist Theory and Inference in Archaeology

Since about 1960, archaeologists have engaged in, or been subjected to, a barrage of epistemological discussions as to how to transform archaeology into a science with a body of "laws." The wearisome monologues on what is a science, whether the social sciences qualify as sciences, and whether archaeology can be made into a science, became more tempered as disillusionment set in during the mid-1970s. Although this initiation to positivism exacted a high cost on civility, archaeology has indeed evolved into a more pluralistic discipline, with broader horizons, a firmer grasp of how analysis is linked to inference, and developed a more explicit problem formulation.

Unlike the philosophers of science and social theorists that some try to imitate, archaeologists have a substantial body of data at their disposal. Their primary concern remains the elucidation of those data, and deliberate reflection on the premises upon which interpretation is offered. Such premises involve assumptions and working models ("theory"), that facilitate (rather than constitute) "explanation." Some models form testable hypotheses and may be converted into generalizable statements or "laws." Other models have intuitive appeal but cannot be verified to general satisfaction, a common problem in the social sciences. Eventually a body of laws and models, based on "behavioral correlates" derived from observed, contemporary contexts, is applied to reconstruct the nature and functioning of past societies.

Since the laws and working models in use commonly relate to lower levels of inference, mainly built directly upon empirical data (the "middle range theory" of Binford, 1989 and earlier works), legitimate questions can be raised whether archaeology has assembled a sufficiently comprehensive body of "theory" to allow high-level reconstruction. To illustrate the nature of this problem, it is essential to examine the several levels of inference integral to such a process.

Low-Level Inference

At the *micro-level*, archaeologists deal with artifacts (*sensu lato*), artifact-patterning, and matters of basic intrasite or intrastructural activity-patterning. Perhaps 95% of the empirical and analytical work of archaeology is expended at this level, and this is what archaeologists do best. From an era of unrealistic confidence in the 1960s that the reconstruction of activity-patterning could be realized within reasonable limits of confidence, archaeologists have learned that such limits must be defined in more elastic terms. Although they refuse to consider that site formation is firstly a geoarchaeological phenomenon (Butzer 1982:chaps.6-7; Rosen 1986), there is increasing acceptance of the disturbing reality that site residues hardly ever record a freeze-frame of representative behavioral activity. That record is normally distorted by natural or cultural processes or both, leading to efforts to filter out that distortion to recover valid behavioral information (e.g., Schiffer 1987).

The problem is not insurmountable if we are prepared to scale back expectations, and emphasize less specific goals: even a disturbed context can be highly informative. Ethnoarchaeological observation and experimental work have similarly shown that artifacts and non-artifactual residues of cultural activities do not allow foolproof inferences as to specific activities or more generalized economic behavior (Gould 1980,1985). Most archaeologists no longer are dogmatic at this level. This has not minimized their research, merely cast their low-level inferences in a more appropriate framework. Others of the social sciences have hesitantly reached the same conclusions: verification in the social sciences is fundamentally different than in the physical or natural sciences, where the number of variables can be controlled and experiments can be replicated at will. For that reason, "laws" in the social sciences rarely are absolute or universal.

Mid-Level Inference

At the *meso-level*, archaeologists attempt to move from the particular to the general, employing site-specific data and related, basic inferences to reconstruct socioeconomic patterns. Here archaeology is on less stable ground and there is need for a healthy level of skepticism. The Espadan examples illustrate that least-effort and other "rational" inferences may be misleading or absolutely wrong. This should not be surprising, since contemporary research in the social sciences contains numerous examples that "satisficing" solutions are more common that "optimal" ones, and that people choose between alternatives and that their strategies vary over time. As anticipated by Clarke (1968:79), the principle of Occam's razor is unacceptable in sorting out inferences: the most parsimonious explanation was seldom right in the Espadan, and "reality" was always more complex.

For many authors, applications of the scattered and inhomogeneous body of what is sometimes called "middle range" theory can be extended upward to include synchronic, socioeconomic reconstructions at the meso-level. A particularly poor assumption in such chains of argumentation is invocation of the principle of uniformitarianism. This geological tool applies only to physical and chemical processes, not to human behavior. The assumptions according to which contemporary examples are chosen may also be wrong (Denbow and Wilmsen 1986). Going beyond the level of linking artifacts and artifact-patterning with direct human activity also underestimates the critical matter of cognitive behavior, alternative choices, and multiple means to achieve a particular goal. From probabilities this moves inference into the realm of possibilities. Despite the disclaimer that even in its heyday, positivistic reconstruction strove only to reduce uncertainty (Watson 1986), occasional qualifiers went largely unnoticed amid the overbearing rhetoric. While the potential margin of error in Palaeolithic archaeology may be acceptably small to some, that is certainly not the case in younger time ranges. Here the Egyptian and Espadan examples show that probabilities of distortion or gross error are a more appropriate frame of reference than "levels" of uncertainty.

At the meso-level we are ultimately left with little but models or hypotheses, that do not appear to be testable in prehistoric contexts, at least not without affirming the consequent. There is a deep chasm between a reasonable interpretation of micro-archaeological activity patterns, on the one hand, and socioeconomic reconstruction, on the other, one that is not bridged by any "robust" laws. Inferring degrees of social differentiation from archaeological data is one thing. Identifying a particular kind of social hierarchy is quite another.

The basic difficulty that archaeology has with verifiability at the meso-level is that complex, synchronic "structures" cannot be reliably inferred from a limited number of simpler components. (Re)constructing the whole requires a new set of laws that must in turn be verified. In the synchronic social sciences this may well be practicable, since the difference is primarily one of integration and scale. In a diachronic context, the interrelationships of a much larger number of variables are not covered by the original, micro-level laws.

The grounds for skepticism increase exponentially when, at the same meso-level, some archaeologists move even beyond socioeconomic reconstruction into the sociocultural or sociopolitical realm. This may be warranted in situations just beyond the ethnographic "present," or in protohistorical contexts. At greater distance in the past, anthropological preconceptions as well as Western biases must be circumvented before even the premises for an explicit model are decided upon.

There are also inherent difficulties as to how culture traits, cultures, and ethnic (self-)identification are linked. Archaeologists commonly assert that material culture is not equivalent to culture, but in practice they do not necessarily adhere to that maxim. In Africa, Hodder (1981) discovered that ethnic boundaries do not coincide with those of material culture or particular spatial macro-patterns. Indeed, even in modern Europe neither ethnic nor linguistic boundaries are tied to readily discernible differences in material culture, and correlations of specific economic strategies or sociocultural (e.g., kinship or inheritance) patterns with ethnic identity only apply in local areas, and cannot be generalized to other segments of such boundaries. In this perspective, the interpretation of large-scale archaeological "components" becomes especially problematical, in default of primary attention to multiple criteria sensitive to identity consciousness.

Protestations to the contrary, too many archaeologists are still willing to ascribe particular social hierarchies and inferred political structures to such "components," e.g. to the Hopewellian or the Mississippian. This is an implicit expression of belief that an area with shared material culture has sociocultural meaning. In general, the problems of cultural or political reconstruction have not yet been adequately addressed or resolved.

Macro-Interpretation

At the *macro-level*, in the leap from "reconstruction" to the interpretation of process and change, all the inherent problems of hierarchical inference are compounded. This can indeed be minimized by selecting partial problems and limiting the number of variables. Process and change can then be attacked with some success, particularly in arenas with a strong empirical base. Examples include Paleolithic subsistence change (Klein 1978,1988), agricultural transitions (Ford 1985) or intensification (Sherratt 1983; Barker 1985), and cycles of settlement expansion and contraction (Adams 1981; Van Andel and Runnels 1987).

Cultural evolution and the origin of complex societies represent the greatest and most difficult challenge for archaeology. Evolutionary concepts are indispensable to the historically oriented social sciences, and underlie most medium- and long-range adaptive processes (Kirch 1980a) in, for example, cultural ecology. The problem in applying archaeology to cultural evolution begins with the premise of evolutionary classification. A group of attributes is directly, or by seriation of sociocultural adaptations, applied to delineate a succession of stages (e.g., the "big-man" collectivity, simple or complex chiefdom, archaic state) to arbitrarily subdivide what is a complex, systemic continuum. Such a directional and ahistorical scheme is evidently based on the normative concepts of ecological succession. It is inappropriate because it assumes, incorrectly, that cultural and social phenomena always vary in tandem, and that socioeconomic functions are linear and hierarchical (see also Richerson and Boyd 1978).

The problem is, then, compounded when "evolutionist" archaeologists accept these preconceptions as reality, and then coerce their data to fit them, regardless of whether models derived from ethnographic case studies on obscure Pacific islands are foisted on peoples with very different cultural traditions. Given the "loaded" assumptions that evolutionary stages are "real" and that they serve as an explanatory tool (rather than as a deductive typology), hypotheses can become self-fulfilling (e.g., Johnson and Earle 1987). Tautological arguments of this kind do archaeology a disservice.

Cultural evolution is not only a valid, but a central concern of archaeology. However, the specific problems and selected methods for investigation must be directly based on the nature and capabilities of the archaeological record, with its objective reality but its less-than-lucid meaning. Archaeology identifies particular sequences in a particular time and place, providing a wealth of empirical data. Cumulatively such studies contribute a wealth of detail that suggests the broad outline of processes, for which one or more explanatory models can be offered. Sometimes such models can be verified reasonably well, particularly if estimation is substituted for prediction, using internal and ratio levels for data sets, so as to examine proportional relationships. The general validity of models may also be corroborated or contradicted by evidence from parallel studies elsewhere. Archaeology can therefore contribute substantially to macro-problems of process and change. But the questions it raises must be formulated and testable within the context and constraints of its data and methods.

Accepting such limitations with good grace would not cripple archaeology, but instead open it up to a more active interchange with a wider intellectual arena. The problems that archaeology has in dealing with high-level inference are not unique, but shared by the other social sciences (or the field of ecology). All are constrained by the vast differences of scale between the basic units of analysis and the complexities of society or societal behavior. None of the social sciences is equipped to tackle macro-problems singlehandedly, but each can contribute a particular kind of insight to what should be envisaged as a cross-disciplinary forum for productive discussion.

Process and Change in the Archaeological Record

The purpose of the preceding analysis and critique of archaeological interpretative methodology was to identify epistemological limits rather than question the contributions and effectiveness of archaeology as a social science. It is disappointing that the historical elucidation of process in cognitive terms cannot be effectively realized in archaeology at this time.

But if a broader, interdisciplinary perspective is accepted, it may well be possible to study process in certain, specific terms. If data are organized and properly manipulated at the nominal, ordinal, interval, and ratio levels, it may be possible to identify or delineate critical factors and variables at different levels of analysis or inference, and to examine and test potential relationships in terms of historical, adaptive, or evolutionary change. The human ecosystem framework, as defined earlier, provides an ideal context in which to both identify and formulate significant questions. That approach also invites the subsequent incorporation of comparative, cognitive data sets.

During the last 30 years the cumulative result of much painstaking archaeological work has revolutionized our appreciation of process and change in the prehistoric past. It is appropriate to briefly outline a selection of macro-level examples that illustrate some of the contributions of archaeology to the formulation and resolution of broader, interdisciplinary questions. Each example is first presented as a "proposition," based on empirical data or verifiable inference, and is followed by discussion or speculation that elucidates the potential implications of the proposition. The organizational framework is eco-systemic and evolutionary.

Stasis and Change in the Paleolithic Record

Three basic themes can be singled out for the later stages of the Paleolithic: the accelerating change of material culture, demographic growth, and diversification of subsistence activities.

Proposition One. During the course of the West European Mousterian, about 115,000-40,000 B.P., there is minimal evidence of synchronic stylistic variability in lithic artifacts and little diachronic, stylistic or technological change; instead, the basic pattern is one of repetitive interchange of several associations of lithic types (facies) over a phenomenal time span of 75,000 years. With the subsequent Upper Paleolithic, more distinctive assemblages appear, that include diagnostic tools, that vary as to both style and inferred function. Such assemblages are regularly replaced after about 2500 years, with lag times of 800 to 1300 years from place to place (Laville et al. 1980; Butzer 1986). By 20,000 B.P. stylistic variation is pronounced and becomes regionally defined (Straus 1983).

Discussion of Proposition One. Assuming that the lack of directional change of the Mousterian technocomplex over 75,000 years is a valid reflection of Mousterian and Neanderthal cultural capacities, it is difficult to comprehend in terms of recent cultural systems. Such a classic case of homeostatic equilibrium suggests different cognitive, communicative, or neural capacities. On the other hand, the succession of material innovations and stylistic shifts characteristic of the Upper Paleolithic implies accelerating change and a dynamic equilibrium mode. The tempo of replacement is similar to that of ethnic/linguistic groups in the European protohistoric and historical record. How distinct the various Upper Paleolithic components were in adaptive, social, or cultural terms remains speculative. But one can posit an incremental increase in experiential information, magnified by an increasing number of permutations and recombinations of exchanged information. It appears that anatomically modern people introduced a more familiar, "human" dynamism of culture to Western Europe.

Proposition Two. Site frequency per unit time increases exponentially towards the end of the Pleistocene, both in Europe and Africa. In northern Spain the number of identified archaeological sites increases from 0.2 per 1000 years about 75,000 B.P., to 1.4 about 28,000 B.P., 8.8 about 19,500 B.P., 9.5 about 13,500 B.P., 14.3 about 10,000 B.P., and 25.7 about 8000 B.P. (Butzer 1986). At the same time site areas and horizon thickness also increase substantially. In a more comprehensive survey of a small basin in southern Africa, this ratio is 1.0 about 350,000 B.P., 9.7 about 100,000 B.P., 312.5 about 12,000 B.P., 753.8 about 4000 B.P., and 4800 about 1000 B.P. (Sampson 1985; Butzer 1988b). Numbers of artifacts and lithic debitage per site also increase exponentially. This suggests some 30 to 70 times as many sites about 10,000 years in age compared with those preserved from about 75,000 years ago. The increase in site frequency begins at different times in different regions: about 22,000 B.P. in northern Spain, 18,000 B.P. in the Nile Valley, 13,000 B.P. in Northwest Africa, 12,000 B.P. in southern Africa, 14,000 B.P. in northeastern Siberia and Japan, 11,500 B.P. in North America, and 5,000 B.P. in Australia (Butzer n.d.2).

Discussion of Proposition Two. The steep increase in site frequency per unit time probably reflects exponential population growth, more than it does better site preservation or visibility in younger time ranges. During Mousterian/Middle Palaeolithic times, population levels appear to have been very low, oscillating in a steady-state equilibrium. Shortly after introduction/development of Upper Palaeolithic-type technologies (beginning between 40,000 and 20,000 B.P.), population began to increase (dynamic equilibrium), followed by a take-off to a notably higher equilibrium level at some point between 22,000 and 5,000 B.P., depending on the area. The threshold of rapid expansion coincides with the appearance of large projectile points (spear or javelin heads) and micro-blades (probably used as transverse cutting-edges on arrow tips or antler points). These technological innovations suggest effective, offensive weapons that should have dramatically improved hunting efficiency. This in turn may explain the demographic pattern: improved food procurement would theoretically have allowed a spurt of population growth, followed by slower but continuing growth as scheduling and other strategies were adjusted (Butzer n.d.2).

Proposition Three. In southern Africa the selection of game and the age-profiles of species taken ("catastrophic" versus attritional) suggest that Middle Stone Age people (about 125,000-80,000 B.P.) were relatively ineffective hunters, concentrating on very young bovids; only in the case of eland, a large and docile bovid, is there a large proportion of prime-age adults, such as would result by driving a small herd over a cliff (Klein 1976, 1989). Later Stone Age peoples (especially after 12,000 B.P.) actively fished, collected marine invertebrates, or hunted flying birds, took a high proportion of dangerous warthog/bushpig, and exploited a wide variety of plants foods, including roots and bulbs (Klein 1976; Deacon 1984). In northern Spain, nimble-footed ibex and chamois as well as ferocious wild boar were already taken in Mousterian times, and the subsequent increase in ibex mainly reflects more frequent, seasonal use of mountain sites. However, Upper Palaeolithic levels show that herds of red deer could be taken at once, and fishing gear such as harpoons become important. After 10,000 B.P. shell middens indicate regular exploitation of marine, estuarine, and riverine resources, while grinding stones are interpreted as integral to plant food preparation (see Butzer 1986). The North Spanish evidence demonstrates that a greater range of environmental and food types was exploited towards the end of the Pleistocene (Clark and Yi 1983), and similar trends are suggested by other regional studies.

Discussion of Proposition Three. Beyond the general improvement of hunting skills and efficiency towards the end of the Pleistocene, new macro- or micro-environments as well as new types of animal foods appear to have been exploited in a more systematic way. These included montane, aquatic, and marine resources. Plant foods, presumably used in a complementary or seasonal fashion throughout human prehistory, were now searched out and systematically harvested. Not only fruits and nuts but also tubers and seeds were exploited, with a higher investment of labor. In some areas, such as the cold steppe of Eurasia, increased hunting efficiency seems to have led initially to a measure of specialization, but even here the trend eventually shifted to greater niche width and niche variability. No later than the early Holocene, hunter-foragers just about everywhere were adapted to a broad spectrum of available resources within each environmental mosaic, as suggested by increasing spatial variability of the archaeological record. Greater procurement effectiveness, diversification, and demographic growth were probably linked by positive feedbacks, perhaps favoring greater socioeconomic complexity.

Agricultural Transformation

The fundamental shift from hunting-foraging to farming-herding lifeways represents a revolutionary change that began in some areas about 11,000 to 6000 B.P. Basic themes include: the domestication of plants and animals, the dispersal of agricultural traits and complexes, the intensification of agricultural exploitation, and the environmental impacts of agricultural land use.

Proposition Four. "Hard" evidence for successful domestication of plants and animals comes in the form of plant materials or animal bone that show specific morphological divergences from their wild progenitors. The Near Eastern record, based on multiple archaeological sequences, is strongest. At Jericho, the earliest Neolithic level (PPNA) at 10,500-9,200 B.P. has remains of domesticated barley, emmer wheat, and pulses; a similar horizon (Bus Mordeh) at Ali Kosh, beginning about 10,000 B.P., also has domesticated cereals and pulses, while the basal level of Tell Aswad, near Damascus, brings evidence of domesticated peas, lentils, emmer and probably barley (Moore 1982). The subsequent Neolithic levels at these and other sites of the steppe-woodland ecotone, dating about

9,600-8,000 B.P., have bones of one or more genera of domesticated animal (goat, sheep, cattle, and possibly pig), a wide array of domesticated grains (emmer and einkorn wheat, barley), and further vegetables (Harlan 1977), although wild game remained prominent; similar lithic assemblages are also found at contemporary sites patently used as hunting encampments (Aurenche et al. 1981). Only in later Neolithic levels, dating about 8,100-5,700 B.P., is the full range of grains, vegetables, and herd animals consistently verified; initially such sites begin to penetrate the woodlands, subsequently they also appear in the larger alluvial valleys, with the first evidence of improvised irrigation.

In terms of similar, semiarid ecologies, domesticated maize is botanically verified in Tehuacán, Mexico, about 7,000 B.P., while cultivated common and lima beans are found in Guitarrero Cave, Peru, some 700 years earlier, and in Tehuacán and Tamaulipas, Mexico, about a millennium later; domesticated squash was used in the Valley of Oaxaca about 9,400 B.P. (Pickersgill and Heiser 1977; Flannery 1986). Other New World plant domesticates only appear in the "hard" record after 3,000 B.P. Although widely distributed in space, it would seem that plant domestication was an equally long-term process in both Mexico and Peru. Agricultural innovation in humid Southeast Asia is still inadequately documented, but rice may have been in cultivation by 5,400 B.P. in China (Gorman 1977, Harlan 1977).

Discussion of Proposition Four. Relying only on the "hard" criteria, domestication appears as an incremental process in so-called hearth areas, with suitable, indigenous wild forms subjected to deliberate planting or breeding at a very early date. At first there was supplementary cultivation of a few plants, within a broad repertoire of plant foraging and manipulation; in other instances, there was some controlled stockraising within the context of more traditional hunting. At least 3,500 years intervened between the first solid evidence for supplementary cultivation in the Near East and the appearance of a full agricultural repertoire of grains, a range of pulses, and the full complement of four domesticated herd animals. Only with the assembling of this complete agricultural "package" about 8,000 B.P. can one speak of a standardized agrosystem. In Mexico, Peru, and Southeast Asia this process of agricultural transformation may have been more attenuated.

But the process of domestication was even longer than the "hard" evidence demonstrates. Intensified harvesting and possible manipulation of wild seed plants are suggested before 15,000 B.P. by grinding equipment and lithic blades with lustrous "sickle sheen" in the Near East (Henry 1988), while selective hunting and possible control of gazelle, fallow deer, onager, and "wild" forms of goat, sheep or cattle were widely practiced by 12,000 B.P. (Moore 1982). In the central United States, essentially sedentary utilization of rich, riparian resources preceded the appearance of standard domesticates by several millennia, with some measure of controlled use of a variety of minor plant foods (Ford 1985). This suggests several, if not many millennia of manipulation, familiarization, and experimentation with a wide range of potential domesticates, prior to the type of agriculture documented by the "hard" evidence. Only after the positive feedbacks, both human and biotic, between the traditional foraging systems and the new "management" techniques began to co-evolve into a new, interactive system, did the selection of a few of these plants or animals for systematic domestication meet the necessary conditions for morphological change. Domestication was, therefore, a very long and punctated process, co-evolutionary in trajectory, revolutionary only in its ultimate impact.

Proposition Five. The problem of agricultural origins is distinct from that of agricultural dispersal, via stimulus diffusion or migration. Even before the full crystallization of a standardized agrosystem in the Near East, large coastal villages, with evidence of grain and vegetable cultivation, and dependent on domestic stock rather than game, appeared in Greece and on Cyprus and Crete 8,400-7,900 B.P. By 7,800 B.P. hunter-foragers in southern France and eastern and southern Spain kept domesticated sheep, apparently driving them in the course of seasonal hunting rounds (Lewthwaite 1986). About 7,300-6,500 B.P. a culturally distinctive agrosystem was established along the coastal plains of the western Mediterranean from Yugoslavia to Portugal (Cardial Neolithic). Another was established in the Balkans 7,400-7,100 B.P. (Starčevo-Körös), and a third in Central Europe 6,500-5,700 B.P. (Linear Pottery) (see Barker 1985). Initially the domesticates themselves were introductions, rather than derived from in situ domestication of local plants or animals. But in some areas where wild prototypes were found, resident hunter-foragers domesticated local stock prior to agricultural penetration (Yugoslavia) or outbreeding produced new

stock or cultivar varieties (Hungary). Subsequent expansion of
agriculture into Northwest Europe, before 5,000 B.P., represented a
selective adaptation of agricultural traits by autochthonous
populations (Barker 1985).

Discussion of Proposition Five. The very rapid dispersal of
agriculture or selected agricultural traits, so well documented for
Europe, presupposes a measure of migration but also identifies
examples of preadaptation. A range of case studies in the Balkans
and the western Mediterranean show that domesticated cattle or sheep
were used by autochthonous hunter-foragers several centuries prior to
the appearance of intrusive groups with distinctive pottery and
cultivars; in the former case, local animals were domesticated, in
the latter, domestic stock were somehow introduced from outside
(Lewthwaite 1985). Excavations in Italy verify the interaction
between small settlements of intrusive farmers with nearby
hunter-foragers, who gradually switched from hunting to stockraising
(Barker 1985).

Complex processes of stimulus diffusion, both local and long
distance, are therefore implicated. Selected agricultural traits
must have seemed highly desirable, despite the need to readapt them
within a different lifeway or to different regional ecologies. In
the case of actual migration, dispersal was nodal, along attractive
segments of the Mediterranean coastline, or from one fertile basin to
the next in Central Europe. Pulse-like surges of migration are
indicated, interrupted by some five centuries of filling in,
ecological adjustment, and perhaps experimentation in forward areas,
characterized by different ecological parameters. The initial delay
in advance of standardized agriculture into the western Mediterranean
may have been predicated on improved navigational skills; the advance
into the Balkan basins certainly required an adjustment to summer
(versus winter) rainfall and, eventually, colder winters. A
significant development in regard to agricultural dispersals has been
the explicit recognition of a complex interplay between: (a)
stimulus diffusion, (b) ecological readjustment (Butzer 1988a), (c)
indigenous innovations, (d) complementary interactions between
resident hunter-foragers and intrusive farmers (Gregg 1988), and (e)
ultimately the "indigenization" of agriculture or agricultural traits
(Kristiansen and Paludan-Müller 1978). A similar co-evolutionary
trajectory is emerging in North America (Ford 1985; Johannessen
1988), and is suggested by ongoing research in subsaharan Africa.

Proposition Six. The standardized agrosystem that emerged in the Near East before 8,000 B.P. and dispersed through much of Europe by 5,000 B.P. was relatively primitive; it only provided plants and meat, while foraged foods remained important. By 3,500 B.C. (about 5,400 B.P., uncalibrated) a new process becomes apparent in the Near East, and within a millennium its impacts were felt as far afield as Britain (Sherratt 1983). Technological innovations such as the plow and wheel were combined with ox traction, and new animals were eventually domesticated to facilitate transport--horse, donkey, and camel. Wool was skeined and woven into cloth, and new breeds of woolly sheep selected. Cows, sheep, and goats were milked, and the milk converted to butter or cheese, which could be stored. Animal manure was applied as fertilizer, while plowing along hillsides deliberately or inadvertently created stepped slope profiles, that may have suggested the principle of terracing. This fundamental change in the relationships between people and animals also tied the planting and herding segments of the economy into an interdependent system. Called the "secondary products revolution" (Sherratt 1983), it represents a part of what agricultural economists and cultural ecologists refer to as "intensification." But it was not limited to animals and animal products.

Polycropping of olive trees with grains produced a new source of oil that did not compete for space, as did flax, and olive groves could also be planted on marginal hillsides. The range of fruit trees was enlarged, vineyards were planted, and grafting was possibly introduced. These innovations, made in Lebanon and Palestine 3,700-3,100 B.C., produced commodities easy to transport and market--olive oil and wine; they also appear to have proved sufficiently important to incorporate into the symbolic realm, and became the hallmark of Mediterranean arboriculture (Stager 1983, 1985; Butzer 1988a). Simultaneously, systematic canal irrigation and urban growth began on the floodplains of Mesopotamia (Adams 1981) and Egypt (Butzer 1976, 1984b), linking agricultural intensification with urbanization. But commercial crops and urban market-economies only spread westward to Spain by the 6th century B.C., and to Northwest Europe--where drainage techniques replaced irrigation--during Roman times. More closely tied to the rapid diffusion of the plow and the wheel were mining and bronze metallurgy.

Discussion of Proposition Six. Intensification reflected a bundle of positive feedbacks that finally integrated all components of a particular agrosystem, triggering a comparatively rapid sequence of socioeconomic adjustments. It can be further argued, but remains to be convincingly tested, that intensification changed dietary, demographic, and sociocultural strategies more fundamentally than did the initial agricultural transformation. These technological innovations represented labor-intensive shifts, that made large families desirable, and reduced biological selection pressures. This appears to be borne out by increasing settlement sizes and numbers, suggesting higher population densities--despite the increasing probability of epidemic disease and the mutation of livestock viruses to deadly vectors of human disease. Improved technologies, coupled with greater labor investment, imply a greater frequency of cultivation of any one plot. The investment in and dependence upon particular land parcels suggests that more complex rules for the communal use of land were called for, leading to the development of private property in some areas, with similarly complex rules for transgenerational property transmission via arranged marriages (Sherratt 1983). The diversification of domestic tasks central to intensification may also have accelerated the differentiation of age- and gender-specific roles.

Most of the changes implied by intensification were incremental in the sense that they reflected countless personal decisions, rather than changes imposed by a centralized authority. But in some areas intensification was paralleled by and interrelated with a shift to more centralized authority. The differential quality of grave goods or burial structures implies development of self-perpetuating social inequality, without necessarily elucidating the nature of social structures. In large settlements marked by greater social inequality, nutritionally inadequate or unbalanced diets become a real probability. New armaments, the use of horses and chariots in warfare, and destruction horizons within town sites all document an increasing role for warfare, sometimes linked with regional discontinuities it the archaeological record. Mining and metallurgy probably played a pivotal role in this process, since control over critical sources of metal conferred economic power, that could in turn be used to demand food or craft products, such as cloth and pottery. Trade and market institutions would be stimulated as economies became more commercialized, favoring urban growth. Symbolic expression in bronze, stone, or architecture becomes striking and seems to have been critical to the maintenance of polities and hegemonies.

These broad themes are particularly challenging for hypothesis-building and testing. Unfortunately the best tests for nutritional implications of dietary change and of social inequality must await a greater breadth and intensity of research in physical anthropology, blood typing, paleopathology, isotopic analysis, and paleonutrition. Such work has not kept pace with other modes of archaeological investigation. And all too often the critical human skeletal remains are minimally preserved or were disposed of by cremation.

Proposition Seven. Geoarchaeological research has incrementally served to elucidate the impact of agricultural land use on local and regional environments, with potential significance for resource sustainability. Episodes of low-density human settlement, with only rudimentary Neolithic farming, had unanticipated effects on the landscape of Greece and Spain via deforestation and massive soil erosion (Van Andel and Zangger n.d.; Butzer, Mateu, and Scott, n.d.). Subsequent, more intensive, Bronze Age or later farming provoked different intensities of landscape degradation, not necessarily proportional to settlement density, and implicitly conditioned by different agricultural procedures. Agriculture in northern Europe led to local and temporary vegetation disturbance or to long periods with extensive clearance, the intensity of use reflected in variations of lakebed geochemistry (Edwards 1979; Simmons and Tooley 1981; Kristiansen and Paludan-Müller 1978). Similar impacts of pre-European settlement have been identified in the eastern United States (Delcourt et al. 1986) and in the Maya lowlands (Deevey et al. 1979). Intensive or long-term disturbance had different impacts in semiarid or subhumid areas, where groundwater recharge was reduced and stream runoff became more intermittent, increasing periodic flood damage, whereas in cool and humid settings interference favored growth of acid heath (blanket bog), with essentially permanent landscape degradation (Butzer 1982: chap. 8). Other studies demonstrate the permanent ecological damage to Pacific Islands by their growing populations, to the point where decreasing resource productivity resulted in population decline (Kirch 1980b, 1982).

Discussion of Proposition Seven. These examples show that the environment was a major variable in archaeologically discernible processes. Soil, water, and biota are unstable, changing, and often fragile resources that demand careful attention, not only in terms of abstract site-catchment analyses or potential productivity measures, but as a complex arena for concrete study. Was soil fertility sustained during long periods of occupation? How was the water supply affected by secular environmental change and human interference in the hydrological cycle? How was cultivation or herding conditioned by human intervention in the natural biota and by successional changes thereafter? Was land use exploitative or conservationist? Productivity is not a given, but a systemic variable that responds to management, both good and bad.

As a first step, management techniques deserve far more explicit attention by searching for direct and indirect evidence, and by evaluating ethnographic analogs. But discussion must also be taken to the level of systemic energy flows, in regard to different ratios between work input and energy output, how risk and shortfalls were accommodated, and whether productivity increased or decreased with time. Was all the food produced consumed locally, or did sociopolitical structures subtract part of that productivity? Can demographic change be monitored by archaeological controls or nutritional levels by skeletal data? Far less tangible, but equally important, are several cognitive questions. How were subsistence resources perceived and did this perception change with time? Were trial-and-error in management and dietary strategies incorporated as "experience" into long-term cultural behavior through beneficial ritual and symbolic expression? Only by devoting more attention to this full range of questions can we eventually hope to test whether expanding populations, increasingly costly labor investment, and declining productivity may have precipitated ecological crises of fundamental importance. Both the time depth represented by archaeology and the data resolution that it can provide are critical to long-term questions of resource sustainability.

Social Transformation and Maintenance

On the threshold of "high" civilizations, the archaeological record allows identification of two basic and interrelated problems: the emergence of complex societies, and the recurrence of population cycles. The latter is amenable to empirical investigation, but the former is primarily an interpretative matter.

<u>Proposition Eight</u>. In different regions the late prehistoric trajectory suggests increasing social complexity (e.g., Wenke 1984). This encompasses economic diversification, elaboration of institutional structures, and verticalization of socioeconomic and sociopolitical components into functional hierarchies, that serve to organize energy and information pathways. Intensification appears indispensible in sustaining the comparatively dense, nucleated and sedentary populations that appear to be integral to such social evolution or to such corollary processes as urbanization or state formation.

Discussion of Proposition Eight. In place of traditional models that emphasize sociocultural evolution, an alternative model, focusing on spatial integration and hierarchical elaboration, can be suggested. This model is no less speculative than existing ones, but offers a different approach, the components of which are more readily tested. Firstly, intensification should serve to increase the productivity and reliability of resource yields, to favor demographic growth and nucleation of population. Secondly, the unequal distribution of natural and human resources, combined with the relative advantages of some locations over others, may lead to differential increases in settlement size and function (see also Ellen, this volume). As the patterns of settlement "rank" change, a vertical hierarchy of "central places" should begin to emerge, in response to population size, economic function, ceremonial roles, or administrative or military advantages. Thirdly, as this process of verticalization continues, feedbacks would come into play as to (a) economic specialization, resource concentration, and social segregation; (b) increasing demands for raw materials or energy; and (c) regional as well as interregional integration. This human ecosystem model centers on a three-dimensional, sociodemographic and economic integration, but can then be elaborated with historical or ethnographic insights to include sociocultural change. Thus the critical structures of the model are reasonably robust, because they are amenable to archaeological testing, and its value is only enhanced by attempting to accommodate the role and impact of decisions imposed by an empowered elite.

Proposition Nine. Both the archaeological and historical records show that change was not progressive but episodic, and sometimes regressive. Most immediately, archaeologically-verified regional settlement histories, such as in Mesopotamia or the Aegean (e.g. Adams 1981; Van Andel and Runnels 1987), fail to exhibit linear trajectories. Settlement histories vary from one valley or subregion to another, so that composite, regional trends are irregular, rather than linear. By whatever surrogate measure, subregional population trends show oscillating, stepwise progressions, interrupted by periods of stabilization or even decline. Seen in the long-term perspective, whether increasing or ultimately in a steady-state equilibrium, population trajectories display cyclical trends of rapid growth and equally rapid decline, so-called millennial long-waves (Whitmore et al. n.d.).

Discussion of Proposition Nine. Demographic growth is rarely possible without improved technology, a broad social access to resources, or a combination of the two. Decline points to fundamental social, political, or environmental problems. Growth, stability, or decline also suggest different questions about the quality of life. In the case of relatively simple, prehistoric farming societies, we still lack data explaining the common and often long-term settlement breaks between archaeological components. Were they due to sweeps of new epidemic diseases, warfare, or declining productivity of primitive agroecosystems, perhaps even Malthusian "overshoot"?

On the other hand, historical examples from complex societies such as Mesopotamia, Egypt, and Spain (Whitmore et al., n.d.; Butzer 1976, 1980b, 1984a, n.d.1), suggest that demographic long waves ultimately coincided with times of strong and efficient government, pointing to the importance of channeling those centripetal forces that favor the integration of institutions, economic networks, and settlement hierarchies which stimulate higher productivity and sustain higher populations. Periods of decline are tied to times of weak or incompetent government that allow centrifugal forces to impair socioeconomic structures and political institutions, to the point where productivity and population levels decline. Energy and information pathways are critical components of this equation. For example, "decline" may also be precipitated by expansion of an imperial system to beyond its logistical limits, when the friction of distance and decreasing efficiency in managing peripheral resources begin to outweigh the energy pumped into the system by the control of labor, production, and trade. A complementary perspective to this is suggested by usurpation of the productive base of a regional system, dissipating energy into non-sustaining activities such as excessive construction activities, indecisive or unsuccessful warfare, or tribute to an external power. Ecological simplification suggests a useful analog to examine "decline," as a product of multiple feedbacks and inherent thresholds.

Outlook

The themes outlined above represent examples of significant problems in historical social science that interest archaeologists today, and for which archaeological methodologies are particularly suitable. The difficulties of testing the interpretations are formidable, but not insurmountable in the long run. Each excavation and survey directed towards testing a particular hypothesis is set within a matrix of both similar and different variables. Eventually the number of equations (case studies) equals the number of unknowns (variables), and identification (role) of the variables can be resolved. Both the formulation of the problems and the interpretive structures suggested above are eco-systemic, and systems or subsystems in archaeology should automatically incorporate spatial and, by extension, environmental components. The utility of this human ecosystem approach, focused on specific data sets at an appropriate level of analysis, transcends its scientific clarity for the selection, formulation, testing, and interpretation of problems. It avoids biased "loading" factors while further allowing the incorporation of relevant sociocultural variables, such as institutional structures, symbolic interaction, and cognition. In effect, a problem can first be tested with ecological or material variables alone, after which further tests can add other variables incrementally. Post-processual concerns can therefore be incorporated.

The basic point is that the variables form a complex (algebraic) matrix, and that no single project can hope to verify a particular hypothesis. Many projects are indispensable to establish a set of modal parameters, and to explain the exceptions without weakening the ground rules.

Acknowledgments

I am indebted to John Peterson, Robert Ricklis, and Samuel Wilson (University of Texas, Austin) for critical suggestions on an interim draft.

References Cited

Adams, J. and A. Kasakoff
    1984   Ecosystems Over Time: The Study of Migration in "Long-Run" Perspective. *The Ecosystem Concept in Anthropology*, Edited by E.F. Moran, pp. 205-223. Boulder: Westview Press. AAAS Selected Symposium, No. 92.

Adams, R.McC.
    1981   *Heartland of Cities*. Chicago: University of Chicago Press.

Adams, R.N.
    1988   *The Eighth Day: Social Evolution as the Self-Organization of Energy*. Austin: University of Texas Press.

Aurenche, O., J. Cauvin, M.C. Cauvin, L. Copeland, F. Hours, and P. Sanlaville
    1981   Chronologie et organisation de l'espace dans le Proche Orient de 12000 a 5600 av. J.C. *Préhistoire de Levant*, Edited by J. Cauvin and P. Sanlaville, pp. 571-601. Paris: C.N.R.S.

Barker, G.
    1985   *Prehistoric Farming in Europe*. New York: Cambridge University Press.

Barth, F.
    1956   Ecologic relationships of ethnic groups in Swat, North Pakistan. *American Anthropologist* 58:1079-1089.

Bennett, J.W.
    1976   *The Ecological Transition*. London: Pergamon.

Binford, L.R.
    1989   *Debating Archaeology*. New York: Academic Press.

Brookfield, H.C.
1964　Questions on the human frontiers of geography. *Economic Geography* 40:283-303.

Butzer, K.W.
1960　Remarks on the geography of settlement in the Nile Valley during Hellenistic times. *Bulletin, Societe de Geographie d'Egypte* 33:5-36.

1976　*Early Hydraulic Civilization in Egypt: A Study in Cultural Ecology.* Chicago: University of Chicago Press.

1980a　Context in Archaeology: An Alternative Perspective. *Journal of Field Archaeology* 7:417-422.

1980b　Civilizations: Organisms or Systems? *American Scientist* 68:517-523.

1982　*Archaeology as Human Ecology.* New York: Cambridge University Press.

1984a　Long-term Nile Flood Variation and Political discontinuities in Pharaonic Egypt. *From Hunters to Farmers,* Edited by J.D. Clark and S.A. Brandt, pp. 102-112. Berkeley: University of California Press.

1984b　Siedlungsgeographie (Settlement Geography). *Lexikon der Aegyptologie* 38:924-933.

1986　Paleolithic adaptations and settlement in Cantabrian Spain. *Advances in World Archaeology* 5:201-252.

1988a　Diffusion, Adaptation, and Evolution of the Spanish Agrosystem. *The Transfer and Transformation of Ideas and Material Culture,* Edited by P.J. Hugill and D.B. Dickson, pp. 99-109. College Station, TX: Texas A&M University Press.

1988b　A "marginality" Model to Explain Major Spatial and Temporal Gaps in the Old and New World Pleistocene Settlement Records. *Geoarchaeology* 3:193-203.

1989　Cultural Ecology. *Geography in America,* Edited by G.L. Gaile and C.J. Wilmott, pp. 192-208. Columbus, OH: Merrill.

n.d.1.　The Realm of Cultural Ecology: Adaptation and Change in Historical Perspective. *The Earth as Transformed by Human Action,* Edited by B.L. Turner, New York: Cambridge University Press, in press.

n.d.2   An Old World Perspective on Potential mid-Wisconsinian
           Settlement of the Americas. *Peopling of the New
           World: New Problems and Issues,* Edited by T.
           Dillehay and D. Meltzer, Caldwell, NJ: Telford Press,
           in press.
Butzer, K.W., and E.K. Butzer
1989   Historical Archaeology of Medieval Muslim Communities
           in the Sierra of Eastern Spain. *Medieval
           Archaeology,* Edited by C.L. Redman, pp. 217-233.
           Binghamton, NY: SUNY Binghamton, Medieval and
           Renaissance Texts and Studies.
Butzer, K.W., E.K. Butzer, and J.F. Mateu
1986   Medieval Muslim Communities of the Sierra de Espadán,
           Kingdom of Valencia. *Viator: Medieval and
           Renaissance Studies* 17:339-413.
Butzer, K.W., and A. Ferrer
n.d.   Excavaciones en el Castillo de Xinquer (Sierra de
           Espadán). *Saguntum* 24, in press.
Butzer, K.W., J.F. Mateu, and E.K. Butzer
1989   Orígenes de la distribución intercomunitaria del agua
           en la Sierra de Espadán. *Los paisajes de
           agua.*Edited by V.M. Rossello, pp. 223-228.
           Valencia: Universidad de Valencia.
Butzer, K.W., J.F. Mateu, E.K. Butzer, and P. Kraus
1985   Irrigation Agrosystems in Eastern Spain: Roman or
           Islamic origins? *Annals Association of American
           Geographers* 75:479-509.
Butzer, K.W., J.F. Mateu, and L. Scott
n.d.   *Holocene Settlement Histories, Environmental
           Degradation, and Alluvial Response in Eastern
           Spain.* Forthcoming.
Chorley, R.J.
1973   Geography as Human Ecology. *Directions in
           Geography,* Edited by R.J. Chorley, pp. 155-169.
           London: Methuen.
Chorley, R.J. and B.A. Kennedy
1971   *Physical Geography: A Systems Approach.*
           Englewood Cliffs, NJ: Prentice-Hall.
Clark, G.A. and S. Yi
1983   Niche-width Variation in Cantabrian Archaeofaunas: A
           Diachronic Study. *Animals and Archaeology,*
           Edited by J. Clutton-Brock and C. Grigson, pp,
           183-208.

Oxford: British Archaeological Reports, International Series 163.

Clarke, D.L.
1968   *Analytical Archaeology*. London: Methuen.

Clarkson, J.D.
1970   Ecology and Spatial Analysis. *Annals Association of American Geographers* 60:700-716.

Deacon, J.
1984   Later Stone Age People and their Descendants in Southern Africa. *Southern African Prehistory and Environments*, Edited by R.G. Klein, pp. 221-328. Rotterdam/Boston: A.A. Balkema.

Deevey, E.S. and others
1979   Maya Urbanism: Impact on a Tropical Karst Environment. *Science* 206:298-306.

Delcourt, P.A., H.R. Delcourt, P.A. Cridlebaugh, and J. Chapman
1986   Holocene Ethnobotanical and Paleoecological Record of Human Impact on Vegetation in the Little Tennessee River Valley. *Quaternary Research* 25:330-349.

Denbow, J.R., and E.N. Wilmsen
1986   Advent and Course of Pastoralism in the Kalahari. *Science* 234:1509-15.

Denevan, W.M.
1983   Adaptation, Variation, and Cultural Geography. *Professional Geographer* 35:399-407.

Edwards, K.J.
1979   Palynological and Temporal Inference in the Context of Prehistory. *Journal of Archaeological Science* 6:255-270.

Ellen, R.
1982   *Environment, Subsistence and System: The Ecology of Small-Scale Social Formations*. New York: Cambridge University Press.

Flannery, K.V. ed.
1986   *Guila Naquitz: Archaic Foraging and Early Agriculture in Oaxaca, Mexico*. New York: Academic Press.

Ford, R.I. ed.
1985   *Prehistoric Food Production in North America*. Ann Arbor: University of Michigan, Museum of Anthropology, Anthropological Papers No. 75.

Geertz, C.
1963   *Agricultural Involution: The Processes of
        Ecological Change in Indonesia.*   Berkeley:
        University of California Press.
Gorman, C.
1977   A Priori Models and Thai Prehistory: A Reconsideration
        of the Beginnings of Agriculture in Southeastern Asia.
        Origins of Agriculture, Edited by C.A. Reed,
        pp. 321-356. The Hague: Mouton.
Gould, R.A. ed.
1978   *Exploration in Ethno-Archaeology.* Albuquerque,
        NM: University of New Mexico Press.
Gould, R.A.
1980   *Living Archaeology.* New York: Cambridge
        University Press.
1985   The Empiricist Strikes Back: A Reply to Binford.
        American Antiquity 50:638-644.
Gregg, S.A.
1988   *Foragers and Farmers: Population Interaction and
        Agricultural Expansion in Prehistoric Europe.*
        Chicago: University of Chicago Press.
Hall, R.L.
1977   An Anthropocentric Perspective for Eastern United
        States   Prehistory.   *American   Antiquity*   42:
        99-518.
Harlan, J.R.
1977   The Origins of Cereal Cultivation in the Old World.
        Origins of Agriculture, Edited by C.A. Reed,
        pp. 357-384. The Hague: Mouton.
Hawkes, C.
1954   Archaeological Theory and Method: Some Suggestions
        from the Old World. *American Anthropologist*
        56:155-168.
Henry, D.O.
1988   *From Foraging to Agriculture: The Levant at the
        End of the Ice Age.* Philadelphia: University of
        Pennsylvania Press.
Hodder, I.
1982   *Symbols in Action: Ethnoarchaeological Studies of
        Material Culture.* New York: Cambridge University
        Press.

1984   Burials, Houses, Women and Men in the European
            Neolithic. *Ideology, Power and Prehistory*,
            Edited by D. Miller and C. Tilley, pp. 51-68.
            Cambridge: Cambridge University Press.
1987   Society, Economy and Culture: An Ethnographic Case
            Study amongst the Lozi. *Patterns of the Past*,
            Edited by I. Hodder, G.L. Isaac and N. Hammond, pp.
            67-95. Cambridge: Cambridge University Press.
Johannessen, S.
1988   Plant Remains and Culture Change: Are
            Paleoethnobotanical Data better than We Think?
            Current Paleoethnobotany, Edited by C.A.
            Hastorf and V.S. Popper, pp. 145-166. Chicago:
            University of Chicago Press.
Johnson, A.W. and T. Earle
1987   *The Evolution of Human Societies: From Foraging
            Groups to Agrarian State.* Stanford, CA: Stanford
            University Press.
Kirch, P.V.
1980a   The Archaeological Study of Adaptation: Theoretical and
            Methodological Issues. *Advances in Archaeological
            Method and Theory* 3:101-56.
1980b   Polynesian Prehistory: Cultural Adaptation in Island
            Ecosystems. *American Scientist* 68:39-48.
1982   The Impact of the Prehistoric Polynesians on the
            Hawaiian Ecosystem. *Pacific Science* 26:1-14.
Klein, R.G.
1976   The Mammalian Fauna of the Klasies River Mouth Sites,
            Southern Cape Province, South Africa. *South
            African Archaeological Bulletin* 31:75-98.
1988   Reconstructing how Early People Exploited Animals:
            Problems and Prospects. *The Evolution of Human
            Hunting*, M.H. Nitecki and D.V. Nitecki, eds. pp.
            11-45. New York: Plenum.
1989   Why does Skeletal Part Representation differ between
            Smaller and Larger Bovids at Klasies River Mouth and
            other Archaeological Sites?   *Journal of
            Archaeological Science* 16:363-382.
Kristiansen, K. and Paludan-Muller, eds.
1978   *New Directions in Scandinavian Archaeology.*

Copenhagen: National Museum, Studies in Scandinavian Prehistory and Early History 1.

Laville, H., J.P. Rigaud, and J. Sackett
1980   *Rockshelters of the Perigord.* New York: Academic Press.

Leone, M.P.
1986   Symbolic, Structural, and Critical Archaeology. *American Archaeology: Past and Future,* Edited by D.J. Meltzer, D.D. Fowler, J.A. Sabloff, pp. 415-438. Washington: Smithsonian Institutional Press.

Lewthwaite, J.G.
1986   From Menton to Mondego in Three Steps: Application of the Availability Model to the Transition to Food Production in Occitania, Mediterranean Spain and southern Portugal. *Arqueología* (Pórto) 3: 95-119.

Miller, D. and C. Tilley, eds.
1984   *Ideology, Power and Prehistory.* Cambridge: Cambridge University Press.

Moore, A.M.T.
1982   Agricultural origins in the Near East: A Model for the 1980s. *World Archaeology* 14:224-236.

Moran, E.F.
1982   *Human Adaptability: An Introduction to Ecological Anthropology.* Boulder: Westview Press.

Orlove, B.
1980   Ecological Anthropology. *Annual Review of Anthropology* 9:235-73.

Pearson, M.P.
1984   Economic and Ideological Change: Cyclical Growth in the Pre-state Societies of Jutland. *Ideology, Power and Prehistory,* Edited by D. Miller and C. Tilley, pp. 69-92. Cambridge: Cambridge University Press.

Pickersgill, B. and C.B. Heiser
1977   Origins and Distribution of Plants Domesticated in the New World Tropics. *Origins of Agriculture,* Edited by C.A. Reed, pp. 803-836. The Hague: Mouton.

Rappaport, R.A.
1968   *Pigs for the Ancestors: Ritual in the Ecology of a New Guinea People.* New Haven: Yale University Press, 2nd ed. 1984, with epilogue.

Renfrew, C., M. Rowlands, and B. Seagraves, eds.
1982   *Theory and Explanation in Archaeology.* New York: Academic Press.

Richerson, P.J., and R. Boyd
1978   A Dual Inheritance Model of the Human Evolutionary Process. *Journal of Social and Biological Structures* 1:127-54.

Rosen, A.M.
1986   *Cities of Clay: The Geoarchaeology of Tells.* Chicago: University of Chicago Press.

Sampson, C.G.
1985   Atlas of Stone Age Settlement in the Central and Upper Seacow Valley. *National Museum, Bloemfontein, Memoir* 20, 1-116.

Schiffer, M.B.
1987   *Formation Processes of the Archaeological Record.* Albuquerque: University of New Mexico Press.

Sherratt, A.G.
1983   The Secondary Exploitation of Animals in the Old World. *World Archaeology* 15:20-104.

Simmons, I.G. and M.J. Tooley
1981   *The Environment in British Prehistory.* Ithaca NY: Cornell University Press.

Stager, L.E.
1983   The Finest Oil of Samaria. *Journal of Semitic Studies* 28:241-245.
1985   The First Fruits of Civilization. *Palestine in the Bronze and Iron Age*, Edited by J.N. Tubb, pp. 172-188. London: University of London, Institute of Archaeology Monograph.

Steward, J.H.
1955   *The Theory of Culture Change.* Urbana: University of Illinois Press.

Stoddart, D.R.
1965   Geography and the Ecological Approach. *Geography* 50:242-251.

Straus, L.G.
1983   *El Solutrense Vasco-Cantabrico: Una Nueva Perspectiva.* Madrid: Centro de Investigación y Museo de Altamira, Monograph 10.

Trigger, B.
  1980   Archaeology: The Image of the American Indian.
            American Antiquity 45:662-676.
Van Andel, T.H. and C. Runnels
  1987   *Beyond the Acropolis: A Rural Greek Past.*
            Stanford, CA: Stanford University Press.
Van Andel, T. and E. Zangger
  n.d.   Land Use and Soil Erosion in Prehistoric and Historical
            Greece. *Journal of Field Archaeology 17*, in
            press.
Watson, P.J.
  1986   Archaeological interpretation, 1985. *American
            Archaeology: Past and Future*, Edited by D.J.
            Meltzer, D.D. Fowler, J.A. Sabloff, pp. 439-458.
            Washington: Smithsonian Institution Press.
Wenke, R.J.
  1984   *Humankind's First Three Million Years: Patterns in
            Prehistory.* New York: Oxford University Press.
Whitmore, T.M., B.L. Turner, D.L. Johnson, R.W. Kates, and T.R.
Gottschang
  n.d.   What Goes Up Comes Down: Long-Term Population Change
            and Environmental Transformations. *The Earth as
            Transformed by Human Action*, Edited by B.L. Turner,
            Cambridge: Cambridge University Press, in press.
Young, M.J.
  1988   *Signs from the Ancestors: Zuni Cultural Symbolism
            and Perception of Rock Art.*   Albuquerque:
            University   of   New   Mexico   Press.

CHAPTER 5

THE ECOSYSTEM MODEL AND LONG-TERM PREHISTORIC CHANGE: AN EXAMPLE FROM THE ANDES

Christine A. Hastorf

The ecosystem concept, as it is used in anthropology, provides a general framework for dynamically viewing humans in their environment, with a focus on the interaction between the physical world and human society. Becoming popular in the 1960's (Rappaport 1967), it was accepted by many archaeologists into the 1980's as providing an integrative approach to study human culture, structuring major relationships and following diachronic changes yet talking about the complexity of the natural ecology and human society. As Jochim points out in this volume, it was adopted as a heuristic device to involve the whole ecosystem, rather than to focus on a single prime mover as the cause for past cultural behaviors. It was enthusiastically incorporated into many archaeological studies because of its broad, complex vision of culture and its interactive ideal to study change. This allowed for the inclusion of multiple causes in explanations of culture change and evolution (Flannery 1972).

But, as Kirch (1980) points out, archaeologists found that the human ecosystem was too large and too complicated to effectively address in any one study. Investigators began to break up the universe into smaller units of study, linking the material to the behavioral in specific relationships. In doing so, they divided society into subsystems and, in particular they focused on environmental and economic forces; either subsistence, technology, settlement patterns, exchange, climate, and/or landscape (Kirch 1980). While still firmly rooted in ecological theory, these subsystemic studies tended to assume that the environment was the fundamental "context" and starting place for causal links (Tansley

1935; Odum 1953). For archaeology this tied in nicely with the cultural ecological tradition of Julian Steward (1938), and projects investigated environmental constraints such as nutrient capture (Keene 1981), transfer (flow) of energy through the environment and society (Earle and Christenson 1980), flow of information (Flannery 1972), or even social archaeology (Bailey and Sheridan 1981).

After a while, criticism began, claiming that the results did not live up to the idealistic goals of the model. The ecosystem approach was not providing satisfying explanations for past human behavior. Some of the fundamental principles were under fire, such as privileging the physical environment in explanation (Jochim 1979). Further, from studying only a part of the human universe via the ecological approach, critics stated that the studies were reductionistic and static, not actually examining the dynamic processes of change and variation. Instead, human intentionality was overlooked and no attention was being paid to alternative possibilities and contradictions (Hodder 1986). What was left out, the critics claimed, was human action and intent. Humans are not like other components of the ecosystem— they have values, histories, intentions, and consciousness that enter into decisions and, it is because of these that humans act rather than react (Jochim 1979; Shennan n.d.). Therefore humans cannot be reduced to maximizing energy or optimizing calories. Decisions and outcomes are complex and if one is not trying to be reductionistic in the extreme these actions should be included in a model. In addition, there is the significance of historical trajectory to certain actions.

Critics claimed that the ecosystemic approach was incomplete in explaining human actions within society and cultural change. The root of change had been "adaptation" (Kirch 1980). In this model, participants adjust to remain in homeostatic balance. This type of single, goal-oriented, adjustment has received a lot of criticism. What was needed, they said, was a different mechanism for change, a framework for intentional action.

Can these criticisms be accommodated in an ecosystemic approach? Is the ecosystemic approach merely a heuristic, organizing device? Must we throw out the whole model? The ecosystem *should* include the social environment but this is far from easy. It requires the incorporation of fundamentally new concepts in the study of human ecology. Perhaps in the 1990's we will see a wedding of the useful aspects of the ecosystemic approach with the theory of practice (Bourdieu 1977) or of structuration (Giddens 1979).

These theoretical orientations focus on human intention, action, and daily life as the cause of change. Humans are actors, and while they maintain a system, including their inter-relationships with the environment, they do so, not as passive participants in a homeostatic world, but rather as reflexive self-regulating human agents (Giddens 1979). That is, individuals act and react routinely based on habits, but they have the consciousness to evaluate the situation and alter their actions based on changing opinions. These actions are derived from institutionalized structures but intentions and situations can change, and thus actions can change, changing outcomes and, therefore, the society. In short, people have the capacity to cause change through their practical consciousness. This approach provides a framework for change originating from humans and their culture, the aspect considered lacking in the ecosystem approach. But, as Shennan (n.d.) has astutely pointed out, this social theory does not provide a mechanism for action. These, he believes, could come from the concepts of natural and cultural selection (Boyd and Richerson 1985).

Given this suggested refinement of the ecosystemic approach, we return to the ecosystem model and ask what are its useful and important aspects that can continue to benefit anthropological issues such as social organization, and political stratification? What aspects of the model can be applied in archaeological studies?

Complexity in a model still is important. Despite criticisms, the ecosystemic approach, as originally conceived, can provide an interactive picture of past lifeways. If we include the additional theoretical perspectives noted above, fluidity of practice and situated change can be achieved. An aspect of the ecosystemic approach especially appropriate in archaeology, has been the focus at a level of analysis above the individual community, with many studies now modeling regional systems (Jochim 1984; Fish & Fish, this volume).

Rather than focusing on one cause for cultural change, a benefit of the ecosystemic approach, is that any part of society can be the nexus of contradiction and change. Thus, if the ecosystemic approach is not seen as privileging the environment as the major instigator for change, change can be generated from any nexus (Foucault 1980). Towards this goal, archaeologists will be interested in tracking different contexts of cultural change through time, analyzed by retaining complexity of the social and the environmental in a study. The material world should be studied and involved in the explanation, including pollen, soil, and climate, but this also means that we must include the cultural contexts of the people and as best we can, their view of that landscape, to better understand how their decisions were formed. With this kind of reformulated ecosystemic approach, we should be able to move forward in the study of how humans transformed their landscape and how this ties in with social and political change.

As an initial exercise on how this might be done, I have chosen an archaeological example in hopes of showing how the economics of a group changed dynamically during political and social change. I will use the economic and political pre-Hispanic trends of an Andean group, the Sausa, to do so. This example is based on my archaeological research in the central Andes of Peru. It illustrates how I re-conceived both the contextual orientation and the ecosystemic approach in the execution of this research. In so doing, I attempt to present a series of interrelationships and their change through time; including the environment, social organizations, political transformations, and economic variables. To do this, many threads of data are brought in to build a context for understanding the historical trajectory of these processes. The constraints are determined from within. This means that the significant nexes for study are chosen from the important characteristics of the human population, not as Lees and Bates (1984) propose, as always emanating from the environment. By opening up the system to include the reflexive actions of humans in their environment, we can begin to gain our goal of a fuller, more understandable view of a society, while still holding onto the dynamic processual ecosystemic approach.

An archaeological example

Before the Inka's arrival and conquest, we have evidence of localized political centralization throughout the highland regions, while major states developed on the coast. To try to understand these changes as well as their causes, I will present a picture of material changes over a one thousand year period. This will be provided in three temporal phases,based on changes in ceramics, settlement size and location, lithic tools, and maize varieties. Change will be interpreted by comparative artifact frequencies and types, agricultural production data, settlement location, and symbolic evidence. This evidence will then be discussed on a larger, interactive scale proposing social, political and economic changes. By trying to contextualize the material within a temporal perspective, a web of continuous, fluid change will hopefully emerge.

The Wanka, an ethnic polity of central Peru, have been the focus of investigation by the Upper Mantaro Archaeological Research Project since 1977 (for example, Earle et al. 1980, Earle et al. 1987; Hastorf et al. 1989). Previous research was carried out primarily by David Browman (1970) and Ramiro Matos and Jeffery Parsons (1978), although many reports have been published of various research projects (see Earle et al. 1987 for a history of earlier research). As a research team, we are interested in the economic and political changes within the northern population of the valley, today called the Sausa (or Xauxa).

The Sausa have lived in the Upper Mantaro Valley, an intermontane valley in the central Andes, about 250 km east of Lima, for more than 2,000 years. Environmental diversity is not just a characterization of the Andes; it is a constant force influencing the economics, and cultural history of its residents, and it dictates their range of available resources and activities. Frost, rainfall, and topography constrain the production of crops and animals today, just as they did throughout the prehistory studied so far.

The study region is approximately 20 by 20 km in size, surrounded topographically by high peaks and plateaus with elevations ranging from 3300 to 4200 m. Small valleys cleave steep rocky hills and higher rolling plateaus that rise east and west up to the craggy snow-covered peaks of the two parallel Andean Cordillera. The home of the Sausa is bounded on three sides by the high, nonarable *puna*, a treeless, windswept plateau that extends up to the Cordilleran peaks.

The Sausa live in, and use, four main environmental zones that are closely intertwined; the valley lands, the hillsides, the rolling upland plateau, and the higher puna plateau. These land-use zones are separated by different temperatures, soils, and moisture and are planted in different cropping patterns (Hastorf n.d.). Dry farming dominates all zones, but irrigation occurs in the valleys and hillsides near springs and streams. Today, as in the past, the farmers of the region produce the two staple crops of the Andes, maize (*Zea mays*) and potatoes (*Solanum*), as well as several other crops. The valleys and lower hillsides can produce maize up to 3600 m, and tubers grow up to 4100 m (see Seltzer and Hastorf n.d. for details on crop geographies in these microzones). Since the first evidence of human occupation of the area, the Sausa herded camelids especially "llama" (*Lama glama*), and alpaca (*Lama pacos*) permanently in the puna and less constantly among the fallow fields of the arable zones. The people, therefore, have lived in a region that provided a range of different production opportunities. Of special note is the proximity of these zones to each other. A small shift in residence will place a group in a different production zone.

People have been residing in this setting (which has not changed significantly in the last two millenia) since 800 B.C.. Although archaeological investigation of the earliest human occupation in the region is incomplete, we think that when the hunter/gatherer/herder population moved permanently off the puna into highland Andean valleys, they did so to take advantage of the diverse resources, focusing early on the arable land. Legumes are known to have been planted at least 6,000 years ago (Lynch 1980). Although we have no precise data for the domestication of the potato, it is thought that the potato was domesticated in northern Bolivia (Hawkes 1989:495), with the earliest date being 10,000 B.C. at Tres Ventanas cave in Peru. Maize had been cultivated on the coast at least from 2300 B.C.. The varieties adapted to the elevations in the Mantaro region probably developed well after that date (Pearsall 1978). The 13 sites found dating to ca. 800 B.C are located slightly off the valley floors near springs, with easy access to upland grasslands and low slope farmlands (Earle, in Hastorf et al. 1989).

By the beginning of the Early Intermediate Period (ca. 200 B.C.), the archaeological evidence suggests that small farming/herding hamlets were scattered throughout the lower hillsides and valley lands. For approximately one thousand years the region's population grew and built new villages. These are scattered fairly evenly across the landscape, focusing on water sources and easy access to several microzones. Although few sites from this time period have been excavated, the evidence suggests that all sites were taking advantage of the local conditions to produce crops and animals. Of the 64 sites surveyed and tentatively dated to this phase, the average size is 4.4 ha. Some of the sites located on hillsides (6) are clearly surrounded by step-slope terraces that would have allowed them to reclaim the often steep hillsides for farming. Site dating based on the ceramic seriation shows that new sites were forming while earlier settlements continued to be occupied. This evidence of site fissioning suggests a pattern of cultural redundancy and growth without structural change (Graber 1984).

The overwhelming sense in each hamlet's material is of local self-sufficiency. Simple crafts, mostly pottery and stone tools, were produced from local raw materials. Exchange occurred on a small scale. From excavations we know that Sausa tools included projectile points, small mortars and pestles, and a range of scraping tools (Russell, in Hastorf et al. 1989). The ceramic inventories include simple storage jars, grain-toasting jars, and serving bowls, made in a wide variety of clays and tempers. The bowls have few colors and simple design styles.

Of note are the ceramic figurines. In the earliest deposits from this period, we have encountered human (often male) plank figures holding knives and trophy heads. This warrior image is the earliest type of ceramic figurine in the region. The motif is the same as the wide-spread Early Horizon North Coast-Central Andean symbolism of human deities as warriors holding images of power. It suggests that the Sausa were linked symbolically with other Andean polities through the use and reproduction of this image. Later in this period and on into the Middle Horizon, the plank figures are replaced by camelid figurines, many of which are smiling and fat (and perhaps pregnant). Interpretation of these figurines is difficult, but power and fertility are the most obvious conclusions, especially because these are what the Andean people say about their own figurines (*illas*) today (Isbell 1978), and it is also what they associate with them when they see pre-Hispanic figurines in museums (Allen 1988). The camelid figurines drop out in the later Wanka I phase. In addition, sacrifices of six-to-eight month-old camelids (complete bodies with no butchering marks) were found throughout the excavated domestic areas. Lack of butchering marks suggests that the animals were buried without meat removal (Sandefur 1988).

The botanical data indicate that potatoes, maize, and quinoa (*Chenopodium quinoa*) were the common crops grown at the lower elevation sites. Potatoes and quinoa dominate at the hillside sites, although all crops were present at each sampled site (Hastorf n.d.). Faunal data show that the bones of domestic camelids are the most common find, as wild species occur much less frequently. The camelid sacrifices and the regular camelid bones in the domestic refuse suggest that, while there is some hunting, herding is the main source of meat in the diet.

The ceramic assemblage includes basic serving bowls, cooking jars, and storage jars. Although the styles vary throughout the entire prehistoric sequence, the functional types do not (LeCount

1987). Like food production, food processing throughout this time remains uniform. The most common ceramic styles of this time are widespread throughout the region, suggesting regular inter-settlement interaction and similar symbolic reproduction at many production locations. Some copper items of local origin are found in burials. Obsidian is common, although its distribution as a long-distance trade item does not suggest restricted exchange, the pieces are small, mainly formed into small points and chipping debris (Russell in Hastorf et al. 1989).

The Middle Horizon (A.D. 600 - 1000) is named after a pan-Andean phase when the Tiwanaku and Wari polities were active, although in this area it is not chronologically equivalent to other areas in the highlands. Until the 1986 field season we were not able to distinguish this phase from the Early Intermediate Period, so similar were many of the materials. In the central Andes, the Middle Horizon has been identified by the presence of Wari ceramics or rectangular architecture, associated with intrusive political or economic interactions from the major polity to the south (Schreiber, pers. comm.). Because we have found very few Wari artifacts in the Sausa region and no hints of Wari architecture, we can not see direct Wari impact on the region. We therefore originally assigned these strata and sites to a general Early Intermediate/Middle Horizon phase.

From the 1986 site survey and more detailed ceramic seriation, we now believe that sometime around A. D. 500-800 the Sausa built more settlements in the uplands and on hills (16 out of 55 sites are on hillsides). The cause for this settlement shift upslope has not been systematically researched, although competition over valley lands and a growing sense of territoriality and identity are likely explanations. One can see how the new settlements are close to earlier communities and, like the earlier hillside sites, are often surrounded by step-slope terraces. The region's sites tend to cluster in and around the northwest Yanamarca Valley suggesting that this region was a locus of economic and social life. Economically, what is unique about the Yanamarca Valley in comparison to the other nearby tributary valleys is that it is closer to the upland zone. All valleys in the region are productive and have easy access to the puna, suggesting that reasons for this clustering were probably more social-cultural than economic. New settlements being built in the same area suggests a desire of the population to stay close to fields, kin, neighbors; and using modern analogies, near to kin-based earth deities (Bastien 1978). This evidence suggests a society of agrarian people, using their nearby land most intensively, but not being forced to divide into specialists or to use land very far away. The different subpopulations may have developed separate identities but their interactions were still fluid and regional.

A new trend of larger but fewer settlements characterizes the next phase, called the Wanka I (ca. A. D. 800-1350). In general, this implies that groups are not splintering off as they had done before, but are staying together. Of the 36 dated settlements, one third are new and these tend to be near a series of recently abandoned sites, as if the local hamlets decided to aggregate. In addition, these settlements tend to cluster in corners of the valleys, continuing the trend seen in the previous phase of groups of sites being associated spatially. There is no evidence that the deserted regions in between these clusters have become less productive, nor the continually occupied areas more productive. Excavated houses show a continuity with the earlier phases: stone foundations topped by adobe blocks built into circular structures and joined by a patio wall (Hastorf et al. 1989).

What separates this phase artifactually from the earlier phases are new ceramic types with more stylistic variability, and a lack of figurines, a change in agricultural production and processing tools suggesting agricultural intensification. Production and exchange of ceramics is evident but pottery distributions appear to have been more circumscribed, suggesting less region-wide interaction and more localized intra-valley exchange (LeCount 1987). As with the settlement pattern data, we see evidence for increased social clustering both on the landscape and in local exchange relations. Some spatial boundedness has been increasing socially and economically.

The ceramics found on the Wanka I sites are in new shapes with new colors and designs. Overall, the serving bowl sizes increase, suggesting either larger families or gatherings. Although figurines drop out, the camelid motifs continue on the large storage jars as embossed faces around the necks or lugs.

Evidence of new agricultural intensification can be seen in the stone tools. The addition of stone hoes suggests a development of new agricultural methods; perhaps more intensive cropping practices, including digging deeper in the soil, more frequent cultivation, or the expansion of cultivation onto new soils. The place for deep soil cropping is in the poorly-drained but fertile valley bottoms or the rolling uplands. In addition, a new form of grinding implement is found consisting of a large flat rocker grinder (*batane*) that could process much more grain than the earlier mortars and pestles (Russell in Hastorf et al. 1989). New larger kerneled races of maize also enter the botanical record (Johannessen and Hastorf 1989), suggesting increased maize yield and interest in its manipulation. More maize in the diet, as a bulk carbohydrate, is documented in the isotopically analyzed encrustations of short-necked cooking jars (Hastorf and DeNiro 1985). The isotopic evidence from the interior encrustations suggests increased maize boiling, possibly because of a dietary shift with increased emphasis on agricultural products, especially maize (Hastorf and DeNiro 1985). Increased interest in maize is culturally very provocative because it is not always the highest yielding crop as it is the most sensitive to the cold regional climate, yet people increased their processing of it. It alludes to an increased interest in the symbolic aspect of the crop in the highlands, one of ritual importance and sacredness, as maize beer has been one of the important elements of ritual, sharing, and offering (Murra 1960, 1980; Skar 1981).

The mix of agricultural crops present in the archaeological deposits, however, is basically the same as in earlier times with a slight increase of maize (Hastorf 1983:252). Camelids continued to be buried, butchered and eaten, with fewer wild species (Sandefur, in Hastorf et al. 1989). Compared to earlier levels, the frequency of worked bone tools in these Wanka I deposits increases, suggesting increased in-house processing of plants, wool, and hides. Local metals are present at the same density as in earlier periods. In terms of long-distance trade items, there is inconsequential evidence of obsidian, although in the 1986 Wanka I excavations at Pancan, we found one piece of *Spondylus* shell from Ecuador.

In sum, the Wanka I evidence suggests a general intensification of local household activity and production with little inter-regional trade. Either households were becoming more bounded, completing more work within their walls or more work was occurring throughout the population. At the same time, we see an increase in discrete local interactions in the increasingly restricted ceramic distributions. We infer little pressure on resources, as settlements are located near prime valley and hillside agricultural and pasture land. There is no physical evidence of inter-site tensions (defensive walls), and the few warrior symbols that existed earlier drop out of the assemblage, perhaps changing to the more contemporary pan-Andean animal and geometric designs.

This phase allows us to view new social relations as group dynamics shifted. The evidence suggests a greater sense of group identity linked to the local landscape with an increased separation between subvalley populations. In addition we have increased visibility of in-house production and processing. Could it be because houses are more bounded and identifiable, or because of increased interpersonal exchanges and gifts in the form of gatherings and feasts? While the second proposal is an analogy based on ethnography, all of the material fits such an idea. If so, the inhabitants are intensifying their interactions to include more food exchange serving larger gatherings that go beyond the individual households.

In the last cultural phase of endogenous development, the Wanka II, we see the artifactual evidence for what I am calling *incipient inequality* in the Sausa population. During this Wanka II phase (A.D. 1350-1460), which was curtailed by the Inka invasion, we have a marked change in many of the material correlates of behavior I have presented thus far. Production and exchange evidence from excavated households differs from the earlier phases. Evidence for increased production costs can be seen in all artifactual classes. In addition, the architectural data suggest heightened political differences within and between sites.

The demographic distribution changed, with aggregation into 23 Wanka II settlements. The population moved onto massive sites in defensive knolltop locations, overlooking the uplands and the valleys. The ratio of hillside/upland sites to valley sites goes from .09 to .29 to .25 in the earlier periods. In the Wanka II time the ratio is .70. None of the Wanka II sites are located in the flat lands of the valleys, all are in protected canyons or on hillside knolls. Given the earlier trend of agricultural intensification in the valleys and our empirical information that upland fields would have been less productive than the lower fields, we know that the switch could not have been driven simply by the need to increase agricultural production efficiency. Further, we are aware that after this move upslope, the climate became slightly colder, thus making the higher fields even less desirable (Seltzer and Hastorf n.d.).

In the Wanka II period we see an escalation of village fusion that began in the Wanka I; villages that were previously separate joined together with other villages in these larger and often new settlements. This population aggregation would produce increased daily interpersonal interaction causing new strains on social relations between residents. Everyone had to share the same local resources, such as water, fuel, and narrow paths, as few had nearby corrals for their animals as had been common previously.

These knolltop settlements are quite well constructed from locally quarried limestone. The regularly occurring artifacts and hearths in the household compounds imply a range of daily activities. The main changes from the earlier adobe brick compounds are the stonewalls and their denser packing. Carbon dating, settlement distribution, and ceramic seriation substantiate that the settlements were not simply retreats during time of war but were occupied by a large population throughout the 110 or so years of the Wanka II phase.

The site sizes suggest both a settlement hierarchy and a political hierarchy. Whereas the largest sites in the Wanka I phase were 8 to 10 ha, with perhaps 1,000-2,000 inhabitants, there were two large Wanka II centers, Tunánmarca (25 ha) and Hatúnmarca (74 ha), that dominated this northern settlement system. From detailed mapping of the standing architecture, we have estimated the sites' populations to range from 10,000 to 16,600 for Hatúnmarca and 8,000 - 13,400 for Tunánmarca (Earle et al. 1987:11).

These two centers each had a constellation of smaller villages surrounding them. Affiliation of each of these smaller villages with one of the larger centers is evident from the notably different ceramic frequencies (LeBlanc 1981) and chert distributions (Russell 1988) between the two site clusters. These distributions suggest restricted exchange networks within each alliance, an escalation of the earlier trends seen in the Wanka I ceramic distributions. The build up of these spatial boundaries seen in the artifact distributions suggests competition between social groups (Bradley and Hodder 1979; Hodder 1979), more visible in the Wanka II times.

The Wanka I collections displayed evidence of bigger serving bowl sizes that I associate with larger gatherings. These larger bowls continue in the Wanka II, increasing in frequency and stylistic variety. At both the household and the alliance level this evidence suggests increased social interaction.

More internal tension and new forms of defense can be inferred from the settlements' construction. The sites' teetering positions on the top of rocky limestone outcrops indicate that the locations were chosen for their defensibility, as one or two defensive walls encircle the habitation area of each knolltop site (Earle et al. 1987:19). Of particular interest are the few, highly fortified site entrances at each settlement. They give a sense not only of keeping the enemy out, but also of monitoring the inhabitants' movements entering and leaving the settlements. Moreover, the thousands of people crowded within the walls would have created new dynamics that should have required new forms of social control, and call for new ways to unify the population. Managerial positions would have become a necessity to maintain these cities.

We have evidence for social change in the physical consolidation and site plans, but what form of political change and leadership developed in this region? Here we can gain entry by looking at the early Colonial documents and ethnographies to learn of Andean political structures. Suggestions of the importance of warfare and associated leadership positions in the Sausa come from historic documents recounting interviews with five Wanka leaders in the early Colonial period. The Inka had consolidated the Sausa within the Wanka ethnic group as an administration unit. The interviews contained specific questions about the pre-Inka period and its political organization (Toledo 1940 [1570]; LeBlanc 1981:335). Although the informants were reporting about the time of their great-grandfathers, their answers are surprisingly consistent.

The historic Wanka leaders talked of *sinchicuna*, or war leaders. There is repeated mention of community protection and defense, with concrete statements about the *sinchicuna's* ability to maintain their power through success in war. The warriors, the informant's report, were the bravest Indians chosen as "captains" to lead the Sausa into battle against other Sausa (Toledo [1570] 1940:27). When victorious, the war leaders received captured land and women. Exactly who the women were is unclear but see (Toledo [1570] 1940:23,31). This reward system suggests that the leaders had some ability to appropriate goods. Further, it was through these wars that leaders (and probably their followers) gained access to land and labor (perhaps women and/or slaves). Access to land and labor means increased capacity to produce more goods. Therefore, differences in agricultural products and artifactual frequencies might be linked to this political behavior.

The Wanka informants imply that the criteria for war leader were based on effective personal action. Occasionally, when the son of a *sinchi* demonstrated appropriate qualities, he could become the next leader (LeBlanc 1981:343). This suggests that the positions were limited and contested, and that there was some desire to consolidate power within kin groups. The documents have evidence that these leaders occasionally acted on internal political issues, but that their decision-making authority was limited.

New political structures are suggested also by the unique site organization at the two central settlements. These sites are unique not only because they are massive, forbidding fortresses that required organization to be built, but also because they have open "public" areas within their walls. At each of the two major settlements there is a central division created by two walled-in unoccupied areas and two large central patios with several square structures attached. This new form of architectural space suggests new public activities that could have involved the entire community at one time, whether this meant large political gatherings, social work areas, or ritual feast days. These leaders of war and community maintenance might have managed the bringing together of the different communities by building on the pan-Andean social ideas of dual division and balance, illustrated in the two large central plazas (Silverblatt 1987; Hastorf n.d.); and also in the idea of exchange and reciprocity, thus requiring that the leaders gain more goods in order to host work parties (with maize beer and coca) and to give away to one's followers (Murra 1980).

Both direct and indirect evidence provide a sense of political rank in the Wanka II times. But why did the changes seen at the onset of Wanka II occur then? What prompted this transition to new and larger social organizations and political developments? We can see in the Wanka I and Wanka II evidence of a trajectory towards group formation, both in larger group size and in more spatially discrete locations. Additional evidence suggests that it was a combination of outside changes in the pan-Andean power structure as well as an increase in internal boundedness and tensions over local territories and resources.

By about A.D. 950 the Wari hegemony had collapsed, causing political ripples and power vacuums throughout the central Andes in the known areas of Wari domination and beyond (Schreiber 1987). The Sausa probably were not exempt from being affected by these events. We might assume that altered inter-regional relationships heightened the sense of political insecurity at home and encouraged more firm local alliances. The more discrete interactions seen in the artifact distributions suggest an increase in territoriality and concern for agricultural production. In the Andes, people's residences are linked to political units, land, and earth deities, suggesting that as groups became more identifiable and distinct from their neighbors, their political and social structures also became more elaborate.

How are these changes illustrated in the economic base? From architectural study and subsequent excavation, we claim to identify economic differences between Wanka II habitations. Residential zones are divided up into walled patio compounds that include one or more circular structures surrounding an open space. These are connected by narrow winding pathways linking all compounds to the center of the settlement.

We have identified differences in construction between the house walls and in size between the enclosed patio spaces. Using higher quality architecture, proximity to the central area, patio area, and number of circular structures, we identified and predicted the "richer" households as different from the "commoner" compounds. We assumed that these larger and better made households were associated with wealth and access to labor, and thus we have labeled this group of households the *economic elite* (Earle et al. 1987; Costin and Earle 1989). What are the material differences between these patios, how much wealth did the "elite" have, did they control production of economic goods, what might they have done with the goods, and how did this wealth associate with the social and political events that were going on?

From the Wanka II ceramics, we can tell that both ordinary and luxury items were produced and used at the settlements. We have evidence of local exchange restricted within each alliance, increased specialization, and differential access to certain types of wares. Ceramic specialization is inferred from the distribution of pottery wasters (malformed during firing; Costin 1986). Analysis of the wasters shows that certain ceramic types were produced at individual settlements. In the northern alliance group, for example, the center, Tunanmarca, produced cooking pots; and one satellite, Umpamalca, produced serving bowls and storage vessels. The waster distribution within the patios suggests that it was the commoners who actually produced most of the pottery. However, in this northern site cluster, all types of pottery were found at all of the patios and settlements (LeBlanc 1981; Earle et al. 1987).

The "elite" households had less evidence of production but more ceramics per unit area, more varieties, more fancy types, more labor-intensive types, and more non-local ceramics. Further, they also yielded more fragments of large jars, which we assume were used for storing crops and for producing *chicha*, the maize beer that was (and still is) essential for social gatherings, work parties, and rituals (Murra 1960; Skar 1981). Direct evidence of other specific wealth production exists but is rare. In the "elite" patios we found a greater concentration of spindle whorls used to spin wool (Costin 1984). From ethnohistoric accounts, we know that cloth was an important Andean wealth item, produced to exchange and display status. The "elite" patios also yielded more evidence of metal production in ore, mold spill, and lugs (Earle et al. 1987:74). In general, metal was rare. When it did occur it tended to be in jewelry or cosmetic implements.

The "elite" emphasis on producing status items: *chicha* beer, cloth, and metal, while trading for "utilitarian" goods such as ceramics and lithic blades, suggests they had interests in activities besides economics; that is, they were producing for a prestige network not for control markets as we see today. Yet, the "elite" did not have control of production. While there are different artifact frequencies between the two statuses, all types of artifacts were present and used in both patio groups. Although the wealthy had more of everything, especially the higher quality ceramics, they still produced and procured their own food. The commoner households had the same range of products as the elites just less of the more elaborate types. This suggests that independent specialists produced and exchanged the pottery (Costin 1986). Household members, servants, or slaves might have been working in the rich compounds producing cloth or metal, but at the time of abandonment (as far as our data show), these tasks were not monopolized or controlled by the "elites".

We see the same production and distribution pattern in the stone tools. We found evidence of site-level specialization in blade production: a dense presence of cores and primary debitage in commoner compounds at Umpamalca. Although initial production took place at this one site, used tools are found in every patio and site sampled (Earle et al. 1987). These blades indicate not only that distribution was open, but that all households, *including* the "elites", used lithic tools in their homes.

Evidence of prehistoric intensive land-use reveals the economic effect of population aggregation in the Wanka II period. When the Sausa relocated into the uplands, they adopted new land-use strategies. The uplands are not as continually fertile as the valleys and are less reliable. In order to maintain adequate yields for the much larger aggregated population, it was necessary for many inhabitants to farm fields that were farther away from the settlement than had been from the smaller Wanka I settlements and to develop intensive farming methods near the upland settlements. Evidence of an irrigation canal that carried water to the northern alliance settlements can be traced across the landscape (Parsons 1978; Hastorf and Earle 1985). Stone-walled terraces are located within and nearby the Wanka II settlements. These intensive strategies resulted in more expensive agriculture (more labor per unit area) but not necessarily more productive agriculture (harvest security is lower in the uplands than in the valleys, suggesting irrigation in the uplands was mainly to secure regular yields, not necessarily to increase them).

The Wanka II agricultural data also reveal the effect of relocation. There is a major shift in what crops were grown. While maize appears in 50 percent of all excavated Wanka I soil samples, it is in only 13 percent of the samples at the upland Wanka II sites. Potatoes are present in 25 percent of the Wanka I samples increasing to 35 percent in the Wanka II samples (Hastorf 1983:252). This crop production reorientation can be partially explained by the shift in settlement location. The arable land around Wanka I valley sites could accommodate all of the Andean crops, including maize. With the relocation, the nearest arable land could not produce maize, but instead would have yielded potatoes, Andean tubers, quinoa, and lupine (*Lupinus mutabilis*). The Wanka II people would have had to travel farther to farm land that could produce maize, making that crop much more expensive. Many Wanka II inhabitants probably had to trade for their maize.

Striking economic contrasts between the "elite" and commoner patios exist in the botanical data. Overall, crops are more dense in the elite samples. The greatest disparity is in the frequency of maize. Forty three percent of the elite samples have maize present while it is in only 12 percent of the commoner samples. Maize is traditionally a very important crop in the Andean world. Today, when it cannot be produced, it becomes a necessary trade good. The maize data from the Wanka II excavations provide clear evidence of differential access to land or trade.

The "elites" either controlled the maize-growing lands, received maize as "gifts," or had greater access to the exchange networks with southern Mantaro Valley residents who could have grown it. We know that one high quality ceramic type, andesite ware, came from the southern Mantaro Valley (Costin 1986:487). "Elite" patios had twice as much andesite ware as commoner patios. Maize could have been traded into the elite households along with or in this andesite ware.

As in the earlier periods, the Wanka II domestic animal bone counts were dominated by llama and alpaca. Much less wild bone was present than in the earlier occupations (Earle et al. 1987; Sandefur 1988). The upland locations placed the populations closer to the camelid grazing lands, and meat may have increased in the diet as a result. Stable isotope analysis completed on Wanka II human skeletons does not suggest that meat was a dominant part of Sausa diet. As expected, "elite" patios have denser frequencies of animal bone than commoner patios.

In this Wanka II phase we see an economic hierarchy in production of subsistence goods (tools and crops), but also in production and access to special status goods. While there were many social changes to the residents' lives with the move uphill, it also effected the economics of the group. The demographic shift reflects the agricultural production changes, but tool, decorative material, and status item frequencies (including maize) suggest political and social transformations. Some households did amass more than others, mustering a labor force to build larger compounds and produce certain goods. The new "economic elites" did not control all production however. They produced more high status items, but had not consolidated the production of most goods.

The Wanka II data show economic, political, and social change. Change through time is evident in material distributions, settlement patterns, architectural organization, production and exchange of goods, and the creation and elaboration of managerial positions. But in the Wanka II phase, it is still not known if the "economic elite" were the political or the managerial leaders. New positions of power and authority had been created in all domains of society but how centralized were they?

In the Wanka II we see the onset of specialized production and distribution of certain items, architectural space, agricultural production, exchange, and decision making. The decision-making evidence is most clear within each alliance. We know that the restructuring of daily life produced new constraints and expenses. The unprecedented settlement size and associated increases in interactions created new institutions for group cohesion and intergroup negotiation. At the same time, this produced increased tensions and factionalism, seen in the site walls, divisions, and controlled entrances.

Given today's Andean ideology of hierarchy in deities (Allen 1988) and the nested political organization that existed throughout the Andes on the eve of the Spanish conquest, we might expect the possibility of social ranking among the Sausa (Hastorf n.d.). The ease of the Inka conquest throughout Andean groups is often attributed to stratification already in place among the conquered people.

Over time, creating a perpetual war threat and subsequent need for protection, the war leaders could have extended their power to organize military actions into power over social relations. To convince people of the benefits of the more costly upland habitation, political leaders first had to convince villagers to join into larger social units and then to organize work parties to build the new settlements. But, as a decision-making body these leaders were not totally specialized in the Wanka I times. A restructuring of the decision-making positions through success in war could have allowed the leaders to accumulate goods and women, whereas previously they could have only traded for these items. Through these channels political and social elite might have become an economic elite. Yet they never became very centralized. Sausa leaders of one alliance never had the political or military power to take over the competing alliance.

The Sausa data suggest that various new positions were developing sometime during the Middle Horizon and the Late Intermediate periods. The presence of these social, political, and economic hierarchies is evident by the end of the Late Intermediate Period, but this was always tempered by the other group of Andean principles—balance and exchange (Bastien 1978; Isbell 1978). Once the social compromise was made, changes in access to goods, productive land, and specialization occurred as unintended consequences. Coercion, warfare, and social stress were the motives for action, but balance and reciprocity caused a leveling of positions. It appears that the underlying causes of change were not simple economic constraints, but were based on a larger, more fluid changing dynamic. The landscape was changing as people were relocated. They still could have been fighting over land, but it would have been over specific locations and over certain soils, not over general insufficiency. Internal conflict and consensus between factions were operating, applying existing Andean ideologies and identities such as warfare and stratification. This creates a complex yet active view of change.

Conclusions

For the Sausa, agricultural intensification was not simply a product of environmental constraints and the need to increase production to feed more people, as some have proposed (Carneiro 1970). It can be due to increased interest in social gatherings, or may be in reaction to demographic shifts due to internal and external social pressures, or it can be due to a legacy of earth spirits inhabiting certain microzones, restraining people from moving away from an area, forcing them to intensify. The same type of multiple argument can be proposed for political change; what is seen in the sequence may tie into regional hierarchies and Andean influences that are brought in or may be unintended consequences of a group that operates on the Andean concepts of balance, reciprocity, and complementarity, giving up some autonomy in the face of aggressing forces.

Several themes seem to have been important in the Sausa "ecosystem" over time: 1) Warleaders became the nexus for overt power and authority as their society became more circumscribed, 2) Arable land was the medium for joining the people together, as crops and stores had to be protected, but technologies had to be developed to intensify production with aggregation, 3) Maize became an important symbol of social interaction, increasing in use, interest, and importance over time, and 4) Social gatherings seemed to expand out from figurines' burial to eating in family households and even designated civic space. These large spaces also reflect the Andean concepts of exchange, reciprocity, duality, and balance.

In this archaeological example I have tried to track and continually weave together a range of data over time to illustrate what an expanded contextual-ecosystemic approach might look like. This picture is not an unidimensional energy flow diagram but has cultural dimensions added all the time. This type of presentation however requires both a great deal of data collection and analysis, and carefully interpreted material. Such a complex cultural vision is beyond classical forms of systems theory and reductionistic boxes and arrows. In the broadest sense, this kind of analysis places *human behavior* at the core of the analysis.

The changes that are evident in the material culture can be viewed as more than human reactions. We can see glimpses of human action, potential causes, and transferred meanings in the evidence.

We can follow the major sources of energy; land, labor, agriculture, animal herds, etc, and see how they intersect with political and social dynamics; sometimes displaying social motivations, sometimes economic ones, sometimes environmental constraints.

While there have been many fair critiques of the ecosystemic approach, by including social action, political forces, and a sense of time (and history), the critics may be silenced and the effective environmental aspects can be retained in future archaeological inquiry. Further, it illustrates how political transformations affect subsistence production and the development of social hierarchy. The usefulness of a long-term perspective can be seen in this example. We see a fluid but complex interaction, each part providing evidence of cultural change with changing emphases each time. As the theory of social action shows us, the same event at a different time can cause a different action, due to new circumstances combined with old habits. Archaeological research is constantly uncovering this variation, the same events are performed again and again, yet over time alterations occur and new traditions are formed. These changes out of the old can be in any domain, maybe it is a new form of recultivation, a new type of peace-making compromise, or the decision not to trade ceramics with the folks down the valley. It is up to the archaeologist to bring these data together and form an explanation about the past, privileging some actions at one time while emphasizing others at another.

Acknowledgments

I would like to thank Emilio Moran for challenging me to attempt this stimulating problem. The material reported on was collected in field work that began in 1977 by the Upper Mantaro Archaeological Research Project, directed by Timothy Earle, Terence D'Altroy and myself, although many others have been involved over the years especially Cathy Scott, Lisa LeCount, Cathy Costin, Glenn Russell and Elsie Sandefur. The field work was funded by the National Science Foundation grant number BNS 82-03723, a Fulbright-Hayes Doctoral Dissertation Grant, and by the College of Liberal Arts at the University of Minnestoa. National Science Foundation Grant BNS 84-51369 funded much of the artifact analysis and the 1986 field season.

References Cited

Allen, Catherine J.
1988   *The hold life has, coca and cultural identity in an Andean community* Washington D. C.: Smithsonian Institution Press.
Bailey, G. and A. Sheridan
1981   Introduction, In *Economic Archaeology*, edited by G. Bailey and A. Sheridan, British Archaeological Reports International Series, Oxford, 96:1-13.
Bastien, Joseph
1978   *Mountain of the Condor, metaphor and ritual in an Andean ayllu.* American Ethnological Society Monograph 64, Washington D.C.
Bourdieu, P.
1977   *Outline of a theory of practice*, Cambridge: Cambridge University Press.
Boyd, Robert and Peter Richerson
1985   *Culture and the evolutionary process*, Chicago: University of Chicago Press.
Bradley, Richard, and Ian Hodder
1979   British prehistory: an integrated view. *Man* 14:93-104.
Browman, David
1970   *Early Peruvian peasants: the culture history of a central highland valley.* Unpublished Ph. D. dissertation, Harvard University.
Carneiro, Robert L.
1970   A theory of the origin of the state. *Science* 169:733-738.
Costin, Cathy L.
1984   The organization and intensity of spinning and cloth production among the late prehispanic Huanca. Paper presented at the 24th annual meeting of the Institute for Andean Studies, Berkeley, Ca.
1986   *From chiefdom to empire state: ceramic economy among the prehispanic Wanka of highland Peru.* Ph.D. dissertation, Anthropology Department, University of California, Los Angeles.

Costin, Cathy and T. Earle
1989   Status Distinction, and Legitimization of Power as Reflected in Changing Patterns of Consumption in Late Pre-Hispanic Peru. *American Antiquity.* 54 (4): 691-714.

Earle, Timothy K. and Andrew L. Christenson eds.
1980   *Modeling change in prehistoric subsistence economies.* New York: Academic Press.

Earle, Timothy, Terence D'Altroy, Catherine LeBlanc, Christine Hastorf, and Terry LeVine
1980   Changing settlement pattern in the Yanamarca Valley, Peru. *Journal of New World Archaeology* 4:1:1-93.

Earle, Timothy, Terence D'Altroy, Christine Hastorf, Catherine Scott, Cathy Costin, Glenn Russell, and Elsie Sandefur
1987   *The effects of Inka conquest on the Wanka domestic economy.* Institute of Archaeology, University of California, Los Angeles.

Flannery, Kent V.
1972   The cultural evolution of civilization. *Annual Review of Ecology and Systematics* 3:399-426.

Foucault, M.
1980   *Power/Knowledge*, edited by C. Gordon, Hassocks: Harvester Press.

Giddens, Anthony
1979   *Central problems in social theory.* London: Macmillan.

Graber, Robert B.
1984   Circumscription as the cause of social growth: a mathematical interpretation. Paper given at 83rd annual meeting American Anthropological Association, Denver.

Hastorf, Christine A.
1983   *Prehistoric agricultural intensification and political development in the Jauja region of central Peru.* Ph.D. Dissertation, Anthropology Department, University of California, Los Angeles.
n.d.   *Resources in Power: Agriculture and the onset of political inequality before the Inka*, New York: Cambridge University Press.

Hastorf, Christine, and Michael DeNiro
   1985   A new isotopic method used to reconstruct prehistoric
           food plants and cooking practices. *Nature*
           315:489-491.
Hastorf, Christine, and Timothy Earle
   1985   Intensive agriculture and the geography of political
           change in the Upper Mantaro region of central Peru. In
           *Prehistoric intensive agriculture in the
           tropics*, edited by I. Farrington. British
           Archaeological Reports, International Series
           232:569-595.
Hastorf, Christine, Timothy Earle, H. E. Wright Jr., Lisa LeCount,
Glenn Russell, and Elsie Sandefur
   1989   Settlement Archaeology in the Jauja region of Peru:
           Evidence from the Early Intermediate Period through the
           Late Intermediate Period: A report on the 1986 field
           season. *Andean Past* 2:81-129.
Hawkes, J. G.
   1989   The domestication fo roots and tubers in the American
           tropics. In *Foraging and Farming*, edited by D.
           Harris and G.C. Hillman, London: Unwin Hyman,
           pp.481-503.
Hodder, Ian
   1979   Social and economic stress and material culture
           patterning. *American Antiquity* 44:446-454.
   1986   *Reading the Past*, Cambridge: Cambridge
           University Press.
Isbell, Billie Jean
   1978   *To defend ourselves: ecology and ritual in an
           Andean Village*, Austin: Institute of Latin American
           Studies, University of Texas.
Jochim, Michael
   1979   Breaking down the system: Recent ecological approaches
           in Archaeology. *Advances in Archaeological Method
           and Theory*, edited by M. Schiffer, New York:
           Academic Press 2:77-117.
   1984   The ecosystem concept in archaeology. *The ecosystem
           concept in anthropology* edited by Emilio F. Moran,
           Washington D.C.: American Association for the
           Advancement of Science, pp. 87-102.

Johannessen, Sissel and Christine Hastorf
1989   Corn and culture in central Andean prehistory, *Science* 244:690-692.

Keene, Arthur
1981   *Prehistoric foraging in a temperate forest: a linear programming model.* New York: Academic Press.

Kirch, Patrick V.
1980   The archaeological study of adaptation: Theoretical and methodological issues. *Advances in Archaeological Method and Theory*, edited by M. Schiffer, New York: Academic Press, 3:101-156.

LeBlanc, Catherine J.
1981   *Late prehispanic Huanca settlement pattern in the Yanamarca Valley, Peru.* Ph.D. Dissertation, Anthropology Department, University of California, Los Angeles.

LeCount, Lisa J.
1987   *Towards defining and explaining functional variation in Sausa ceramics from the Upper Mantaro Valley, Peru.* M.A. thesis, Department of Anthropology, University of California, Los Angeles.

Lees, Susan H. and Daniel G. Bates
1984   Environmental events and the ecology of cumulative change. *The ecosystem concept in Anthropology*, edited by Emilio F. Moran, Washington D.C.: American Association for the Advancement of Science, pp.133-159.

Lynch, Thomas
1980   *Guiterrero Cave: Early man in the Andes* New York: Academic Press.

Murra, John V.
1960   Rite and crop of the Inka state, *Culture and History*, edited by S. Diamond, New York: Columbia University Press, pp. 393-407.
1980   *The economic organization of the Inca state.* Greenwich, Conn.: JAI Press. Orig. publ. 1965.

Odum, E. P.
1953   *Fundamentals of ecology*, Philadelphia: W. B. Saunders Co.

Parsons, Jeffery
        El complejo hidráulico de Tunánmarca: canales, acueductos y reservorios. *El hombre y la cultura*

andina: actas y trabajos del III Congreso, tomo II, edited by R. Matos M, pp. 556-566. Lima: Universidad Nacional Mayor de San Marcos.

Parsons, Jeffery and Ramiro Matos Mendieta
1978   Asentamientos prehispánicos en el Mantaro, Peru: informe preliminar. *El hombre y la cultura andina: actas y trabajos del III Congreso, tomo II*, edited by R. Matos M, pp. 539-555. Lima: Universidad Nacional Mayor de San Marcos.

Pearsall, Deborah
1978   Phytolith analysis of archaeological soils: evidence for maize cultivation in Formative Ecuador *Science* 199:177-178.

Rappaport, Roy A.
1967   *Pigs for the ancestors.* New Haven: Yale University Press.

Russell, Glenn
1988   *The impact of Inka policy on the domestic economy of the Wanks in the Central Highlands of Peru: Stone tool production and use.* Ph.D. dissertation, University of California, Los Angeles.

Sandefur, Elsie
1988   *Andean Zooarchaeology: Animal use and the Inka Conquest of the Upper Mantaro Valley,* Ph.D. dissertation, University of California, Los Angeles.

Schreiber, Katerina J.
1987   From state to empire: the expansion of Wari outside the Ayacucho Basin. *The origins and development of the Andean State*, edited by J. Haas, S. Pozorski and T. Pozorski, Cambridge: Cambridge University Press. pp. 91-96.

Seltzer, Geoffrey and Christine Hastorf
n.d.   Climatic change and its effect on prehispanic agriculture in the Central Peruvian Andes. *Journal of Field Archaeology* (in press)

Shennan, Stephen
1990   Evolution, adaptation and the study of prehistoric change. *Between Past and Present: Issues in Contemporary Archaelogical Discourse.* Edited by R. Preucell, Carbondale: Southern Illinois University Press.

Silverblatt, Irene
  1987   *Moon, sun and witches*, Princeton: Princeton
          University Press.
Skar, Sarah L.
  1981   Andean women and the concept of space/time. *Women
          and space*, edited by S. Ardener, London: Croom
          Helm, pp. 35-49.
Steward, Julian H.
  1938   *Basin-Plateau aboriginal sociopolitical groups.*
          Washington D. C.: Bureau of American Ethnology,
          Bulletin 120.
Tansley, A. G.
  1935   *Introduction to plant ecology.* London: George
          Allen and Unwin, Ltd.
Toledo, Francisco de
  1940   [1570] Informacion hecha por orden de Don Francisco de
          Toledo en su visita de las provincias del Per, en la
          que declaran indios ancianos sobre el derecho de los
          caciques y sobre el gobierno que tenn aquellos pueblos
          antes que los Incas los conquistasen. *Don Francisco
          de Toledo, supremo organizador del Per, su vida, su
          obra 1515-1582.* Vol. II, edited by R. Levillier,
          pp. 14-37. Buenes Aires: Espasa-Calpe.
Ugent, D., S. Pozorski, and T. Pozorski
  1982   Archaeological potato tuber remains from the Casma
          valley of Peru. *Economic Botany* 36:82-192.

CHAPTER 6

AN ARCHAEOLOGICAL ASSESSMENT
OF ECOSYSTEMS IN THE TUCSON
BASIN OF SOUTHERN ARIZONA

Suzanne K. Fish
Paul R. Fish

Introduction

The ecosystem concept in archaeology has been aptly characterized as an heuristic device (Moran 1984:10; Jochim 1984:87). Compared to biological and human ecological studies of living populations, archaeological application necessarily entails a more generalized approach. Precision diverges from levels attainable in contemporary investigations because past environmental and cultural variables cannot be directly measured. Reconstruction is an intervening step to analysis. A more restricted range of variables can be identified or observed, and their significance in past ecosystems is inferential. Interrelationships among populations and other variables must be tested indirectly.

In spite of the operational difficulties, ecological orientations are pervasive in current archaeology, fostered by concern with the physical environment of prehistoric populations. A benefit of explicit ecosystem analysis is a rigorous examination of underlying assumptions about interrelationships among cultural and environmental factors in past systems. As a generalized frame of reference for organizing and understanding data (Netting 1984:231-232), ecosystem concepts may even be invoked *ex post facto* in archaeology, if relevant observations have been gathered. Indeed, the process of identifying archaeological systems may legislate such a sequence because data revealing system shape are either acquired simultaneously with detail for particular components, or in reverse order.

Definition of meaningful boundaries is a preoccupation of ecosystem studies (Moran 1984, this volume; Ellen 1984, this volume; Butzer, this volume). Some level of system familiarity is requisite for drawing even the most preliminary boundaries from which to proceed to more detailed examination. The problem is particularly acute for archaeological systems because evidence for components is selectively preserved. The present case study of Hohokam remains in the Tucson Basin of southern Arizona illustrates the operational requirements of ecosystem studies in archaeology and the interplay between data collection and analysis.

The Hohokam

Cultigens and an agricultural lifestyle were adopted in the Sonoran Desert of south-central Arizona (Figure 6.1) by about 1000 B.C. Decorated pottery styles that define the Hohokam archaeological culture appeared in the first few centuries A.D. Cultural identity was maintained thereafter for over 1000 years, until or just preceding the 16th century. Throughout this interval, the Hohokam lived in pithouse structures of wattle-and-daub construction, with the addition of adobe surface structures in the later centuries of the sequence. Elements of ceremonial expression and public architecture such as ballcourts and mounds of earthen construction were shared with Mesoamerican cultures to the south, although in less elaborated forms.

Archaeological study of the Hohokam has been concentrated in the contiguous lower valleys of the Salt and Gila Rivers. This area surrounding Phoenix, Arizona, termed the Hohokam core, is marked by massive irrigation networks totalling more than 400 km of primary canals (Masse 1980) and the highest population densities, estimated in the range of 8 to 25 persons/km^2 (Haury 1976:356; Doyel, in press; Fish and Fish, in press). Other regions of the Hohokam sphere lacked similar perennial rivers with large upland watersheds outside the desert. Farming in these other areas was accomplished through smaller-scale irrigation and a variety of techniques tapping intermittent and ephemeral watercourses. Until recently, archaeological study in non-core regions was limited. Settlement patterns among more dispersed populations were virtually unknown, and it was widely held that expressions of Hohokam high culture were attenuated or absent (See Fish 1989; Doyel 1980a; McGuire and Schiffer 1982).

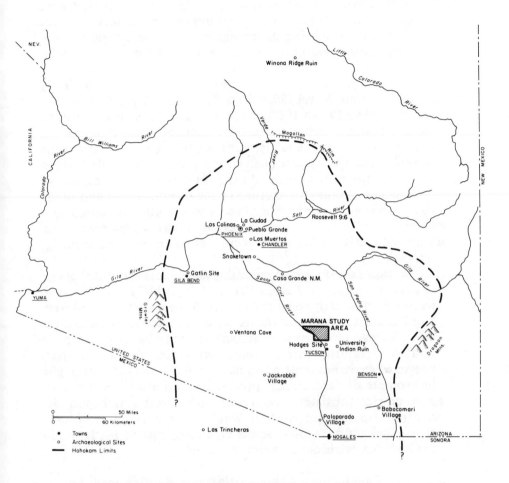

Figure 6.1 Important locations within the Hohokam region of southern Arizona.

In the Phoenix Basin, irrigation systems and larger sites were mapped in the early part of this century before intensive modern development, permitting the recognition of interdependent site clusters along a shared canal network. Units of more closely interrelated sites along the same network have been inferred from the presence of prominent sites with ballcourts and mounds at fairly regular intervals (Doyel 1980b; Wilcox and Sternberg 1983; Gregory and Nials 1985; Crown 1987). Ballcourts are the common form of public architecture during the earlier Hohokam sequence, superseded in later times by mounds. Integrative functions for settlements without public architecture in an adjacent radius are ascribed to the sites with these structures. At the start of the present study, instances of public architecture in the Tucson Basin and other noncore regions implied similar functions of multi-site integration. However, configurations of related settlement surrounding preeminent sites with this architecture were unknown in areas lacking irrigation.

No basis existed for even estimating magnitudes and distributions of populations in the Tucson Basin. The goal of research was to recover a sufficiently comprehensive settlement pattern to address such issues and to document the nature of the underlying subsistence base. Topographic diversity across the linear basin would have created variability in subsistence strategies, suggesting that integrated territories centered on mounds and ballcourt sites might incorporate differentiated productive activities. In order to characterize subsistence diversity and investigate economic relationships among sites, it would be necessary to record archaeological distributions across broad expanses of the basin, in conjunction with environmental data.

Populations, Communities, and Replication

Ecosystem analysis necessitates the partitioning of a continuous reality. Boundary definition of a unit appropriate for study is based on a discernible level of systemic closure, wherein interactions among components are more strongly oriented within than without. For some interactions, human populations may overcome physical separation by long intervening distances, and concepts such as "nested zones" or "graded boundaries" (Ellen 1984: 195-198) may be necessary for different realms of transaction or systemic inclusiveness; the role of extralocal materials in an archaeological system is largely evaluated through quantities and contexts of extralocal occurrence. Nevertheless, following the general orientation of Rappaport (1968), an analytically relevant ecosystem for an archaeological population may be defined by the distributional extent of remains associated with its basic sustenance and reproduction. The integrity of an ecosystem defined in this manner may be indicated by spatial separation from other contemporary systems which also possess internal capacities for satisfaction of subsistence needs.

Segments of the productive potential inherent in regional environment are bounded and integrated through cultural principles by human populations. According to the preceding definition of ecosystem, replications of such culturally-prescribed segments would be independent for the bulk of subsistence needs and would therefore encompass the resources and interactions fundamental to population support and continuity. Archaeological criteria for such a segment or unit would include: (1) environmental opportunity for internal production of most subsistence materials known to have been used; (2) remains of such productive activities; and (3) evidence for territorial cohesiveness and segregation from similar contemporary units. Among the Hohokam, boundary criteria for ecosystems of this nature do not seem best satisfied by single sites or small groups of sites in areas of largescale irrigation or in those emphasizing alternative technologies. The productive unit replicated across the landscape is a community of economically interdependent sites, whose identity and integration is symbolized by shared participation in public functions associated with ballcourts and platform mounds at a central site.

Tucson Basin Case Study

Environmental Setting

As the drainage basin of a major river, the Tucson Basin is a typical focus of regional research in archaeology. Mountains rimming the basin and dividing it from other drainage systems form physical barriers promoting a degree of both natural and cultural closure. Annual rainfall, between 225 and 300 mm, supports a relatively diverse desert vegetation dominated by shrubs, small, leguminous trees, and cacti, including arborescent forms. Somewhat more than half the rainfall occurs in summer as thunderstorms with diameters of less than 5 km. Rapid runoff and high evaporation rates in this season increase the need to concentrate direct rainfall for successful cropping.

The northern basin is the setting of the present study. Topographic diversity is mirrored to either side of the Santa Cruz River between the floodplain and mountains at the basin rim. The river runs near the base of the low Tucson Mountains (maximum elevation = 825 m) on the west, creating a foreshortened valley slope. The more massive Tortolita Mountains (maximum elevation = 1375 m) rise at a greater distance from the river to the east, and the eastern valley slope occupies a majority of the basin interior. The Santa Cruz is intermittent in the study area, but surface flow persists for extended periods near the end of the Tucson Mountains. Here, igneous intrusions of bedrock prevent water infiltration into deep valley fill, creating locally high water tables that might be tapped by shallow wells. Permanent or extended water sources also occur as springs and seeps in drainages near the base of the Tortolita Mountains.

Definition of Communities

A longterm program of regional survey between 1981 and 1986 eventually encompassed a study area of 2100 sq km (Figure 6.2). The basis of the present discussion is a contiguous survey block of 390 sq km near the town of Marana (S.Fish et al. 1989a). This area was covered by surveyors at intervals of 30 meters, permitting the identification of small, isolated loci of extractive activities and evidence of agricultural fields, as well as the more substantial remains of habitation sites. The resulting array of over 700 sites and thousands of small artifact scatters reflects only minor preservation bias. Ancillary geomorphological studies reveal a low degree of site burial for the Hohokam sequence, and historic land disturbance is confined to limited segments near the Santa Cruz River.

The elements of settlement pattern constitute the basic variables of archaelogical ecosystem analysis, but could be segregated into meaningful spatial and interactive units only through examination of distributions at a regional scale. The nature of such configurations and their territorial extent was unanticipated, emerging only after the initial stages of data acquisition. Boundary definition, in this case, proceeded from higher to lower levels. Broader territorial

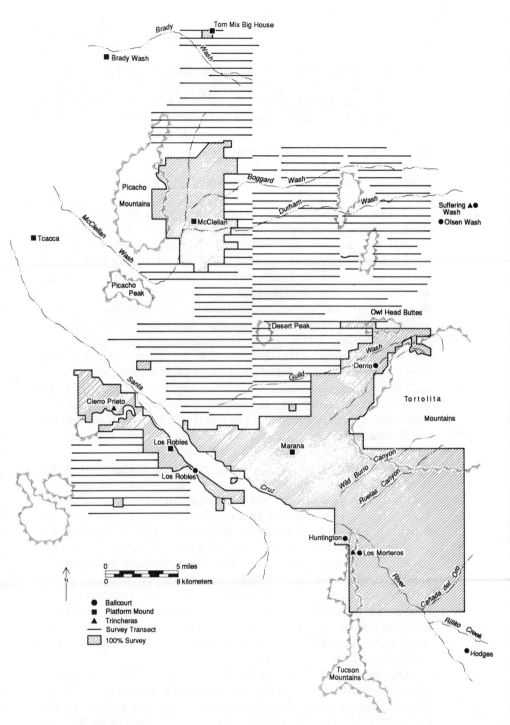

Figure 6.2 Survey areas and important site locations in the northern Tucson Basin.

integration was recognized through settlement clustering and the
positioning of central sites with public architecture as integrative
nodes. Within these multi-site communities, local zones of direct
environmental interaction for inhabitants of individual sites were
then discerned through patterning of residence, productive activity,
and environmental variables.

Community Dynamics and Evolution

Archaeology provides the opportunity for viewing dynamics of
human ecosystems over time. The temporal precision at which this can
be realized rests on the ability to discriminate the chronological
association of relevant remains. The shortest time increments,
corresponding to the finest subdivisions of ceramic style, are
difficult to isolate in Hohokam settlement patterns. Restricted
occurrence of water for seasonal agriculture and permanent domestic
supplies in the desert basins encouraged continuous reoccupation of
optimal locales. Later occupations disturbed earlier ones, and
ceramics of sequential phases are mixed on the surface.
Non-residential sites reflecting hunting and gathering activities or
field cultivation commonly contain small numbers of widely dispersed
artifacts, few of which may be temporally diagnostic. Ecosystems of
a generally synchronous nature are therefore inclusively defined for
the Preclassic period (circa A.D. 700 to 1050) and the early part of
the Classic period (circa. A.D. 1050 to 1300).

Change over time in cultural ecosystems is apparent in the
northern Tucson Basin, in spite of locational constraints imposed by
the desert environment and an essentially static productive
technology over the period in question. However, there are also
aspects of long-term continuity in patterns of land use, which begin
with settlements of the earliest farmers. Sites of all periods
parallel the Santa Cruz River. This concentration includes both
sites at the edges of the floodplain itself and sites situated on the
adjacent lower edge of the valley slope. A second concentration is
found along the flanks of the Tortolita Mountains. Preferred
locations within river and mountain flank bands of settlement are
attested by numerous sites with superimposed occupations and
clusterings of discrete ones.

During the earliest phases (A.D. 300 to 700) defined by ceramic
styles, scattered sites occur in both bands of settlement, but do not
form cohesive clusters. Linkages among sets of sites are unclear for

this time. Later in the Preclassic as site numbers increase, two loose clusterings of residential sites can be identified in the immediate study area near Marana (Figure 6.3). These clusters satisfy the criteria of broad self sufficiency in subsistence for communities as units of ecosystem analysis. Both the cluster along the Santa Cruz river at the end of the Tucson Mountains and the one to the north on the upper valley slope of the Tortolita Mountains contain focal sites with ballcourts, suggesting independently integrated territorial units. Both incorporate permanent sources of potable water, diversity in productive activities, and a range of site types. Major habitation sites in each group appear to represent equally permanent occupations, as indicated by comparable assemblages of items difficult to transport such as grinding stones, and exotic goods such as shell ornaments. Each community is surrounded by areas lacking substantial habitation sites and with sharply reduced densities of other classes of remains.

The two Preclassic communities are of equivalent territorial scale. The Tortolita flank community covers 57 sq km and the one along the Santa Cruz covers 70 sq km. Oval, earthen banked ballcourts at the central sites are widely assumed to represent ball game arenas as in Mesoamerica, although use for ceremonies or dances has also been suggested. Competitions could have involved players from sites within the community or from different communities. Occasions of games likely would have served secondary purposes of exchange and social interaction (Wilcox and Sternberg 1983). Preclassic ballcourt communities are replicated across the Tucson Basin, with intervening areas of minimal settlement. For example, the ballcourt site of a second riverine community lies 16 km to the south of the one in the study area. Yet another ballcourt 19 km southeast along the mountain flank is similarly located with respect to its study area counterpart.

Ecosystem boundaries are transformed in the early Classic Period by merger of the two previous communities into a single territorial unit encompassing 146 sq km (Figure 6.4). Settlement and societal reorganization at that time appears to have been accompanied by additions to population from outside the preceding communities. Residence and land use expanded into the former intervening area. A mound constructed at a large, new site in this area designates the focal point of the Classic community.

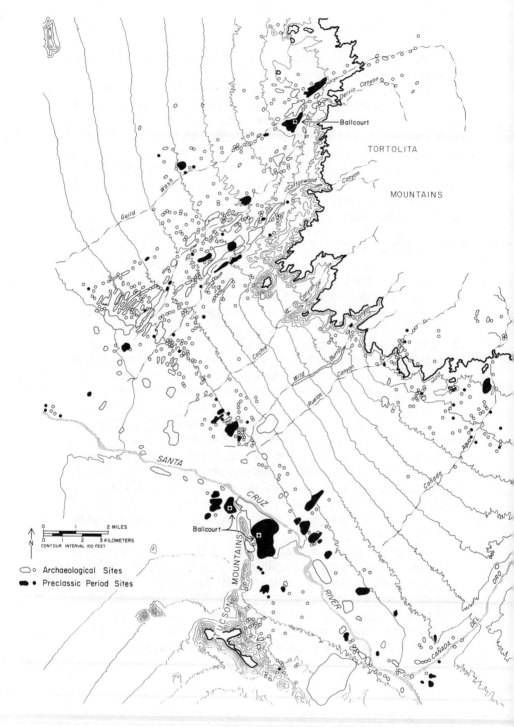

Figure 6.3 Preclassic settlement pattern in the northern Tucson Basin.

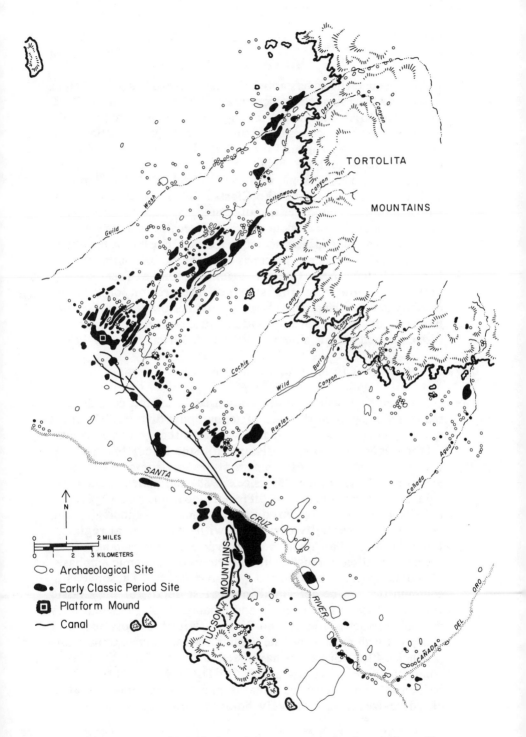

Figure 6.4 Classic Period settlement pattern in the northern Tucson Basin.

The mound site covers approximately 1,000,000 square meters. Inhabitants lived in adobe structures enclosed by compound walls of the same material. The mound, serving as a platform for rooms on the top, and adjacent structures of presumably specialized function were also enclosed by a compound wall. In the Preclassic, observances of ceremony and ritual are thought to have been dispersed because objects such as figurines and censors are widely distributed among and within sites. During the Classic Period, such objects of individual use are no longer manufactured; evidence for ceremonial events is concentrated at mound loci. Differentiated personnel are associated with the residences and other rooms of mound compounds, in contrast to the unenclosed precincts of Preclassic ballcourts without conjoined structures. Centralization of ritual at focal sites is believed to indicate the stronger integration of Classic communities. Site hierarchies based on architectural elaboration and patterns of consumption evident in artifacts can also be distinguished at this time (S. Fish et al. 1989a; Howard 1987).

Zonal Patterns in the Classic Community

Gatherable resources in the Tucson Basin exhibit considerable spatial redundancy in the face of topographic diversity. There is species overlap across sequential elevations, although densities may differ. Short trans-valley distances in the linear basin allow convenient access to those items that are environmentally more localized. Charred plant (Miksicek 1988) and pollen (Fish 1988) evidence for crops raised in different locations also registers a redundant list. However, productive capacities, technologies, and risks are differentially correlated with hydrological regimes. Environmental variability in productive potential seems to have been further magnified through cultural patterns of economic emphasis.

Differentiated production within northern Tucson Basin communities can be described in greatest detail for the Classic Period. Individuals, households, and inhabitants of particular sites did not interact routinely and directly with the ecosystem as a whole, but with more immediate subdivisions and discrete components (Ellen 1984:168). Six zones have been defined within the Classic community (Figures 6.5 and 6.6). Differential patterns of residential and productive remains, and environmental variables characterize these relatively homogeneous zones.

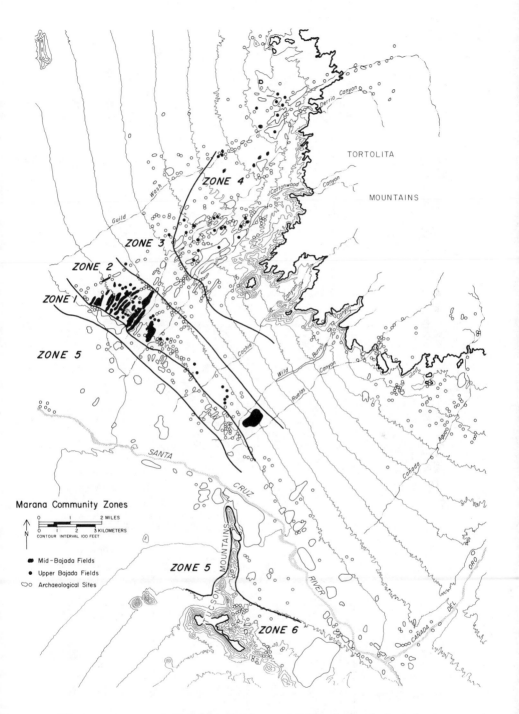

Figure 6.5 Zonal divisions in the early Classic Period settlement pattern.

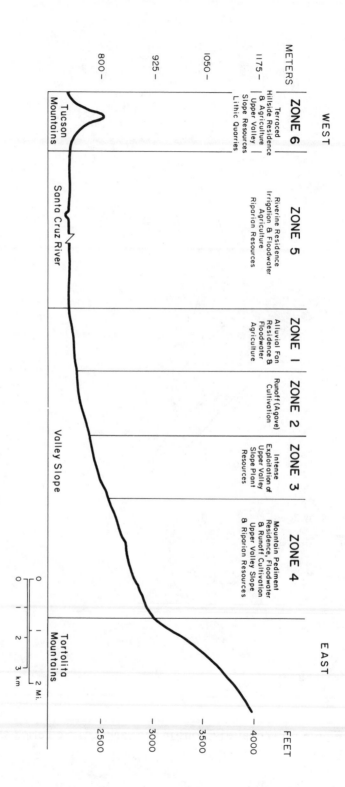

Figure 6.6 Idealized cross-section of the northern Tucson Basin showing zonal relationships of the early Classic Period Community.

Zone 1. Zone 1 residential settlements, including the mound site, occur in a more or less continuous band along the lower edge of the Tortolita valley slope. Coalescing alluvial fans create an active depositional environment. The density of remains in this zone is likely the result of opportunities for floodwater farming in an area receiving runoff from the full expanse of the valley slope.

The most desirable floodwater farming situations are concentrated in the southern third of the zone. More shallow, easily diverted channels carry water after storms to moisten fields and fertilize them with organic detritus (Nabhan 1986). Permanent water is nearby in high riverine water tables. This is the segment of Zone 1 with earliest settlement, predating Classic period reorganization in the study area. Later expansion to the north and away from the river is correlated with a new source of domestic water made available through canal construction from the river to the mound vicinity.

Zone 2. Zone 2 is uphill from the mound and lower valley slope sites. This zone lacks any sources of drinking water. Agricultural water from flow in drainages is less abundant and more difficult to divert than in zones above and below. Residential sites are virtually absent.

Dominant remains consist of huge complexes of agricultural features dependent on surface runoff. Low, rounded piles of rock are the prominent feature type, accompanied by cobble terrace alignments, checkdams, and roasting pits. Perennial crops were planted in the rock piles, which act as water-conserving mulches and create barriers to rodent predation of crop bases and roots (S. Fish et al. 1985: 110). Large rock pile fields, each measuring 10 to 50 ha in size, cover more than 485 ha. A zonal total of 42,000 rock piles and 120,000 m of linear terrace alignments is estimated. This magnitude represents a dramatic increase in the Classic Period over a few earlier instances of small fields.

The principal crop of Zone 2 fields was the agave or century plant (*Agave* sp.) This drought-adapted succulent is widely cultivated in Mesoamerica, where it is called "maguey." Although wild stands were known to have been extensively exploited in the southwestern United States, recognition of agave as a major Hohokam cultigen is recent. Carbohydrates stored in the plant base over its lifetime for a sole and final flowering episode provide a sweet and nutritious food after pit-roasting. Stiff leaves produce a fiber for cordage crafts and coarse textiles.

Scattered stone tools in rock pile fields consist predominantly of knife and scraper types ethnographically related to agave harvesting and processing. Huge--up to 35 m diameter--roasting "areas" in all large fields appear to have been communally reused over many years. Excavated pits have yielded charred agave. Zone 2 fields occur on ridges between secondary drainages on gentle mid-valley slopes. Corn pollen recovered from sediments of the small drainage bottoms suggests occasional cultivation of this annual crop in seasons of particularly favorable rainfall. Such opportunistic supplements to dependable harvests of drought-resistant agave through interspersed plantings are also found in contemporary Mexican fields (e.g. Wilken 1979; Johnson 1977; West 1968).

Zone 3. A third zone is located in the middle portion of the eastern valley slope. A few artifact scatters that may represent isolated or temporary structures are dispersed in optimal situations for water diversion and utilization. However, like Zone 2, Zone 3 is characterized by a scarcity of substantial habitation remains and the occurrence of unique and specialized sites.

Sites marked by abundant pottery and few or no other artifact types extend up to one km in length and tend to be arranged linearly along the ridgetops. The obviously specialized function of these sites has not been directly confirmed, but locations coincide with high densities of saguaro cacti. Longterm seasonal gathering by large groups is probable. Rock rings, thought to have served as supports for large conical-bottomed baskets, are usually associated with procurement of saguaro fruits (Goodyear 1975; Raab 1973). These rings are the only surface features. In this waterless zone, pottery accumulations would result from containers for imported drinking supplies and for resource processing.

Zone 4. Zone 3 settlement patterns intergrade with a fourth zonal type nearest the mountains and between three major drainages. Unlike the lower zones underlain by deep colluvial basin fill, Zone 4 corresponds with mountain pediment, where shallow bedrock prevents deep percolation of water originating on the Tortolita slopes. A relatively high and accessible water table is therefore maintained in the drainages. The proliferation of large and small habitation sites undoubtedly reflects this availability. The three major drainages and secondary ones all appear to have supported cultivation. Features such as terraces, rock piles, and checkdams are numerous in Zone 4 in conjunction with large sites, small sites, and isolated structures. Agricultural complexes are never located at a distance from habitations as in Zone 2, nor do they approach the size of large fields common to that zone.

Zone 5. The floodplain and terraces of the Santa Cruz River constitute Zone 5. Large and small residential sites of all periods occur on both sides of the river as it parallels the mountains in response to the elevated water table and more persistent surface flow. Two of the largest, most continuously inhabited sites in the community are located on either side of the mountain terminus.

From the southern boundary of the study area to the end of the Tucson Mountains, the river channel and floodplain are concisely delimited. To the north, the river flows more infrequently and for briefer durations. Terraces become poorly defined and the considerably broadened floodplain is subject to flooding after major precipitation. Residential sites are smaller and overall densities are reduced.

The highly foreshortened valley slope between the Tucson Mountains and the Santa Cruz River compresses zonal topography equivalent to that of the eastern valley slope. Inhabitants of southern Zone 5 undoubtedly diverted flow from Tucson Mountain watersheds; some sites at the western floodplain edge seem oriented toward such floodwater situations. Riverine canals undoubtedly account for the denser populations and consistently favored locales, however.

Zone 6. The Tucson Mountains form a low chain with a maximum study area height less than 130 m above the floodplain. These dark volcanic hills that form Zone 6 are covered with a variety of cacti and leguminous trees. Inhabitants of the river edge therefore had immediate access to these characteristic resources of upper valley slopes.

Terraces and walls of dry-laid masonry occur on Tucson Mountain hillslopes in both large and small groups. The largest concentration consists of over 250 terraces. Some excavated terraces have yielded evidence of agriculture while others contained foundations for what appears to have been permanent habitations (S. Fish et al. 1984; Downum 1986).

Intervening Areas. The most concentrated extractive activities took place in zones within community boundaries. Widely scattered artifacts in the areas lacking settlement between communities attest to more occasional use for hunting, gathering, or collection of raw materials. Intervening territory may have buffered potential conflict between communities, but it is likely that it was also essential for acquisition of resources exhausted by longterm use near habitation sites. In particular, hunting of larger game and firewood gathering in the desert environment may have required travel beyond community boundaries.

In the most productive zones along the river and the edge of the uplands, spaces between adjacent communities are less extensive and sharply defined. Broad expanses of the middle valley slopes at appreciable distances from community settlements contain virtually no indications of exploitive visits, in spite of abundant seasonal resources such as cacti. The burden of water transport likely discouraged use of this otherwise available bounty. During the prolonged summers when temperatures routinely exceed 100° F (37.7 ° C), the weight of water for extractive trips of any distance or duration may have been prohibitive.

Community as Ecosystem

According to the definition of ecosystem in this study, inhabitants of sites in the productively differentiated zones of the Classic community must be interlinked by more than shared ritual or political identity. Economic relationships basic to community subsistence and continuity cannot be documented directly from the archaeological record, but must rest on cumulative, inferential evidence. Diversity in zonal resources, and agricultural opportunity, become significant to the degree that economic integration can be demonstrated.

Under average or better circumstances, the zones of long-term, dense habitation along the river and mountain flanks might have been largely independent for both cultivated and noncultivated subsistence needs. Indeed, the viability of separate community units was apparent during the Preclassic Period. However, community continuity requires strategies for successfully circumventing intervals of poorer-than-average conditions for desert farmers. The ability to counteract frequent and unpredictable adversity is affected by the ratio of population to stored or alternative resources. Higher levels of population in the Classic Period may have negated the options of the smaller, independent Preclassic communities. The magnitude of contrast between earlier and later populations can be appreciated by a comparison of cumulative totals for areas within sites devoted to residential function: 2,300,000 square meters in Preclassic sites compared to 6,100,000 square meters in the Classic ones.

The two Preclassic communities faced reciprocal environmental hazards common to the deserts of the southwestern United States (Lightfoot and Plog 1984; Abruzzi 1989). Storms that flooded riverine zones and destroyed canal intakes might ensure bountiful harvests on valley slopes. Conversely, rains sufficient for irrigated crops and posing no floodplain threat might produce scanty harvests in upper fields. Seasonal limits to cultivation exhibit a similar dual pattern. Although predictable yields per area would be highest on the irrigated floodplain, pronounced cold air drainage and temperature inversions in desert basins seriously inhibit winter crops on valley floors. Winter rains could be used for an early crop on upper valley slopes free of frost.

Strategies of diversification were followed historically by many Southwestern groups to counteract environmental threats. In the Classic Period community, diversification was shifted further above the level of individuals, sites, and zones to the broad productive spectrum of the expanded boundaries. Localized risks were defused to the extent that fortunes and resources were shared within the larger entity. Overlapping biota among zones and a parallel suite of staples at sites yielding botanical evidence might imply limited routine exchange. Yet, differing magnitudes, emphases, and organization of production within the community suggest resource circulation.

Specialization and Exchange

Fine-scale examination supports subsistence exchange even within single zones. The most complete intra-zone data come from six excavated sites to the south of the mound in Zone 1 (Rice 1987). Artifacts, facilities, and botanical remains emphasize agave production at the northernmost of these sites, which is located nearest the large Zone 2 rock pile fields. Potential for floodwater farming is lowest at this site, with correspondingly few indications of adjacent cultivation of corn or other annual crops. Grinding stones document substantial corn consumption at the site, but it is doubtful that much of it was grown locally.

Other patterns among the six excavated sites lack obvious environmental correlates. For example, the same northernmost site with poor land for corn had best access to plentiful saguaro. Nevertheless, botanical remains do not show that saguaro gathering was intensified. A second site with adjacent floodwater fields yielded evidence for the most intensive saguaro processing found. This site similarly emphasized cholla cactus, a uniformly available resource. To the extent that production and consumption can be discriminated from archaeological remains, differential productive emphasis and exchange appear to have both broadened and homogenized consumption at individual Zone 1 sites (Fish and Donaldson, in press).

The strongest expression of subsistence specialization is the class of remains unequivocally linked to agave production. Specialization in this case is coincident with agricultural expansion onto marginal lands in the vicinity of high Classic population densities. Zone 2 rock pile fields represent a means to support the inhabitants of the mound vicinity, a location lacking situations for floodwater farming comparable to other segments of Zone 1. Based on quantities of stone agricultural features and a plant-to-feature ratio, the minimal estimate of annual yield is 40.8 metric tons of edible product. This amount would supply the total caloric needs of 155 people and protein for 110, or an appreciable fraction of these dietary requirements for a much larger population (FAO/WHO 1973).

Agave specialization above the mound has implications for the organization of production as well as its extent. Huge roasting areas that entailed communal harvest and processing schedules occur only in large Zone 2 fields. Massive complexes with thousands of contiguous stone features necessitated coordination in apportioning plots and maintaining boundaries. These fields, separated spatially from habitations, must have created greater need for land tenure consensus than small fields in proximity to residences, as in Zone 4. Finally, large Zone 2 fields required greater labor input per unit of production than the ones adjacent to dwellings. Fields at a distance from living areas demand an additional investment in travel and transport (Chisholm 1970; Bentley 1987;42-44), and reduce the ease of monitoring. Concentrated labor pools with limited alternatives may have encouraged pursuit of this option. Water supplies could not support intensified cropping of corn and other high-moisture annuals near the mound. In an arid land variation on agricultural intensification, as defined by Boserup (1965), total production was increased by expansion onto marginal land rather than by more intensive cultivation of existing fields.

Specialized production of raw materials is often accompanied by specialization in manufactured products. Locations of agave crafts need not duplicate those of primary production, but dense populations with poorer agricultural land might be drawn toward both activities. The artifactual correlates of fiber processing are prominent at the six excavated Zone 1 sites. Stone knives for fiber extraction and spindle whorls of a type suitable for agave products were recovered at each site. The estimated minimum annual yield of 3.72 metric tons of agave fiber from Zone 2 fields (S.Fish et al. 1985: 112; S. Fish et al. 1989b) should have provided a surplus of conveniently portable goods for exchange, perhaps both within and outside community boundaries.

As with subsistence resources, part of the basis for community exchange in other items is differential opportunity. Chipped stone materials occur in the volcanic rock of the Tucson Mountains rather than in the granitic Tortolita range. Circulated raw materials and finished products from these localized sources were essential for everyday extractive and processing equipment. Almost 90 percent of stone tools in sites of the Tortolita slopes were made from raw materials requiring exchange. The abundance of circulated items is commensurate with specialized acquisition by residents near the sources and reciprocal exchange on the part of consumers elsewhere in the community.

Exchange outside the Classic community was necessary for other materials equally vital for subsistence activities. Knives used in agave production and the majority of grinding implements originated to the north and west. Quarries for both have been identified in another contemporary community 30 km from the mound site. Importation of finished products is indicated by manufacturing debris at sites near those sources but not within the community under study. Some community surplus must have been offered in return.

Nonutilitarian raw materials and manufactured products also circulated widely. Many categories such as shell, turquoise, red pipestone, imported ceramics, and copper originated outside the community, often at considerable distances. Some exotic materials such as shell were worked into finished forms at a variety of locations in the community, while working of other materials, such as red pipestone for ornaments, has been documented only at the mound site. It is probable that both long distance and internal exchange of these items also involved subsistence resources, but its extent is unknown.

Several lines of evidence suggest a close relationship between resource circulation, hierarchical structure, and principles of political or religious integration in the Classic community. Routine use of nonlocal raw materials and a degree of differentiated and specialized subsistence production created a widespread base for exchange. Geographical centrality of the mound site within the community bestowed an advantage for performers and regulators of transfer activities. In fact, few other advantages can be cited for the newly settled location, with no natural source of domestic water and secondary agricultural potential. As specialized producers from marginal land, inhabitants of the mound site had high stakes in a comprehensive and dependable system of exchange. Strong association of nonlocal ceramics with this site, and the three others marked by compound architecture, suggests the prominence of long distance exchange. Rice (1987) has proposed that Classic community exchange resembled redistribution in a chiefdom context. However, additional possibilities such as periodic market-like gatherings, ritually organized exchange among relatively equal social units as in the Pueblos (Ford 1972), or some mixture of these according to resource type are consistent with the evidence available.

Conclusions

In archaeological ecosystem studies, where relationships among variables cannot be elicited from actors or directly observed, correlations between environmental factors and cultural remains provide a starting point. Such correlations reveal combinations of environmental opportunity and productive activity that underlie the interactions of human populations in an ecosystemic context. Settlement patterns at a regional scale were crucial in the present study for an understanding of ecosystem dynamics beyond this elemental level and for the designation of boundaries. Regional replication also confirms the multi-site community as an appropriate unit, or level of analysis, in ecosystem studies.

Although patterns of environmental correlation can be distinguished within the Classic community, broader consideration of the community and the northern Tucson Basin underscores their insufficiency for predicting many aspects of prehistoric economic behavior. Choices by community inhabitants among a range of economic alternatives seem to have shaped both synchronic settlement expression of ecosystem interactions and ecosystem dynamics over time.

With present archaeological data, the role of economic decisions can best be evaluated at the level of individual sites. For instance, potential investment in rock pile fields on middle valley slopes was virtually unlimited. Topographic and hydrologic variables similar to those of the largest fields are duplicated widely in proximity to the Classic community. As judged by the relative lushness of natural vegetation across the middle slope, segments with somewhat greater moisture availability were used less intensively than drier ones. The extent of rock pile fields in the vicinity of particular sites can best be understood as the outcome of investment decisions that took into account other kinds of agricultural opportunity and population densities.

The distribution of Zone 2 rock pile fields in relation to Zone 1 habitation sites can be seen in Figure 4. The largest fields and the greatest total acreage occur above Zone 1 sites with highest population densities and poorest potential for floodwater farming. These fields are located in a slightly drier portion of the valley slope and at a greater distance from habitations. Even though it likely entailed higher labor input per unit return, expansion of slope cultivation was economically advantageous under local circumstances. To the south in Zone 1, less dense population and small sites occur in a more promising area for floodwater farming, adjacent to better vegetated middle slopes. Nevertheless, inhabitants of these sites constructed fewer and mostly small rock pile fields. Further south still, fields are absent on slopes opposite heavy settlement along the river at the Tucson Mountains. The agricultural labor of river dwellers could evidently be absorbed more effectively in irrigated production.

Agave cultivation in rock pile fields also exemplifies the primarily cultural dynamic of ecosystem change from Preclassic to Classic configurations. Botanical and geomorphological evidence reveal no significant environmental change during this time. The requirements of higher populations were met by expanded production on marginal land, without new technologies or new cultigens; scattered, small fields producing agave are documented in the Preclassic Period. Innovation occurs only in culturally determined variables of cultivation, such as the extent, distance from residence, and communal nature of the rock pile fields.

With regard to the critical factor of domestic water, location of the preeminent mound site and new, dense Classic settlement can also be seen as the result of cultural rather than environmental criteria. Similar canal technology was implemented along the river during the Preclassic (Bernard-Shaw 1988), but construction of the 10 km canal to the mound site coincides with development of the Classic community. Thus, unlike earlier sites, the focal site of this period was dependent on distant water sources and its canal lifeline was vulnerable to interruption by human and natural forces.

The polemic between culture and environment as prime cause of social patterns and diversity has formed the basis for intellectual development of an ecological anthropology (Moran 1984:3). Archaeological investigations provide an opportunity to assess the interplay of culture and environment through time and to test conclusions derived from synchronic perspectives. Not only must tendencies to assume ecosystem homeostasis be overcome in archaeological studies, but differential roles for culture and environment can be evaluated in particular cases of change.

Ecosystems that include human components are subject to uniquely cultural sources of instability through changing perceptions of alternative behaviors. Hohokam settlement in the northern Tucson Basin illustrates the extent to which the ecosystem of a human population may be culturally defined even under restrictive conditions of an arid environment, Neolithic economy, and moderate level of political complexity. Classic sociopolitical integration encompassed greater environmental diversity within expanded community boundaries. The Classic Period community differed from its Preclassic precursors in overall numbers, localized densities, and geographic locations of population. Desert production was intensified in large part through expanded cultivation of a drought-adapted crop on marginal land. Notwithstanding an essentially static technology, Hohokam populations generated the motivations and possessed the ability to bring about ecosystem change.

References Cited

Abruzzi, William S.
    1989   Ecology, resource redistribution, and Mormon settlement in northeastern Arizona. *American Anthropologist* 91:642-655.
Bentley, Jeffrey W.
    1987   Economic and ecological approaches to land fragmentation: in defense of a much-maligned

phenomenon. *Annual Review of Anthropology* 16:31-67.

Bernard-Shaw, Mary
  1988   Hohokam canal systems and late Archaic wells: the evidence from the Los Morteros site. *Recent Research on Tucson Basin Prehistory*, edited by William Doelle and Paul Fish, pp. 153-173. Tucson: Institute for American Research Anthropological Paper 10.

Boserup, Ester
  1965   *The Conditions of Agricultural Growth: The Economics of Agrarian Change under Population Pressure*. Chicago: Aldine.

Chisholm, M.
  1970   *Rural Settlement and Land Use: An Essay in Location*. Chicago: Aldine.

Crown, Patricia L.
  1987   Classic Period Hohokam settlement and land use in the Casa Grande Ruins area. *Journal of Field Archaeology*. 14:147-162.

Downum, Christian E.
  1986   The occupational use of hill space in the Tucson Basin: evidence from Linda Vista Hill. *The Kiva* 51:219-232.

Doyel, David E.
  1980a   The prehistoric Hohokam of the Arizona desert. *American Scientist* 67:544-554.
  1980b   Hohokam social organization and the Sedentary to Classic transition. *Current Issues in Hohokam Prehistory*, edited by David Doyel and Fred Plog, pp. 23-40. Arizona State University Anthropological Research Paper 23.
  In Press Hohokam cultural evolution in the Phoenix Basin. *Changing Views on Hohokam Archaeology*, edited by George Gumerman. Albuquerque: The University of New Mexico Press.

Ellen, Roy.F.
  1984   Trade, environment, and the reproduction of local systems in the Moluccas. *The Ecosystem Concept in Anthropology*, edited by Emilio Moran, pp. 163-204. AAAS Selected Symposium 92. Boulder: Westview Press.

FAO/WHO
    1973   Energy and protein requirement: report of a joint
              FAO/WHO ad hoc expert committee. *World Health
              Organization Technical Report Series* 522.
Fish, Paul R.
    1989   The Hohokam: 1000 years of prehistory in the Sonoran
              Desert. *Dynamics of Southwestern Prehistory*,
              edited by George Gumerman and Linda Cordell, pp.
              19-64. Washington, D.C.: Smithsonian Institution
              Press.
Fish, Paul R. and Suzanne K. Fish
    In Press Hohokam   political   and   social   ogranization.
              Changing views on Hohokam Archaeology, edited
              by George Gumerman. Albuquerque: The University of New
              Mexico Press.
Fish, Suzanne K.
    1988   Environment and subsistence: the pollen evidence.
              Recent Research on Tucson Basin Prehistory,
              edited by William H. Doelle and Paul R. Fish, pp.
              31-38. Tucson: Institute for American Research
              Anthropological Papers 10.
Fish, Suzanne K., Paul R. Fish, and Christian E. Downum
    1984   Hohokam terraces and agricultural production in the
              Tucson Basin. *Prehistoric Agricultural Strategies
              in the Southwest*, edited by Suzanne K. Fish and
              Paul R. Fish, pp. 55-71. Tempe: Arizona State
              University Anthropological Research Paper 33.
Fish, Suzanne K., Paul R. Fish and John Madsen
    1989a   Differentiation and integration in a Tucson Basin
              Classic   Period   Hohokam   community.   *The
              Sociopolitical Structure of Prehistoric Southwestern
              Societies*, edited by Steadman Upham, Kent
              Lightfoot, and Robert Jewett, pp. 237-267. Boulder:
              Westview Press.
    1989b   Analyzing prehistoric agriculture: a Hohokam example.
              *The Archaeology of Regions: The Case for Full
              Coverage Survey*, edited by Suzanne Fish and Stephen
              Kowalewski, pp. 189-219. Washington, D.C.: Smithsonian
              Institution Press.
Fish, Suzanne K. Paul R. Fish, Charles Miksicek, and John Madsen
    1985   Prehistoric agave cultivation in southern Arizona.
              Desert Plants 7:107-112.

Fish, Suzanne K. and Marcia Donaldson
In Press Production and consumption in the archaeological
        record: a Hohokam example. *The Kiva.*
Ford, Richard I.
1972   An ecological perspective on the Eastern Pueblo.
        New Perspectives on the Pueblos, edited by
        Alfonso Ortiz, pp. 1-17. Albuquerque: University of
        New Mexico Press.
Goodyear, Albert C.
1975   *Hecla II and III: An Interpretive Study of
        Archaeological Remains from the Lakeshore Project,
        Papago Indian Reservation, South-Central Arizona.*
        Tempe: Arizona State University Anthropological
        Research Paper 9.
Gregory, David A. and Fred L. Nials
1985   Observations concerning the distribution of Classic
        Period Hohokam platform mounds. *Proceedings of the
        1983 Hohokam Symposium*, edited by A.E. Dittert, Jr.
        and D. E. Dove, pp. 373-388. Phoenix: Arizona
        Archaeological Society Occasional Paper 2.
Haury, Emil
1976   *The Hohokam: Desert Farmers and Craftsmen.*
        Tucson: The University of Arizona Press.
Howard, Jerry B.
1987   The Lehi Canal System: organization of a Classic Period
        community. *The Hohokam Village: Site Structure and
        Organization*, edited by David E. Doyel, pp.
        211-222. Glenwood, Colorado: Southwestern and Rocky
        Mountain Division of the American Association for the
        Advancement of Science.
Jochim, Michael
1984   The ecosystem concept in archaeology. *The
        Ecosystem Concept in Anthropology*, edited by Emilio
        Moran, pp. 87-102. AAAS Selected Symposium 92.
        Boulder: Westview Press.
Johnson, Kirsten
1977   Disintegration of a traditional resource use: the Otomi
        of the Mezquital Valley, Hidalgo, Mexico. *Economic
        Geography* 53: 364-367.
Lightfoot, Kent G. and Fred Plog
1984   Intensification along the north side of the Mogollon

Rim. *Prehistoric Agricultural Strategies in the Southwest*, edited by Suzanne K. Fish and Paul R. Fish, pp. 179-195. Tempe: Arizona State University Anthropological Research Paper 33.

Masse, W. Bruce
   1981   Prehistoric irrigation systems in the Salt River Valley, Arizona. *Science* 214:408-415.

McGuire, Randall H. and Michael B. Schiffer
   1982   On the threshold of civilization: the Hohokam of southern Arizona. *Archaeology* 35: 22-29.

Miksicek, Charles H.
   1988   Rethinking Hohokam paleoethnobotanical assemblages: a progress report from the Tucson Basin. *Recent Research on Tucson Basin Prehistory*, edited by William H. Doelle and Paul R. Fish, pp. 47-56. Tucson: Institute for American Research Anthropological Papers 10.

Moran, Emilio F.
   1984   Limitations and advances in ecosystems research. *The Ecosystem Concept in Anthropology*, edited by Emilio F. Moran, pp. 3-32. AAAS Selected Symposium 92. Boulder: Westview Press.

Nabhan, Gary P.
   1986   *Ak-chin* "arroyo mouth" and the environmental setting of the Papago Indian fields in the Sonoran Desert. *Applied Geography* 6:61-75.

Netting, Robert McC.
   1984   Reflections on an Alpine village as ecosystem. *The Ecosystem Concept in Anthropology*, edited by Emilio F. Moran, pp. 225-235. AAAS Selected Symposium 92. Boulder: Westview Press.

Raab, L. Mark
   1973   AZ AA:5:2, a prehistoric cactus camp in Papagueria. *Journal of the Arizona Academy of Science* 8:1-9.

Rappaport, Roy A.
   1968   *Pigs for the Ancestors: Ritual in the Ecology of a New Guinea People*. New Haven: Yale University Press.

Rice, Glen E.
   1987   *Studies in the Hohokam Community of Marana*.

Arizona State University Anthropological Field Studies 15.

West, R.
1968   Population densities and agricultural practices in Pre-Columbian Mexico, with emphasis on semi-terracing. *Verhandlungen des XXXVIII Internationalen Amerikanistenkongresses*, Band II, pp. 361-369. Munich.

Wilcox, David and Charles Sternberg
1983   *Hohokam Ballcourts and Their Interpretation.* Tucson: Arizona State Museum Archaeological Series 150.

Wilken, G.C.
1970   The ecology of gathering in a Mexican farming region. *Economic Botany* 24:286-295.

PART III
ECOSYSTEMIC APPROACHES
IN CULTURAL ANTHROPOLOGY:
RESEARCH DESIGN AND PRACTICE

CHAPTER 7

TRADE, ENVIRONMENT, AND THE REPRODUCTION OF LOCAL SYSTEMS IN THE MOLUCCAS

Roy F. Ellen

Introduction

Theoretical objections to the ecosystem concept as applied to our understanding of social relations in human populations fall into two broad categories. The first stem from the very special character of the coded information which circulates in addition to matter, energy and other forms of non-cultural messages. The second arise from a fundamental ambiguity in the intentions of those biologists who originally developed and employed the concept. For, on the one hand, the notion of an ecosystem has permitted an emphasis on the relationship between variables otherwise considered separately, while on the other it has allowed the *separation* of a discrete universe for the purpose of analysis.

Yet while these objections are valid, and criticisms of the particular use of the ecosystem concept and its dangerous excesses often trenchant, it is difficult to deny the advantages which it has brought to scholarship. Some of these are worth repeating here. In empirical terms the new techniques and data associated with it have enabled us to correct mistaken notions about often unfamiliar subsistence regimes. In theoretical terms it has stressed the necessity for holism while focussing on *specific* relationships between human populations and features of their environment, and directed attention toward the existence of ramifications of particular relationships. It has shifted the emphasis away from the vulgar correlations of environmentalism, possibilism and cultural ecology toward more specific and integrated studies, and avoided simple-minded determinism by its stress not only on reciprocal causation, but also on complex networks of mutual causality. It has focused on the organization and properties of systems, on the degrees and forms of stability which they may attain, and on the mechanisms which regulate their functioning and determine their evolution. By focussing on populations rather than on groups defined ethnically or socially it has provided a much more satisfactory basis for establishing conditions of analytical closure and for the exploration of the validity of the concept of adaptation.[1] More generally, it has provided a framework for description and analysis which accords due recognition to the complex and varied interactions of environmental and cultural variables. In so doing it has also paved the way for the introduction and development of further systems concepts, such as the idea of trophic exchanges, and has revived an interest (through parallelism) with Marxist dialectical theories of systems. While it has not escaped entirely from the constraints of disciplinary chauvinism, it has at least moderated them.

In this paper I restrict myself to the problem posed by the necessity for system closure, a matter which is not solely of interest to those concerned with ecological issues. I wish to examine the ways in which boundaries can be delineated (in both empirical and theoretical terms) for a series of populations interacting at different levels of intensity. These populations are situated in the south and central Moluccan islands of eastern Indonesia (Fig. 7.1). By showing how boundaries may be established so that such populations can be seen as parts of systems on different scales of inclusiveness, I want to argue for the development of concepts of graded boundaries and a notion of system (and ecosystem) defined less in terms of absolute and discrete boundaries than in terms of centers, peripheries and structural focus. What I have to say is highly compressed, provisional and short on supporting data. As with an earlier piece (Ellen 1979), it is basically a theoretical exercise undertaken in connection with some long-term research on inter-island trade which has yet to be completed.

On Delineating Boundaries

It is becoming ever more difficult to argue that even the most isolated human population and its immediate environment can be treated as an unproblematic self-reproducing closed system (Bennett 1976:256; Friedman 1976; Langton 1973:133-135). Such systems are too

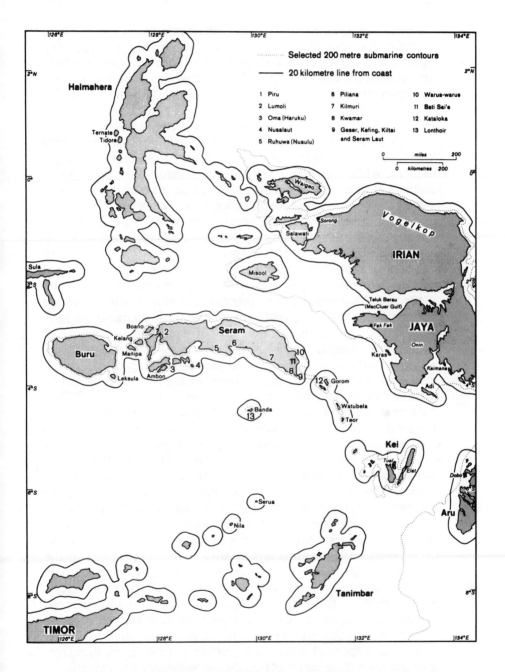

Figure 7.1 Some topographic aspects of the Moluccan region of eastern Indonesia, indicating localities mentioned in the text.

empirically complicated to specify in terms of patterns and numbers of connections or intensity of flow. Moreover, it is rare for material exchanges suddenly to discontinue at a border, while systems are generally identified through subjective judgments as to what processes are critical to their functioning. Thus, it is usually necessary to demarcate or adjust boundaries according to research interests, models employed, and the patterns in emerging data.

However, it is empirically evident that some human systems *do* maintain a degree of integrity and are relatively well defined, while it is both practically and intuitively a relatively simple matter to isolate 'systems' which make sense of the data and which may be used in the investigation of a wide range of different problems. Those who most vociferously argue that we recognize the openness of human systems must at the same time employ some notion of closure themselves when determining what universe of variables to examine. If the search for "systemness" is to remain an important issue in the analysis of human social relations (and it is difficult to see how this could ever be otherwise), then it is better that this be done explicitly. The arbitrary and analytic character of most boundaries, together with the reality of exogenous intrusions of varying types, magnitudes and origins, should not compel us to accept a sloppy notion of receding causation, whereby the critical determinants always seem to lie outside the local system subjected to detailed analysis (cf. Mason and Langeheim 1957; Newcomer 1972:5). Neither should we infer the character and consequences of external influences entirely from specific local effects. That a small glass bead is recovered archeologically in highland Papua New Guinea may well serve to represent that absolute frontier of the Portuguese mercantilist world system of the sixteenth century, but the systemic consequences of its appearance, compared with those of all other inflows of material, energy and information, were probably negligible. It may nevertheless be evidence for a kind of kick-on effect, of changes which have reverberated through a series of overlapping local systems geographically connecting the site in the New Guinea highlands with some putative point of origin in Portuguese adventurist activity in the Moluccas. Whatever the local consequences of this may have been in terms of Moluccan social formations, for highland New Guinea these will have been transformed into shifts in exchange and political relations, of which the occurrence of the bead is quite independent. The problem, therefore, is to recognize those empirical conditions necessary for assuming a particular degree of closure in the analysis of any one system, and to specify the magnitude and character of exogenous links in terms of their consequences for the effective reproduction of local social formations.

In order to do this it is useful to underline three axiomatic characteristics of human ecological systems. Few would now deny these, but in the past they have often been understressed or ignored in those studies to which the label 'ecological functionalism' is sometimes applied. The first is that subsistence areas associated with particular local human populations (and therefore ecological systems) generally overlap. Even among relatively self-sufficient groups where food production is technologically simple and where conditions for observing the functioning of the most isolated systems are best met, spatial distribution is preferably represented as a wide network of demes (Weiner 1964:401-2), where different groups and individuals are engaged in the mutual transfer of resources (personnel, genes, energy, materials and value). The subsistence and reproductive bases of such populations (whether designated ecologically or sociologically) are therefore not independent, and their boundaries might better be conceptualized as clines (Rappaport 1968:226).

It is unnecessary to labor this point by repeated ethnographic illustration, but the Nuaulu villages of the south coast of central Seram well exemplify the often complicated geographical relationships between the resource areas of nearby settlements in the Moluccas. These villages are part of a web of settlements of varying size, market integration and ethnolinguistic origin, which have for almost 100 years been in close political association as a result of events and population movements which took place between 1880 and about 1920. The land around Nuaulu villages constitutes a mosaic of cultivated plots: those owned and worked by the villagers themselves, those owned by Nuaulu from other villages, and those owned by Muslim and Christian outsiders (Ellen 1978:82). However, the resource base of the Nuaulu village is by no means limited to the area cultivated. That of the village of Ruhuwa amounts to some 214.75km^2 of forest and swampland at varying altitudes, from which are drawn a wide range of different food and non-food products (*ibid.* 61-64). But this zone also overlaps the resource areas of perhaps 10 other major settlements. Although this kind of overlap is more unusual in upland and inland areas of Seram where population density is lower, the pattern must be quite common in all coastal areas of the Moluccas.

A second relevant characteristic of human ecological systems is that the environment upon which any one population subsists is seldom uniform. Ecosystems are not homogeneous but patterned, often in a way which is critical to an understanding of the functioning of the human populations within them. It has sometimes been useful to portray human groups as participating in several different relatively well-defined zones with varying food-getting potentials (e.g., Dornstreich 1977). Such discontinuities may affect the distribution of resources, yields and the selection of sites for cultivation, as well as emphasizing that human populations do not interact directly with *total* environments but only with particular fragments and species. Such spaces characterized by a particular biotic, geomorphologic and climatic composition are conventionally described as *biotopes*.[2]   Table 7.1 shows that kind of biotopic pattern common for many parts of coastal Seram. This particular example is drawn from Nuaulu settlements in the Amahai subdistrict.[3]

A third characteristic of human ecological systems is the Heraclitean one: that ecosystems and human populations (together with their various social and cultural representations) are in a constant condition of flux, coping with sudden or emergent environmental problems by 'disturbing' rather than by maintaining balance. Emphasis must consequently be placed on types and rates of change. The recent interest in hazards research (Vayda and McCay 1977), the critique of negative feedback models and their replacement by amplification-deviation approaches (e.g., Bennett 1976; Ellen 1979) is ample indication that this point is now well taken.

Some Environmental Conditions for Closure

If closure is always a matter of degree, it becomes important to ascertain, in as precise terms as possible, what qualitative conditions might prevent the completely open flow of energy, materials and information for a particular human population. These can be grouped under three headings: geomorphologic (or physical), biogeographic, and cultural.

Geomorphologic closure is that which prevents the free flow of matter and its images between any two points through non-biological and non-cultural means. This may be achieved through altitude, distance, climate, terrain, seismic conditions, or any combination of these. As blocking mechanisms these rarely act independently of each other, or indeed of forms of closure under the remaining two headings. Thus, in eastern Indonesia, seismic disturbances may affect local weather by triggering-off torrential rain, while more regular patterns of precipitation are closely associated with the distinctive hydrological cycle of mature tropical forest (Kenworthy 1971). Moreover, the type and degree of closure possible in each case will vary. Altitude, distance and terrain must primarily be thought of in terms of spatial closure, while the immediate effects of seismic and climatic disasters are to induce temporal closure for particular localities. On the islands of the Banda arc, settlements have often been destroyed or severely damaged by tidal waves and earthquakes (Admiralty 1944:14, 16, 290; Kennedy 1955:147, 149, 150, 174), while the villages and nutmeg plantations of Banda itself have a long history of periodic destruction through volcanic eruption (Hanna 1978). The illustrations discussed here are confined to distance, terrain and the question of spatial closure.

Terrestrial distance in the Moluccas (whether measured orthographically, topographically or pherically) has important consequences for the closure of local systems. Although also affected by a range of other factors, distance is an important determinant of the distribution of Nuaulu cultivated plots. The relationship between village distance and both the number and total area of plots in hectares for the Ruhuwa Nuaulu is strongly inverse (Pearson's r = -0.89). This is so despite the amount of land under cultivation in 1970 (when this calculation was made) being almost equal for concentric zones of one kilometer width up to three kilometers from the village, and despite over 40 percent of the plots being more than two kilometers from the village. However, beyond the three kilometer radius the disadvantages of distance begin sharply to outweigh any advantages, as well as the disadvantages linked to the spatial constraints imposed by a preference for plots nearer to the settlement (Ellen 1978:133-8). However, the relationship between the amount of activity devoted to the extraction of non-domesticated resources and distance from a settlement does not show this strong correlation (Pearson's r = 0.1544; for data see *ibid.* 63). This is in part due to an unevenness in the distribution of forest resources, and in particular to the tendency of exhausting resources nearer the settlement before moving further afield. Outside of subsistence, the effects of land distances can also be seen in the degree of social interaction, including spatially represented marriage relations (*ibid.* 19-21).

Provided terrain and vegetation do not make routes impassable, land distances can be traversed without technical aids. This is not the case with maritime distances where some form of transport is unavoidable. However, as on land, the difficulties of communication increase in proportion to the distance involved, and short inshore movements are relatively common and straightforward.

In Figure 7.1 a continuous line has been drawn at a distance of 20 kilometers from the coast of all inhabited land-forms, a distance which corresponds approximately to the practical limit imposed by the use of small dug-out outrigger canoes by present-day Moluccans. This limit depends in part on the distance of the horizon, which may vary between 4.18 and 188.3 km, but is generally no more than 60 km. The physical demands of handling a small craft, the location of fishing grounds, the duration of journeys between nearby islands and perceived weather hazards also affect patterns of use. However, as a rule, small outrigger canoes keep well within sight of land. Traditional multiple-oared and long-hulled plank-built boats, as well as sailing boats of imported design, are not nearly so limiting. Their capabilities are well known (McKnight 1980), although it should be stressed that sub-types and individual craft are extremely variable in this respect. Much depends upon such factors as design and size. The dotted line on the same map marks the 200 meter depth contour. Both lines encompass island groups within which we might reasonably expect local settlements to be highly connected. If this is so--and I shall presently demonstrate in some fairly specific ways that it is--then the notion that islands necessarily provide us with convenient laboratories with a high degree of geographical closure (Thompson 1949; Fosberg 1963:5; Bayliss-Smith 1977:12) is misleading. This is so at least as far as human populations on geographically unremote islands are concerned.

The idea that small islands provide us with almost experimental conditions of closure is attractive, but in the Moluccas, and for large parts of insular southeast Asia and Melanesia, short sea-barriers are less limiting for humans than land barriers. The seas of the Indonesian archipelago are mostly warm, shallow and sunlit, with a rich and varied biomass. Although mud, shifting sand and submerged reefs sometimes provide navigational hazards, the seas are not subject to frequent bad storms or strong currents and are well suited for easy communication. Moreover, the main east and

Table 7.1. The relationship of resource areas to food-getting activities among the Nuaulu of south central Seram.

Resource area (or biotope)	Altitude range (m above sea level)	Food-getting activities [1]										Total number of activities
		a	b	c	d	e	f	g	h	i	j	
1. Hamlet site - current	0 - 200				+		+		+			3
2. First year garden	0 - 400	+				+						2
3. Old garden - staple foods exhausted	0 - 200	+		+		+						3
4. Abandoned garden	0 - 200	+	+	+	+		+	+	+			7
5. Sago swamp and domesticated sago groves	0		+	+					+	+		4
6. Other groves	0 - 50			+					+	+		3
7. Mixed secondary forest	50 - 400			+	+		+	+	+			5
8. Mature rain forest	200 - 1000				+				+	+		3
9. Montane rain forest	1000 - 1400									+		1
10. Rivers	0 - 200								+		+	2
11. Streambank	0 - 400			+		+			+			3
12. Littoral	0								+		+	2
13. Sea	0										+	1
14. Total number of resource areas		3	1	3	5	3	1	4	9	6	8	

Note:
1. a, swiddening; b, starch extraction from non-domesticated *Metroxylon* (sago) palms; c, starch extraction from domesticated *Metroxylon* palms; d, silviculture; e, gathering non-domesticated plant foods; f, animal husbandry; g, trapping; h, collecting non-domesticated animal foods; i, hunting; j, fishing. The mode of presentation adopted in this table (reproduced from Ellen 1982:167) broadly follows that employed by Dornstreich (1977: 249-50) in his description of Gadio Enga subsistence.

west monsoon winds are ideal for sailing craft moving between the mainland and various lateral points in the archipelago (See Urry 1981:2).[4] Even for populations at a low level of technical development using primitive vessels, small island groups are often better connected along coasts, across narrows and short stretches of open sea than coasts are with interiors divided by mountains, fast-flowing rivers, dense forest and difficult terrain. Moreover, lowland coastal regions draining large upland interiors are often swampy, and their foreshores covered with mangroves. In parts of the Moluccas (e.g., Nusalaut, parts of Haruku and Seram: see Figure 7.1), islands rise precipitously from the sea, and steep vegetation-covered slopes make coastal land travel awkward. In such areas, and also where forest, swamp, quagmirish paths, rocky and undulating surfaces prevent effective movement on firm land, intervillage travel may be along beaches. This littoral zone is particularly favored in this respect, but the availability of such routes depends on the tides. In the large islands to the west and in the Malay peninsula, estuaries and river networks provide access to the interior, but in the Moluccas such waterways are often shallow, full of rocks, liable to wet season flash floods and generally difficult to navigate.

In the Moluccas the density of network connections has for a long time been infinitely greater along coasts and between small nearby islands than between the coast and the interior, and this has been closely linked to a high interdependency of local populations. By contrast, the most isolated and least dependent populations are those found in the interior and on highland peripheries. But while coastal connectivity within the small island groups of the Moluccas may well be high in terms of both human and non-human flows, as maritime distances increase beyond the 20 kilometer and 200 meter depth frontiers, so communication becomes technically more difficult and geographic barriers more resistant. As distance and sea-depth increase so separated land-masses become ecologically more closed and some potential lines of closure more apparent than others.

The second kind of closure to be considered is that between *biotopes*. Table 7.2 lists the main biotopes which can be usefully distinguished in the Moluccan area in the context of an essentially ethnographic examination. The table also indicates their degree of specialization, susceptibility to hazard and ability to sustain human populations. Six points about this table are worth noting.

Table 7.2. Major biotopes distinguishable in the central and southeastern Moluccan region.

Biotope	Example	Degree of Special-isation [1]	Susceptibility to hazard [2] Man-Induced	Natural
1. Mature *Agathis* rain forest	extensive areas of the interiors of Seram and Buru below 1000 m	1	1	3
2. Montane forest	highland areas of Seram and Buru over 1000 m	2	1	1
3. Mixed secondary forest	extensive areas of Buru; on Seram mainly around human settlements	2	2	4
4. Bamboo scrub	lowland areas of Seram mainly around human settlements	7	3	3
5. Coconut palm and other groveland	extensive coastal areas of Kei Besar (coconut), Banda (nutmeg) and the Ambon group (clove)	7	3	3
6. Swiddens at different stages of development	within 4 km of most settlements on Seram and Buru	5	4	4
7. Irrigated rice fields	rare - but some in transmigration areas of Seram	10	5	5
8. *Metroxylon* palm swamp	lowland riverine areas of Seram such as Nua-Ruatan confluence and Masiwang estuary	10	1	1
9. Mangrove swamp, muddy shores and estuaries	extreme southeast portion of mainland Seram between settlements of Kwamar and Kwaos	10	2	2
10. *Imperata* grassland	coastal areas of Tanimbar and Kei Kecil; patches on all major islands	7	1	1
11. Savannah	southern parts of Aru group	6	2	2
12. Beach	all coastal areas	7	2	1
13. Intertidal (littoral) zone	all coastal areas	6	1	1
14. Lagoons	extensive in Kei archipelago	5	1	1
15. Close inshore waters and surf zone	all coastal areas	5	1	1
16. Banks and offshore shallows	Geser group	5	1	1
17. Cays — small coral islets	Geser-Gorom group; Kei archipelago	5	3	1
18. Coral shoals and reefs	Geser-Gorom group; Kei archipelago	5	1	1

Notes:
1. A notional 10 point scale in which 1 represents the ecologically most generalised biotope and 10 the ecologically most specialised. Degree of specialisation usually reflects species diversity.
2. Two 5 point notional scales in which 1 represents a biotope least subjected to or affected by hazard and 5 that most subjected to or affected by hazard. Man-induced hazard may take the form of either intensification of disturbance or withdrawal of system- maintaining practices.

1. The biotopes can be arranged according to their degree of ecological specialization, ranging from the most generalized (mature tropical rain forest, deep sea) to the most specialized (sago palm swamp, mangrove swamp).

2. The biotopes can be arranged according to their susceptibility to hazard. Generally speaking, more complex biotopes, such as mature rain forest, are the least affected by natural hazards, although the converse is not the case. While simple biotopes, such as sand bars, are certainly extremely fragile, the specialized conditions of the sago palm swamp are such that only the most severe disturbances are likely to alter its overall composition. If we order the biotopes according to their susceptibility to *man-induced* hazard in terms of known Moluccan history, then mature tropical forest must be seen as much more susceptible to hazard due to the continual firing and clearing of swiddens. On the other hand, the already stable conditions of the sago swamp are only enhanced by human husbandry which, while involving the periodic felling of stands, also ensures their vegetative propagation.

3. The boundaries between different biotopes vary in their degree of closure, fixity and clarity. Thus, mixed secondary rain forest merges imperceptibly into mature rain forest as it develops and increases in complexity. The break between sago swamp and mature forest is more abrupt: the circumstances in which one can develop into the other are rare and the overlap in species content negligible.

4. All biotopes listed may be used for human subsistence, although not all to the same degree. Both mature and mixed secondary forest is used by the Nuaulu for a wide variety of purposes (Table 7.1; Ellen 1978:61-80). By contrast, the extensive mangrove swamps of the extreme southeast of Seram are only used by inhabitants of the settlement at Kwamar as a source of firewood, palms (*Nipa fruticans*) and some minor marine foods (but see Dunn and Dunn 1977). However, although the range of products available from mangrove swamps is narrow, the economic significance of collecting firewood is considerable as the fuel depleted offshore islands of Geser and Kefing (Figure 7.1) have long been dependent upon it.

5. Biotopes vary in their ability to support human habitation

and cater for the major part of their subsistence needs. Most of the more specialized biotopes are excluded on this basis, as are those which are entirely maritime. However, no Moluccan settlement is dependent upon only one (or even a few) biotopes for its subsistence. The Nuaulu example presented in Table 7.1 illustrates a particular case, although some idea of how the combination varies may be gained from an examination of Table 7.3 The overall stability of a particular resource base, its susceptibility to fluctuation and the predictability of its processes will depend on the combination of biotopes and the degree of subsistence dependence upon each. Crudely, the greater the number of different biotopes and the higher the ratio of generalized to specialized biotopes, the more stable a resource base is likely to be.

6. The degree to which these biotopes perpetuate themselves varies. All are dependent upon energy and materials from parts of a wider ecosystem. Mature rain forest and other generalized systems can perpetuate themselves over a period of decades and centuries. Mixed secondary forest cannot maintain itself over the longer timescale since it is gradually moving toward primary rain forest. It can be maintained in a more immature condition, however, by periodic human intervention, through selective thinning and clearance. Though specialized, sago swamp is ecologically stable and can reproduce itself effectively over the 15 year life-span of the *Metroxylon* palm. Swiddens are highly unstable associations which are only prevented from changing into mixed secondary forest, grassland or bamboo scrub by constant horticultural attention. Groveland is more stable in this respect.

Resource Dependence and the Introduction of Asymmetry

Having examined some of the environmental preconditions which facilitate varying degrees of closure between Moluccan populations, we next turn to a comparison of these with the extent of cultural closure maintained in particular localities. All contemporary Moluccan populations are parts of wider trading systems. In Table 7.3 I have tried to indicate degree and type of closure for some of

these. We must now investigate to what extent it is possible to
ascertain the degree of functional independence of such populations
at the present time or at some specified period in the past.

In an earlier publication (Ellen 1979a:47-52), I described the
character of the most elementary Moluccan subsistence unit it is
possible to conceive of in the light of available historical and
ethnographic evidence. This, in part, is characterized by a critical
dependence upon *Metroxylon* sago as a source of carbohydrate
with a minimal use of domesticated resources, slight production for
exchange and little reliance on other populations for food and
materials. At the present time this condition is most closely
approached by those populations listed in Table 7.3 as having the
widest spread of food-getting activities over different terrestrial
biotopes. Such populations also come nearest to fulfilling the
maximum conditions of closure possible. Nevertheless, even when
populations are capable of meeting their own food requirements in
overall terms, periodic shortages of particular resources necessitate
a degree of exchange with other similar populations. In many cases
the resource areas of such populations may be almost identical or
overlap extensively, but local demographic, ecological, subsistence
and social variables give rise to imbalances in resource output per
head which to some extent can be evened-out through trade with
immediate neighbors. Assuming a situation of basic exchange equality
between populations, this kind of lateral trade has a mainly
regulatory effect. We can say this without positing any
preconditions for closure. It is enough to assume that all
populations in the universe under observation are connected equally
with each other and that ordinarily imbalances in local production
are adjusted whenever possible by using connections nearest to the
settlement concerned. The boundary of a system is never constant.
No permanent frontier exists, only a gradation moving outward from a
focus along which the limit defined at any one instance fluctuates.
Moreover, such local systems overlap in proportion to the number of
focal settlements.

It is now necessary to look at the ways in which asymmetry is
introduced into trading relations, but before doing so I wish to make
it clear that our working definition of asymmetry must be in terms of
structural significance and not in terms of some convenient economic
or ecological index which does not take into account the varying
importance of particular resources in different localities. Thus,
because a population is a specialized component in a world system

Table 7.3. Spread of food-getting activities for selected Moluccan settlements with some measures of their reproductive independence. [1]

Local settlement and Population [2]	Basic subsistence pattern	1	2	3	4	5	6	7	8	9	10	11	12	13	14	15	16	17	18	Trade dependence [4] a b c d e f	Degree of connectivity [5]
1. Ruhuwa (Nuaulu), south Seram - 100+	sago extraction, hunting and collecting, gathering, swiddening	+	(+)	+	+	+	+		+					+		(+)				1 2 1 2 0 0	1
2. Piliana (Huaulu group), central Seram - 100+	sago extraction, hunting and collecting, gathering, swiddening	+	+	+	+	+	+		+											1 2 1 2 0 0	1
3. Bati Sai'e, inland east Seram - 100+	sago-extraction, hunting and collecting, gathering, swiddening	+		+	+	+	+		+											1 2 1 3 0 0	1
4. Warus-warus, coastal east Seram - 576	sago extraction, swiddening, fishing, cash-cropping			+	+	+	+		+					+		+	+			2 3 2 0 0 0	2
5. Sepa, south Seram - 500	swiddening, cash-cropping, fishing, sago extraction			+	+	+	+		+					+		+				3 3 3 0 0 0	2
6. Oma, Haruku - 2245	swiddening, cash-cropping, fishing, sago extraction				+	+	+		+					+		+				4 4 3 0 0 0	3
7. Lonthoir, Banda - 1500	swiddening, cash-cropping, fishing					+	+							+		+				4 5 4 0 0 0	3
8. Kataloka, Gorom - 1500	swiddening, cash-cropping, fishing, sago-extraction			+	+	+	+		+				+	+		+	+	+	+	4 4 3 0 0 0	4
9. Elat, Kei Besar - 2000+	swiddening, cash-cropping, fishing, sago extraction				+	+	+		+					+		+	+			4 4 3 0 0 0	4
10. Kefing, Geser group - 500	fishing								+					+		+	+			5 5 5 0 5 5	4
11. Gorogos, Gorom group - 100	fishing											+	+			+	+	+	+	5 5 5 0 5 5	3
12. Geser, Geser group - 1230	fishing, trade													+		+	+	+	+	5 5 5 0 5 5	5

Notes:

1. The data upon which these schematic indices are based were collected in 1981, except for Piliana which were collected in 1975. In the cases of Ruhuwa and Sepa, the observations span 12 years from late 1969.

2. Figures based on field data (1975-81) and annual statistical compendia available in relevant *kecamatan* (district) offices. These latter sources provide approximate figures only.

3. Code refers to numbered biotopes in Table 2.

4. Entries in column *a* are based on a notional 5 point scale of *overall* dependence on trade goods, where 1 represents minimal dependency. Entries in columns *b-c* indicate degree of dependence on non-food items, rice, fish, sago and vegetable foods respectively. In each case dependency is assumed to be theoretically measurable by dividing import subsidies (expressed by weight, calories, energy cost, time cost or economic value) by total population.

5. Crude estimates expressed as a 5 point notional scale, where 1 represents minimal connectivity. However, the empirical information is available to construct graphs for local networks of which listed populations are parts. An 'accessibility index' for each interaction (or vertex) can then be calculated using standard techniques of network analysis (Haggett 1969).

does not automatically mean that it is a dependent component or that it need reduce its own generalized condition. For many centuries the highland villages of the larger Moluccan islands served as specialized suppliers of forest products which were of great significance in world markets, and they were able to do this while still maintaining a generalized subsistence base (cf. Dunn 1975). On the other hand, trade which may appear negligible measured by its bulk, calories, in terms of its external cash equivalent or the number of persons involved in its conduct, may involve products which have a high local value. Consequently, its role in an understanding of the functioning of indigenous social formations may be much greater. The value of shell bangles and pre-twentieth century trade porcelain and stoneware has always been much greater to the Nuaulu than to either the craftsmen of Gorom who made bangles or the traders who supplied the dishes. Now that both of these commodities are no longer available, their value has risen concomitantly. The value attributed by the Nuaulu to *kain berang*, a type of red cloth, has similarly been greater than that attributed to it by traders in Ambon from whom it was obtained or the factory owners in Java who manufactured it. However, unlike bangles and porcelain, red cloth has become more available not less. While shortages of the former items have had deflationary consequences, the increasing availability and circulation of red cloth has had inflationary consequences. Each of these categories of object obtain their local value through their role in ceremonial exchanges necessary for the social reproduction of clans, households, ritual houses, in the settlement of disputes, and so on. In some cases the inability to obtain new items at all (for example, *patola* cloths) to replace old ones generates crises for particular households or clans, which may herald a more dramatic breakdown of traditional social organization. The general loss of valuables over time (such as old Chinese porcelain) encourages either an unwillingness to participate in the exchanges at all or the switching to substitute articles (cheap trade plates, for example) which do not carry the symbolic load of traditional items and which in this way lead to a decline in the significance of the exchanges. Our notion of trade asymmetry must therefore always be one which takes into account differential structural significance.

Asymmetry may be introduced into trade relations either exogenously or endogenously. There is no doubt that the creation of asymmetry from exogenous links—that is, through Asian and European demand for forest products, marine products and spices--has been historically crucial. In terms of unequal exchange it accounts for the widest disparities and has had the greatest impact on changes in local social formations. However, I wish to focus here on asymmetry of endogenous origin. It will become clear that in the end we cannot consider one without reference to the other.

Endogenous asymmetry may arise through consistent shortfalls in particular resources. These may result from ecological disturbance, population fluctuations or socially determined changes in the pattern of extraction or demand. In such situations, local ties of dependency may arise in an otherwise symmetrical model. Asymmetry may also derive from the creation of demands which the local environment has *never* been able to supply, such as stone for the manufacture of implements or shell for ornaments. The historical and contemporary evidence for this kind of local trade in the Moluccas is considerable. *Trochus* shell for armbands is known to have been traded from the Gorom archipelago to highland Seram, while flint implements have been found in locations which suggest overland transport of raw materials (Glover and Ellen 1975) and pottery has been traded widely for many centuries (Ellen and Glover 1974; Spriggs and Miller 1979). At the present time much of the movement of sago into areas where it is deficient, and of fish from coastal to highland settlements, may be attributed to this kind of local endogenously-derived imbalance. In many places it takes the classic 'fish for sago' (coastal-highland) pattern. The creation of this kind of continuing dependency provides permanent structural foci for a local trading network, which may grow symmetrically through the incorporation of other specialist localities. Some of these networks may assume a degree of symmetry through circular linkages, of a kind exemplified by the *kula* in the New Guinea Massim. None of these developments are, of course, independent of other social relationships, which may well determine particular trading partnerships or patterns. A congruence between marital alliance and trading relationships is an obvious possibility. The final putative phase in the evolution of an endogenously-inspired trading network is, where conditions permit, the development of central places which become more important than other localities in the control of exchange within the network. It is at this level of development that exogenous links have proved most critical and have led to the development of more inclusive and highly structured local trading systems.

Networks and Nesting Trading Systems
in the Moluccas

Moluccan trading networks can be reconstructed on the basis of both ethnographic and historical evidence. Figure 7.2 provides an outline taxonomy for those discussed here, indicating the four levels of spatial organization which I have found useful to identify: local, intermediate, regional and long-distance. Differences in scale between different networks may be judged in terms of distance, numbers of linkages, quantities of goods handled, size of populations integrated, and technological infrastructures, such as types of boat.

All networks have some systemic properties, but the degree of "systemness" varies between them. Conceptually, a trading *system* is composed of one or more networks and is characterized by a set of linkages which theoretically permit its reproduction over time through both a geographical and social division of labor. Most systems discussed here have a center and a periphery, connected by radial links. A center exhibits control over a periphery, although the extent of this control tends to diminish as we move further from the center. The center connects the system as a whole with the outside world, yet is dependent itself on the resources of the periphery. Settlements on the periphery which are points of production rather than simply nodes in a trading network, can be described as *termini*. The extent to which higher level systems absorb those at a lower level depends upon the degree and character of external commercial intrusion. Membership of a higher level system may in turn redefine the structure and dynamics of lower level networks.

Local Trading Networks

In the Moluccas local trading networks tend to be confined to single islands, small island groups, and stretches of coast and hinterland. In Figure 7.3, the Gorom (10), Banda (8), and Geser (9) networks clearly illustrate this level of organization. In each case, local imbalances in the production of various subsistence crops are usually evened-out with the area, and different villages within the zone provide certain specialized services, such as pottery manufacture, ironwork, sago production and fishing. The degree to which these zones overlap varies and affects their systemic

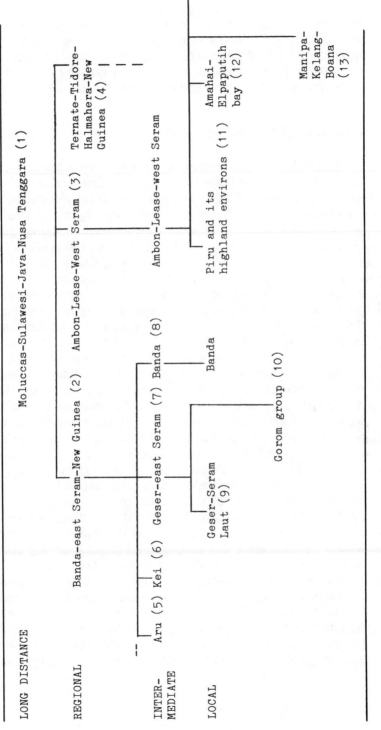

Figure 7.2 A hypothetical taxonomy of trading networks in the Moluccas.

character. The Banda group maintains a high degree of closure in terms of local trade, the Gorom group considerable less. Zones which include stretches of coastline, because of continuous habitation, tend to overlap much more. Some settlements may best be portrayed as simultaneously belonging to two or more local networks with different foci.

Intermediate Trading Networks

At a higher level, the central and southern Moluccan area can be seen as being composed of relatively discrete intermediate networks. At the hub of each of these is a small island (or number of small islands) serving as a redistributive center and point of articulation with the outside world: West Seram (3) is focussed on Ambon and the Lease islands, Aru (5) is focussed on Dobo, East Seram (7) on Geser, and so on (Figures 7.2 and 7.3). The trading centers, or foci, show varying degrees of dependency on their peripheries. The foci are generally very small islands which appear to owe their importance to their geographical centrality in the local trading area and because they provided intrusive traders or refugees with good harbors and safe havens in otherwise hostile territory.[5] Occasionally the reasons may have been environmental. For example, local inhabitants on Kefing claim that it was first inhabited to avoid the mangrove swamps along the southeastern coastline of Seram. But in this case one might ask why not settle at that point on the mainland where the mangrove ends. The real reason is more complicated and is probably largely political and economic in character: that an island among banks and shoals provided ready access to desirable sea products, that land elsewhere was politically inaccessible, that anchorages were convenient and safe, and that Kefing lay in the immediate vicinity of what was already an important trading center (the Seram Laut archipelago). The reasons for settling on such small islands must have been strong since the disadvantages were often severe. Small islands such as Geser, Kiltai and Kefing are flat, tiny and exposed coral islets, subject to frequent flooding, vulnerable to tidal waves, with virtually no possibilities for cultivation and with only brackish drinking water. There is oral and written evidence, as well as that of contemporary administrative boundaries, that in the past local political domains rivalled for control of these small strategic islands (Kolff 1840:287-8). Islands with so little natural wealth, with the possible exception of sea

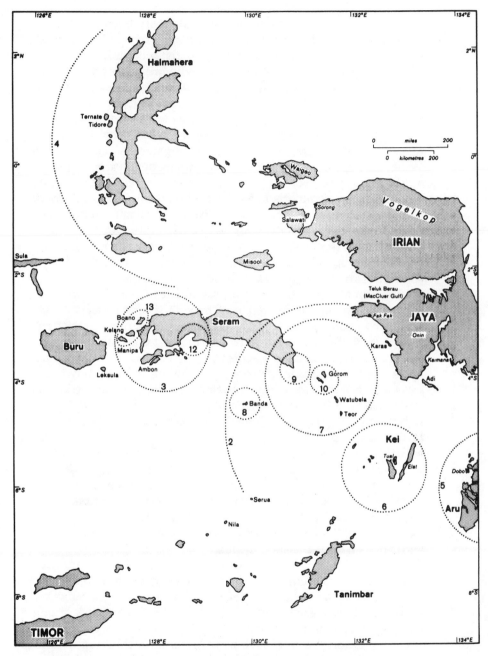

Figure 7.3 Spatial relationship between nesting Moluccan trading
networks listed in Figure 7.2.

produce, must be contrasted with powerful political domains such as Kataloka on Gorom (Gorong, Goram) which also exported locally-grown spices and had opportunities for subsistence cultivation.

This play-off between precarious environmental conditions and supreme economic advantage does not only apply to minute islands whose entire livelihood depends on the transhipment, rather than production, of commodities. We have already mentioned that access to exotic sea produce may have been an important locational feature for some of these centers, at least since the eighteenth century (Urry 1980). The same is true for certain settlements on the periphery, either populated reefs subject to storms (Gorogos) or islands vulnerable to seismic disturbances (Nila) (see Figure 7.1). Banda has had a continuous and violent history of seismic disturbance (Hanna 1978), and much the same may be true of Ternate and Tidore. But in these cases, we are not only dealing with ideal centers for transhipment but with primary locations for the production of nutmeg (Banda) and clove (Ternate/Tidore).

Very small islands with no possibilities for cultivation are entirely dependent on the import of basic foods and materials: rice and manufactured goods from outside the Moluccas and other starch staples (largely sago), vegetables, fruit and constructional and craft materials from other islands in the immediate vicinity. Such needs in a hypothetically pristine world might initially have been met through highly local trade, such as between Geser and Seram Laut. Geser, Kiltai and Kefing are still very dependent on the vegetable produce of nearby Seram Laut, and the economic significance of this island is reflected politically in the division between the polity (now *pemerintah negeri*) of Kelu on Kefing and Kiltai. But long-term requirements for trade objects to exchange for rice and manufactured goods and the increasing scale of such trade and home consumption have led to the growth of a much larger periphery. In the case of Geser, Kiltai and Kefing, it meant the incorporation of the eastern part of mainland Seram and parts of Teluk Berau and the Onin peninsula of New Guinea.

Trade in locally produced commodities falls into two categories: goods which are consumed within the network (or perhaps in adjacent ones) and those which are part of long-distance trade to Sulawesi, Java, or elsewhere. Commodities consumed within the network include sago, pottery, vegetables, fish, red sugar, timber, firewood and alcoholic liquors. Within each zone different localities specialize in different products for trade and the local circulation of these products, either radially through the center or laterally within the periphery, serving to integrate a complex social and economic *system*. Thus, the Banda islands as a whole are self-sufficient in red sugar, firewood, vegetables and fish, although no one settlement within the group is completely self-sufficient. But while the circulation of commodities maintains integral systemic properties within each trade zone, trade also occurs between zones. Thus, timber (and previously large quantities of sago) is generally imported into Banda from southeast Seram and Gorom (Kolff 1840:287), whereas Kei pottery was exported (after 1621) both to Seram and Banda. The circulation of such commodities serves to integrate Kei, Banda and S.E. Seram into a wider intermediate system, both laterally (between localities of more-or-less equal political and economic significance) and radially (focussing on secondary and primary centers).

Trade in commodities of the second kind--nutmeg, clove, exotic forest products such as edible birds' nests, trepang, pearls, tortoise-shell, and so on--only serve to integrate center with periphery (Seram Laut with Geser, Geser with Banda, etc.) or various points on the periphery *through* the center, since the demand is not local and is only mediated through traders operating from central places.

Termini are of two kinds. First of all there are those which are generally self-sufficient but which are the source of products desired elsewhere, in local centers, other parts of Indonesia, Asia and Europe. Prior to 1900 all interior mountain villages would have fallen into this category Thus, Lumoli in West Seram has traditionally exported *damar* resin (*Agathis alba*) to Piru, while Kilmuri on the southeast coast of Seram has traditionally exported sago to Geser. Other highland termini could have been involved in the export of timber, rattan, medicinal barks (*kayu manis, kayu lawan*), dried meat, antler, bird plumage and wild clove (*cengkeh hutan*). Because such termini were self-sufficient in a material sense, is not to deny that the objects which they received in exchange (red cloth, *patola, ikat*, porcelain) represented a form of congealed value very necessary for the maintenance of local exchange and ritual cycles (Ellen 1979; Elmberg 1968). I have already made a point of emphasizing this. The second kind of terminus is one whose very existence is dependent on a wider trading network, secondary rather than pristine. An example of this would be the Gorogos reef, a fishing settlement exporting fish,

trepang, *lola* and *batu laga* via Kataloka and Geser.
Despite having to import all of its vegetable food, starch staples,
and timber, Gorogos has grown from a temporary to a permanent
settlement in the course of the last 20 years.

Secondary and primary centers can also be classified into those
which are basically self-sufficient and those which are utterly
dependent upon others. Examples are found in Kataloka (on Gorom) and
Geser respectively. They can also be distinguished according to
whether or not they are also centers for spice production. It now
appears that spice cultivation at such centers may have been
secondary to their emergence as centers of trade, as local trading
communities dealing in spices produced elsewhere attempted to
exercise more control over the production of the commodity. This at
least appears to be the case for the east Seram network and Banda.
We know that for many centuries wild nutmeg (*pala panjang*)
had been coming from the forests of New Guinea via the trading
polities of east Seram. It is possible that the emergence of Gorom
as an important trading and political center was linked to the
planting of nutmeg (*pala panjang* and *pala bundar*)
brought from New Guinea. In turn, it is equally possible that the
first nutmeg grown on Banda (*pala bundar*) was brought from
Gorom. Thus a wider Banda-focused system (Figure 7.2) may have been
a rational outgrowth and extension of an east Seram system (Figure
7.2), linked to the planting of nutmeg in the Banda archipelago, the
growth of the islands as a general primary trading center for
incoming traders from the west and the consequent re-focussing of
other radial supply links. But in any case, the principal Banda-east
Seram artery remained for good geographical, navigational and
economic reasons.

Regional Trading Networks

There is some historical evidence for the existence of three
reasonably distinct regional trading nexuses located in the
present-day Indonesian province of the Moluccas and adjacent areas of
the New Guinea coast (Figures 7.2 and 7.3). Each is focused on a
densely-populated spice-producing center, dependent upon an extensive
sago-producing periphery. The three spice-producing foci are
Ternate/Tidore, Ambon-Lease and Banda; the major sago-producing
peripheries are, primarily and respectively, Halmahera, west Seram
and east Seram-Kei-Aru-New Guinea.[6] These trading nexuses

did not all emerge at the same time, and the conditions of their emergence were in each case significantly different. However, by the early part of the seventeenth century their simultaneous existence is quite well documented, while the consequences of the interaction of external factors with their internal dynamics are evident over the subsequent 300 year period. In the past they have shown strong systemic properties.

I shall confine myself here to the Ambon and Banda systems. However, these two systems differ in one important structural respect quite apart from the many detailed historical and ethnographic differences between them. The Ambon system (Figure 7.3) is a network without obvious intermediate-level components, despite certain sub-systemic local tendencies (e.g., in the Elpaputih bay and Boano-Kelang-Manipa areas). On the other hand, the Banda system (Figure 7.3) is composed of at least four separate and identifiable intermediate networks plus some less clearly attached networks. The four major intermediate level components are the Banda, Kei and Aru groups and the network connecting the Geser-Teor chain with the mainland of east Seram. The attachments are the New Guinea (Irianese) coast from Teluk Berau southeastward to Mimika and the extreme southeastern islands focused on Tanimbar. The systemic properties of the Banda regional nexus derive almost entirely from intrusive long distance trade. The incorporation of various intermediate networks has depended on the importance of their products outside the Moluccas and, to a lesser extent, demand elsewhere in the system. With the drop in external demand for their products, some intermediate (and local) networks regain something of their earlier autonomy. Thus, strong local sub-systemic properties have continually threatened the integrity of the system as a whole.

Connected with this important structural difference between the Ambon and Banda systems is the question of distance, and the appropriate boat technology in each case. While in the past the Ambon system could reproduce itself internally by relying on small coastal craft, the Banda system involved major sea-crossings. Historically, this has often meant the employment of much larger vessels, often those belonging to intrusive trading groups such as the Buginese and Butonese.

It would be a mistake, however, to see the emergence of trading patterns in the Moluccas as simply the working out of some game-theoretic or market rationality under ideal Von Thünen and Christaller-Lösch conditions. Trade relations within any network

were always constrained as well as facilitated by other social
relations, such as inter-village *pela* alliances (in the
Ambon-West Seram zone, or their functional equivalents elsewhere),
marital alliances and ties of political subordination and control.
Although it is difficult at this stage to make even intelligent
guesses regarding the political organization of trade prior to the
European period, the larger systems clearly brought together
political entities of very different kinds: semi-dependent tribal
groups (such as the Nuaulu), slave-based *kerajaan* (such as
Kataloka) and class based traditional societies (such as Kei). From
the seventeenth century onward, the number of different kinds of
political entity and modes of production increased to include Dutch
East India Company-run estates (Banda) and mercantilist control of
peasant producers (Ambon group). It is quite clear that the Dutch
period resulted not simply in changes in traditional trading
patterns, but in some cases in their consolidation and the
accentuation of some of their more distinctive properties. Thus, the
geographical division of resources in the Banda archipelago after
1621 was administratively re-defined: firewood could only be taken
from Gunung Api, and it was mandatory that vegetable gardens and
nutmeg groves be located in different places. Clove cultivation was
prohibited in large parts of Seram (e.g., the Huamoal peninsula) in
order to maintain the Dutch monopoly, encouraging these localities to
develop other trading specialities. But although Banda under the
East India Company represented an entirely unique kind of social and
cultural formation, the expansion and concentration of spice
production not only connected Moluccan centers more closely with the
world system but, paradoxically, made them more dependent on local
trading links as well. The growth and specialization of centers of
spice production and trade as the result of colonial policy had the
effect of producing a more distinct and complex local division of
labor and network of exchange relations. As land under spices and
population increased, so also did the local trade in sago, root
crops, vegetables and other products necessary to supply deficient
spice-producing areas. When centers already important for trade were
adopted as convenient administrative centers by the Dutch, bringing
an increasing number of retail enterprises to cater for a growing
number of wage earners and bureaucrats, trade routes became
increasingly centralized, radiating to and from the administrative
center.

Degree of Closure and Causality

Let me recapitulate the main points made in the previous section. Moluccan trading patterns may be conveniently represented as a series of nesting (and sometimes overlapping) zones of varying degrees of "systemness". Each local population may participate in up to four levels of trading organization: local, intermediate, regional and long-distance. Participation is defined minimally in terms of the destination of a population's exports and the origins of its imports. However, the structural significance of participation at any one level depends also on the types, proportions, range, value and volume of goods traded. For example, it may be important to know the scale of the trade in food products or the local significance of the trade in 'valuables.' Collectively, such factors determine the degree and kind of reproductive independence a local social formation maintains, and maintains at different levels of systemic inclusiveness. Table 7.4 illustrates how these patterns of participation at different levels of organization vary for six historically specified populations and between different kinds of functional components in wider systems. The patterns defined in these terms can help us to decide where we may legitimately draw boundaries in particular analyses, and what factors we can reasonably hold constant. Depending on the purpose of the analysis and the populations specified, we may decide to emphasize boundaries at the local, intermediate or regional level; but in all cases we must remember that boundaries are necessarily graded and determined by the structural foci of the systems they encompass.

We are now in a position to reinforce a number of observations about closure and the spatial character of causality:

1. We may agree with Faris (1975:239) that external causes only become effective through internal ones. Thus, growth in the spice trade accelerated local trade in sago, other food products, and basic resources. In turn, this changing pattern in purely local trade had repercussions elsewhere in the system. One cannot understand the spice trade in the Moluccas or the maintenance of trading centers without first tackling the dynamics of local trade. Even though the spice trade itself did not extend to the periphery, local trade in these areas did respond to its kick-on effects. For example, the increase in the east Seram sago trade was a

Table 7.4 Reproductive dependence of selected Moluccan populations at various levels of systemic inclusiveness.

| Local population [1] | Incoming goods classified by level of origin [2] | | | | Main exports |
	local	intermediate	regional	long-distance	and re-exports
1. Ruhuwa (Nuaulu), south Seram, 1969-81	11. vegetable shortfalls, fish, certain	3. pottery, red sugar		1. *cloth*, *porcelain*, manufactured goods, salt	cloves, copra
2. Bati Sai'e inland east Seram, 1981	8. *fish*, pottery, ironwork	6. <--	2. pottery	1. *cloth*, manufactured goods	cloves
3. Bandaneira, Banda group, 1797 [3]	8. vegetable shortfalls, red sugar		2. *timber*, thatch, coconut oil, *sago*, slaves, pottery, *trepang*, pearls, yams	1. *cloth*, rice, ironwork, porcelain	nutmeg, trepang, pearls
4. Kefing, Geser group, 1825 [4]	9. *sago, all vegetable foods*	7. sago	2. edible bird nests, tortoise-shell, cloves, nutmeg, pottery	1. firearms, gunpowder, porcelain, cannon, *cloth*	trepang, edible bird nests, tortoise-shell, cloves, nutmeg
5. Gorogos, Gorom group, 1981	10. *sago, all vegetable foods, timber*, thatch, ironwork	7. sago, timber	2. pottery	1. *cloth*, manufactured goods, *rice*	trepang, fish
6. Geser, Geser group, 1981	9. sago, *all vegetable*, foods, timber, thatch	7. sago pottery	2. pottery, tortoise-shell, *trepang*, nutmeg, cloves	1. *cloth*, *manufactured goods, rice*	trepang fish

Notes:
1. In terms of system components: 1 and 2 are generalised termini, 3 and 6 are primary centres, and 4 and 5 are specialised termini. Numbers refer to named networks listed in Figure 7.2.
2. Those import items most crucial in structural terms are written in italics.
3. Source: Miller 1980.
4. Source: Kolff 1840.

response to the growth of a secondary trading center in Geser, which itself was the outcome of nutmeg cultivation and commercial activity in the Banda archipelago. Similarly, changes in local exchange relationships in the Vogelkop of western New Guinea (Elmberg 1968) following the introduction of increasing quantities of cloth were an indirect spin-off of the central Moluccan trade in spices and international demand for maritime products, such as *trepang* and pearls, in the southeastern islands.

2   Although poorly connected systems may depend upon other levels of organization for their reproduction, this is not to reduce their overall dynamic to events at another level (Waddington 1970:180-3). While extrinsic factors may trigger changes in purely local trading arrangements, extrinsic factors alone are insufficient to explain the structural character of specific causation. One should no more expect to understand the spice trade in the Moluccas or the maintenance of trading centers without first tackling the dynamics of local trade than one should expect to understand local trade in isolation from external factors.

3.   Few (if any) modern populations are engaged in trade which is confined to local networks. Where this may have been the case historically, local patterns of trade had consequences for those populations participating in higher levels of organization. Moreover, even if a population does not handle materials produced within a system of which it is part, since objects must pass through any intermediate level, that trade must have some effect on participating populations.

4.   While being part of a system does not in itself indicate dependence on that system, certain populations could not exist except as parts of wider systems. In some cases this is physically true, for example, in the case of Geser or Gorogos. The ethnography of Melanesia is full of examples of small offshore islands apparently developing as centers in endogenously pristine systems. But although this may provide us with a possible explanation for the origin and development of places such as Geser as centers, their subsequent development is no doubt closely bound-up with a role in regional systems and long-distance trade. Gorogos, on the other hand, we know to have developed in the last 20

years *precisely* to supply marine products in demand in the Asiatic and world systems.

Conclusion

Since ethnographers will continue to make assumptions about the boundaries of the systems which they investigate and about the degree of closure which they attribute to them, it is necessary to know with the greatest possible accuracy what criteria can be used to best establish the degree of reproductive autonomy of local populations (both ecologically and socially) under particular ethnographic conditions. It is also important to know to what extent legitimate and useful boundaries may be drawn at different levels of spatial or temporal inclusiveness, since where boundaries are drawn determines the relationship between causes and effects.

I have shown that in order to ascertain the degree of dependence of particular Moluccan populations on wider systems we require information about:

(a) degree of physical isolation;

(b) spread of subsistence base;

(c) involvement in trade analyzed according to the destination of exports, the origin of imports and the general web of relations which connect local populations to wider systems; and information on the proportions, range, value and volume of goods traded.

But although this provides us with the basic preconditions for making realistic statements about the character and degree of reproductive independence, it is insufficient. It is not enough to know the *extent* to which a population is dependent or simply the items upon which this dependence lies; it is also important to understand the spatial position and function of a population in more inclusive local and regional systems. Once this information is available (and it implies a general understanding of the overall structural focus of superordinate systems), it then becomes possible to go beyond the explanation of a particular local change or pattern with reference to some general 'external' event(s) and specify in more detail how a cause has been transmitted from an imputed point of origin and transformed through intermediate variables, with varying and multiple consequences, to produce a particular local effect. It is then also possible to ascertain whether the population in which the effect is realized is simply a passive recipient of a set of consequences or whether it is itself an essential agent in the process which led to its being realized. I suspect that in most cases we will have to conclude that an effect is produced by an nexus of dependent factors which have their separate origins at different degrees of proximity.

Acknowledgments

I am grateful to the British Institute in Southeast Asia and the University of Kent at Canterbury who made possible the presentation of this paper at the January 1982 meeting of the American Association for the Advancement of Science in Washington, D.C. The empirical data referred to is derived largely from the pilot phase of a long-term project on 'Change and the social organization of regional trading networks in the Moluccas, eastern Indonesia,' financed by the British Academy and the University of Kent at Canterbury under the auspices of *Lembaga Ilmu Pengetahuan Indonesia* (the Indonesian Academy of Sciences). I would like to thank Emilio Moran, C.W. Watson, and Nikki Goward for their comments on an earlier draft of this paper. Jane Shepherd has drawn the maps.

Notes

[1]A notion of closure can only be effectively applied to discrete objects in a delineable field. Collectivities defined by cultural representations and particular social relationships do not easily meet this condition, even less so their reifications as 'cultures' and 'societies'.

[2]Allee and Schmidt 1951; Coe and Flannery 1967. The term *biotope* is convenient in that it avoids some of the rather special theoretical and structural implications associated with *ecosystem*. However, I would not wish to reject entirely the notion of ecosystem or to suggest that biotopes do not also possess systemic properties.

[3]For a more detailed account of general environmental variation in the Nuaulu area see Ellen 1978:5-10, 22-6, 33, 43-6, 61-89, 108-18, 130-60.

[4]It was such considerations which led Coedes (1944:2) to develop his concept of an internal archipelagic sea as a powerful formative force in the history of southeast Asia, a notion which has been returned to in more recent writing on regional prehistory

(Bellwood 1978: Urry 1981). In many ways, this role is similar to that attributed to the Mediterranean by Braudel (1949).

[5]For some comparative Melanesian studies of the role of small islands as crucial nodes in wider trading networks, see the following: Allen 1977a, 1977b (Motupure island in the vicinity of Port Moresby), Brookfield and Hart 1971:314-34 (general), Harding 1967 (Siassi islands in the Vitiaz straits), Harris 1979 (Torres Strait islands), Malinowski 1922 and Lauer 1970 (Tubetube, Amphletts and other islands in the *kula* area), Sahlins 1972 (general) and Schwartz 1963 (Admiralty archipelago). Such systems might be regarded as pristine in evolutionary terms. Similar systems in which the nodal points have their origin as settlements for exogeneous trade are also well known if not analyzed. Consider, for example, the role of Zanzibar, Pemba, and the Comores in Arabic commerce along the east coast of Africa, or Fernando Po in the Spanish-consolidated trading links across the Bight of Biafra.

[6]In an earlier paper (Ellen 1979:70) I have described the peripheries of the Ambon and Banda systems as 'Seram and Aru-Kei-New Guinea' respectively.

References Cited

Admiralty
    1944   Netherlands East Indies 1. Naval Intelligence.
Allee, W.C. and K.P. Schmidt
    1951   *Ecological Animal Geography.* 2nd ed. New York: Wiley.
Allen, J.
    1977a   Fishing for Wallabies: Trade as a Mechanism for Social Interaction, Integration and Elaboration on the Central Papuan Coast. *The Evolution of Social Systems.* Edited by J. Friedman and M.J. Rowlands, London: Duckworth.
    1977b   Sea Traffic, Trade and Expanding Horizons. *Sunda and Sahul: Prehistoric Studies in Southeast Asia, Melanesia and Australia.* Edited by J. Allen, J. Golson, R. Jones, London: Academic Press.
Bayliss-Smith, T.P.
    1977   Human Ecology and Island Populations: The Problem of Change. *Subsistence and Survival: Rural Ecology*

in the Pacific. Edited by T.P. Bayliss-Smith and
R.G.A. Feachem, London: Academic Press.

Bellwood, P.
1978 *Man's Conquest of the Pacific.* Auckland:
Collins.

Bennett, J.W.
1976 *The Ecological Transition: Cultural Anthropology
and Human Adaptation.* New York: Pergamon Press.

Braudel, Fernand
1949 *La Méditerranée et le Monde Méditerranéen a
l'Epoque do Philippe II.* Paris: Librarie Armand
Colin.

Brookfield, M.C. with D. Hart
1971 *Melanesia: A Geographical Interpretation of an
Island World.* London: Methuen.

Coe, M.D. and K. Flannery
1967 *Early Cultures and Human Ecology in South Coastal
Guatemala.* Smithsonian Contributions to
Anthropology 5. Washington, D.C.: Smithsonian
Institution.

Coedes, G.
1944 The Empire of the South Seas. *Journal of the
Thailand Research Society* 35(1):1-15.

Dornstreich, M.
1977 The Ecological Description and Analysis of Tropical
Subsistence Patterns: An Example from New Guinea.
*Subsistence and Survival: Rural Ecology in the
Pacific.* Edited by T.P. Bayliss-Smith and R.G.A.
Feachem, London: Academic Press.

Dunn, F.L.
1975 *Rainforest Collectors and Traders: A Study of
Resource Utilization in Modern and Ancient Malaya.*
Kuala Lumpur: Monographs of the Malaysian Branch of
the Royal Asiatic Society. No. 5.

Dunn, F.L. and D.F. Dunn
1977 Maritime Adaptations and Exploitation of Marine
Resources in Sundaic Southeast Asian Prehistory.
Modern Quarternary Research in Southeast Asia.
G.J. Bartstra, et. al. Rotterdam: A.A. Balkema.

Ellen, R.
1978 *Nuaulu Settlement and Ecology: An Approach to the*

Environmental Relations of an Eastern Indonesian Community. (Verhandelingen van het Koninklijk Instituut voor Taal-, Land en Volkenkunde 83) The Hague: Martinus Nijhoff.

1979   Sago Subsistence and the Trade in Spices: A Provisional Model of Ecological Succession and Imbalance in Moluccan History. *Social and Ecological Systems.* Edited by P. Burnham and R.F. Ellen, (Association of Social Anthropologists Monograph 18) London: Academic Press.

1982   *Environment, Subsistence and System.* Cambridge: Cambridge University Press.

Ellen, R.F. and I.C. Glover
1974   Pottery Manufacture and Trade in the Central Moluccas, Indonesia: The Modern Situation and the Historical Implications. *Man* (N.S.) 9(3):353-379.

Elmberg, J.-E.
1968   *Balance and Circulation: Aspects of Tradition and Change among the Mejprat of Irian Barat.* Stockholm: Ethnographical Museum Monograph Series No. 12.

Faris, J.C.
1975   Social Evolution, Population and Production. *Population, Ecology and Social Evolution.* Edited by S. Polgar, The Hague: Mouton.

Fosberg, F.R. (ed.)
1963   *Man's Place in the Island Ecosystem.* Honolulu: Bernice P. Bishop Museum.

Friedman, J.
1976   Marxist Theory and Systems of Total Reproduction. Part I: Negative. *Critique of Anthropology* 2(7):3-16.

Glover, I.C. and R.F. Ellen
1975   Ethnographic and Archaeological Aspects of a Flaked Stone Collection from Seram, Eastern Indonesia. *Asian Perspectives* 18(1):51-61.

Haggett, P. and R. Chorley
1969   *Network Analysis in Geography.* London: Edward Arnold.

Hanna, W.
1978   *Indonesian Banda: Colonialism and Its Aftermath in*

the Nutmeg Islands. Philadelphia: Institute for the Study of Human Issues.

Harding, T.G.
1967 *Voyagers of the Vitiaz Strait: A Study of a New Guinea Trade System.* Seattle: University of Washington Press.

Harris, D.R.
1979 Foragers and Farmers in the Western Torres Strait Islands: An Historical Analysis of Economic, Demographic and Spatial Differentiation. *Social and Ecological Systems.* Edited by P. Burnham and Roy Ellen, (Association of Social Anthropologists Monograph 18) London: Academic Press.

Kennedy, Raymond
1955 *Field Notes on Indonesia 1949-1950.* Harold C. Conklin, ed. New Haven, Connecticut: Human Relations Area Files.

Kenworthy, J.B.
1971 Water and Nutrient Cycling in a Tropical Rain Forest *The Water Relations of Malasian Forests; Being the Transactions of the First Aberdeen-Hull Symposium on Malasian Ecology Hull 1970.* Edited by John Roger Flenley, University of Hull, Department of Geography: misc. ser. 11.

Kolff, D.H.
1840 *Voyages of the Dutch Brig of War Dourga - Through the Southern and Little-Known Parts of the Moluccan Archipelago and Along the Previously Unknown Southern Coast of New Guinea .. 1825-1826.* London: Madden & Co.

Langton, J.
1973 Potentialities and Problems of Adopting a Systems Approach to the Study of Change in Human Geography. *Progress in Geography: International Review of Current Research* 4:125-79.

Lauer, P.K.
1970 Amphlett Islands Pottery Trade and the Kula. *Mankind* 7(3):165-76.

MacKnight, C.C.
1980 The Study of Praus in the Indonesian Archipelago. *The Great Circle* 2(2):117-128.

Malinowski, B.
  1922   *Argonauts of the Western Pacific.*   London:
          Routledge and Kegan Paul.
Mason, M.L. and J.H. Langenheim
  1957   Language Analysis and the Concept Environment.
          Ecology 38:325-340.
Miller, W.G.
  1980   An Account of Trade Patterns in the Banda Sea in 1797,
          from an Unpublished Manuscript in the India Office
          Library. *Indonesia Circle* 23:41-57.
Newcomer, P.H.
  1972   The Nuer are Dinka. *Man* (N.S.) 7(1):5-11.
Rappaport, Roy A.
  1968   *Pigs for the Ancestors: Ritual in the Ecology of a
          New Guinea People.* New Haven: Yale University
          Press.
Sahlins, M.
  1972   *Stone Age Economics.* London: Aldine.
Schwartz, Theodore
  1963   Systems of Areal Integration: Some Considerations
          Based on the Admiralty Islands of Northern Melanesia.
          Anthropological Forum 1:56-97.
Spriggs, M. and D. Miller
  1979   Ambon-Lease: A Study of Contemporary Pottery Making
          and its Archaeological Relevance. *Pottery and the
          Archaeologist.*   Edited by M. Millet, London:
          Institute of Archaeology Occasional Paper.
Thompson, L.
  1949   The Relations of Man, Animals and Plants in an Island
          Community (Fiji). *American Anthropologist* 51,
          253-67.
Urry, James
  1980   Goods for the Oriental Emporium: The Expansion of
          Trade in the Indonesian Archipelago and Its Impact on
          the Outer Periphery. Unpublished manuscript.
  1981   A View from the West: Inland, Lowland and Islands in
          Indonesian Prehistory. Unpublished paper presented at
          Section 25A-Archaeology, 51st ANZAAS Congress,
          Brisbane, May.
Vayda, A.F. and B.J. McCay
  1977   Problems in the Identification of Environmental

Problems. *Subsistence and Survival: Rural Ecology in the Pacific.* Edited by T.P. Bayliss-Smith and R.G.A. Feachem, London: Academic Press.

Waddington, C.H.
1970 *Towards a Theoretical Biology.* Edinburgh: Edinburgh University Press.

Weiner, J.S.
1964 The Biology of Social Man. *Journal of the Royal Anthropological Institute* 94:230-240.

CHAPTER 8

LINKS AND BOUNDARIES: RECONSIDERING THE ALPINE VILLAGE AS ECOSYSTEM

Robert McC. Netting

In occasionally referring to the Swiss mountain village of Törbel as an "island in the sky" and describing the intricate economic and social means by which its inhabitants over the centuries struck a balance with their alpine environment (Netting 1981), I may well have been guilty of the ecosystematic fallacy. This common anthropological error involves an overemphasis on functional integration, stability, and regulatory mechanisms within the community and a relative neglect of disequilibrium, changes emanating from more inclusive political-economic systems, and instances of evolutionary maladaptation. The solid specificity of terrain and climate may appear as a fundamental set of constraints on human activity, especially when ethnological comparison with communities in similar environments is lacking. The nature of long-term resident field research, our anthropological reverence for an holistic perspective, and the romantic mystique of the self-sufficient, autonomous, emotionally rewarding "little community" all perpetuate our proclivity to learn a lot about a very limited group.

Ecological anthropologists with their commitment to gathering a wider range of non-cultural data and organizing these variables into systems models often focus on small units of interaction for both practical and theoretical reasons. Biologists have told us comfortingly that ecosystem is an accordion concept, applying equally well to a drop of pond water or the entire biosphere. If indeed the ecosystem may be delimited at any magnitude appropriate for a particular investigation, and if natural boundaries are desirable but not essential (Fosberg, personal communication), we can continue to study those convenient social entities where we have always worked--the band, the village, or the tribe. The boundaries are, of course, recognized as artificial or drawn for "heuristic" purposes.

We vigorously deny that we still labor in the expiring vineyard of neo-functionalism (Vayda and McCay 1975; Orlove 1980). Nevertheless, we remain concerned with more or less closed systems whose internal processes are regulated by negative feedback loops and we tend to make common cause with the biologists who stress the idea of ecosystems rather than of natural selection as an organizing concept (Richerson 1977). Distinguishing between local and regional ecosystems (Rappaport 1971a) recognizes part of the problem, but the sheer complexity of charting energy flows in local food production and consumption (Rappaport 1971b) may be so demanding that exchanges of goods, services, and people in the wider network over long periods of time cannot be adequately handled.

The appearance of biological orthodoxy and objective comparability that comes from selecting "populations" rather than "cultures" as units of investigation (Vayda and Rappaport 1968) is often illusory. Lacking species distinctiveness and geographical isolation, questions of where to bound a population as ecosystem may rely on endogamous groups or other ethnic bars to mating whose creation and effective maintenance are based on cultural rules of acceptable marriage (see Butzer, Hastorf, and Ellen, this volume). The fact is that human populations must be typified and analyzed by demographic means, but we necessarily begin with localized, co-resident, on-the-ground groups that identify themselves by name and tell us about the links of kinship, cooperation, and citizenship that bind their members together. Without hard information on such biosocial factors as diet and disease, warfare mortality and migration, and changing age-specific fertility (Buchbinder 1978; Morren 1977; Hassan 1978), ecological approaches to human cultures, populations, or ecosystems, however defined, can hardly rise above the level of explanatory sketches.

Jochim (1981:4) has pointed out that the boundary definition of human ecosystems is even harder than that of biological ones because of the more diverse interactions and because humans establish cultural boundaries that may or may not coincide with any natural ones. It seems to me now that I was led down the garden path of the independent population subsisting on its own resources in a clearly defined geographical area by the extraordinarily definite and enduring congruence between the Swiss folk model of the community and the historic realities of peasant village economy in the Alps. But in the very process of finding out how orderly, effective, self-correcting, and responsive the local system had been in sustaining a self-conscious, corporate population through time in an unremittingly difficult mountain environment, I became aware of the often concealed interdependencies that sustained the system at its points of weakness and rectified its dangerous imbalances. We will consider some of the ways in which a culturally and materially defined local community ecosystem survived by means of significant economic and demographic flows back and forth across its boundaries.

Self-sufficiency and the Market

Anthropologists coming from technologically complex, occupationally specialized, economically interdependent societies may be attracted to groups where the entire labor process from raw natural resource to finished product and consumption is visible and comprehensible. We are devotees of mechanical solidarity for whom small has always been beautiful and ecological homeostasis in traditional societies is assumed until proven otherwise. It was with a sense of considerable satisfaction that I settled on the community of Törbel in the Visp valley of Valais Canton as a site for field research. Törbel seemed by all accounts a representative alpine village whose peasants had lived since at least the eleventh century A.D. on the returns of agropastoral subsistence pursuits carried on within their own demarcated territory. A contiguous area of 1545 ha. (including 967 ha. of field, meadow, and pasture, 455 ha. of forest, and 123 ha. of unproductive land in 1924) sloped from the Mattervispa river at 770 meters above sea level to the peak of Augsbordhorn at 2972 meters (Netting 1981:2). Various altitudinal zones were used for vineyards, hay meadows, grainfields, gardens, and summer grazing grounds (Netting 1972), so that almost the entire supply of wine, rye bread, dairy products, potatoes, vegetables, and meat were locally produced. Wood for building and fuel, slate for roofs, and wool for textiles also came from village lands, and the mountain stream powered grist mills, a saw mill, and a fulling mill. Every farm family had access to the various types of land and other means of production such as barns and livestock (Netting 1981:10-41). Climatic fluctuations were cushioned by the scatter of individually-owned fields with various degrees of slope and sun exposure (Netting 1976), an extensive system of meadow irrigation (Netting 1974), and effective techniques for the storage of hay, grain, bread, cheese, and dried meat. The provision of adequate food, clothing, and shelter from within the village territory by subsistence methods that changed little from at least the fourteenth century does not appear to have degraded the local environment. Soil fertility was maintained by manuring. Terracing and uphill transport of earth limited erosion, and carefully controlled timber cutting prevented deforestation (Netting 1981:46, 67). The localized ecosystem seemed to epitomize a well articulated, self-sustaining interdependence of physical environment, subsistence techniques, and human population.

Both the Swiss inhabitants of Törbel and the outsider ecologist are inclined to stress a peasant past in which comparative independence and autarchy distinguished the community from other European rural societies. But economic isolation was probably never the rule. A path through the Törbel hamlets of Burgen and Feld is part of the ancient trail connecting the Rhone valley with Italy. Roman coins found on the Theodul Pass indicate early connections, and a hill fort in Zeneggan near Törbel dates to the Iron Age (Netting 1981:8). Trails of this kind were perhaps less important than the Simplon or St. Bernard routes in linking northern Europe with the Mediterranean, but the alpine traffic could only be interrupted and the passes closed when a disaster like the bubonic plague caused medieval Valais to quarantine itself. Mountain agriculture is hardly imaginable without axes and other woodworking tools, hoes, and caldrons for cheese making. Metals have always necessarily been imported. Salt for the dairy cattle and for preserving cheese had also to be brought in from a distance. Indeed, the trade in salt from either France or Italy was a cantonal monopoly around which the late medieval international relations of the mountain districts revolved (Dubois 1965). Sixteenth century documents record an annual wagonload of salt delivered to Törbel, and in the nineteenth century it was estimated that an ordinary household with cattle used 70 kg of salt per year (Franscini 1848:139).

From the time of the earliest parchment documents relating to the community, land sales have required money, and both churchly tithes and personal debts with interest are given cash values. Törbel had to participate in the market, but it is not completely clear how this was done. To this day, Törbel has a reputation for raising milk cows, and most families sell some breeding or slaughter stock every year. Animals were driven to the lower Rhone valley, across the passes into Italy, or disposed of at the annual markets in Visp and Stalden. Live sheep and goats as well as raw wool and cheese may also have been traded in former times, but the alpine herding pattern was more oriented to subsistence than exchange, and it seems likely that an appreciable surplus was not regularly produced. Cash income may have been dependent more on the export of labor than of agricultural goods. The trans-alpine trade required the seasonal work of drivers, mule skinners, and guards, and Törbel men may have taken such jobs as they did when the tourist industry began in the late nineteenth century. They also served for longer periods as mercenary soldiers in the armies of France, Spain, Naples, and the Pope (Netting 1981:54). If more recent accounts of wage labor outside the community as miners, craftsmen, and waitresses represent past practices, the earnings of such workers were returned in large part to parents or conjugal families in the village. The export of labor power allowed the community to meet its needs for commodities and manufactured goods, pay taxes, and conduct internal exchange on a cash basis. Though the typical farming village was something of a commercial *cul de sac*, its continued existence on the alpine margins has always required active exchange with the capitalist European centers of the world system.

Just how porous is the membrane separating the peasant economy from wider spheres is demonstrated by two processes. In the first place, self-sufficiency in food has been both aided and diminished by contact with external agricultural sources. There is evidence that mountain agriculture was made substantially more productive and dependable by the adoption of the potato in the late eighteenth century (Netting 1981:159-168). On the other hand, the purchase of cheap maize meal from Italian sources in the early 1900's decreased Törbel's reliance on its own rye crop. Construction of a road and daily bus service with valley towns more recently allowed the buying of bread made from imported wheat. The price and convenience of the white loaves spelled the quick demise of local grain *Aecker*, grist mills, and bake ovens.

In the course of our research, it also became clear that a model of village wealth built on inheritance of agricultural resources within a closed system did not adequately represent reality. If the major determinant of an individual's property in farm land and buildings was the holdings of his parents (and the parents of his spouse) to which he was heir, then there should have been an association between father's and son's wealth. Quantitative data from tax valuations between 1851 and 1915 failed to disclose significant correlations between father's and son's property measured by average wealth, maximum wealth, wealth at marriage, or wealth at age 40 (McGuire and Netting 1982). Even controlling for the number

of siblings with whom the partible inheritance was shared brought no
better prediction of the wealth of the son. This considerable
mobility up and down the village wealth spectrum suggested that
inherited resources were less of an influence than we had thought,
and that differences resulting from cash earned outside the village
context and then invested in agricultural property were substantial.
The hard summer work on the mule train postal transport or on railway
construction tapped funds from the national economy that allowed
peasant households to prosper without leaving the land.

Population Growth and Self-Regulation

Ecological anthropologists have found it difficult to resist the
attractions of an hypothesized human ecosystem in which population
was somehow regulated without direct imposition of Malthusian
sanctions, and growth rates, if present, were extremely low. Simple
models of populations maintaining themselves below carrying capacity
have been criticized (Street 1969; Brush 1975; Hayden 1975; Jochim
1981), and theoretical constructs of "group selection" in adjusting
numbers so as to maintain local resource stability have been
questioned (Lewontin 1970; Bates and Lees 1979; Orlove 1980). My
choice of a Swiss village with adequate vital records and our major
effort to reconstruct 300 years of local historical demography were
motivated in part by the wish to determine (1) if population
equilibrium with fixed resources had in fact been achieved and (2)
what was the role of social factors in influencing population
growth. The more extensive the quantitative data to be analyzed, the
more complex and partial become the answers to simple questions. We
have found that the population of Törbel went through periods of
quite marked growth as well as times of relative stability (Netting
1981:90-108). These dynamics have been responsive to a variety of
internal factors, both biological, such as life expectancy,
seasonality of conception, lactation, and nutritional patterns, and
socio-cultural variables, such as age of marriage, celibacy, and
inheritance (Netting 1981:109-158). But at no time could the
fluctuations of local population be understood in isolation from
surrounding populations.

Törbel's demographic distinctiveness from its neighbors was due
less to topographical barriers than to the conscious imposition of
legal and political barriers that successfully barred most
in-migration. Unlike the rapid turnover of population characteristic

of most European rural settlements (Schofield 1970; Gaunt 1977; Macfarlane 1978), Törbel family lines showed remarkable continuity over time, and no new family names took root in the village from before 1700 to 1970 (Netting 1981:70-89). Village citizenship descended in the male line, and local statutes first written down in 1483 prohibited outsiders from enjoying communal rights in the forests or the Alp. Without such resources to supplement privately owned farm lands, alpine agro-pastoralism would not have been possible. Formulation and enforcement of these rules was explicitly an activity of the corporate community (Wiegandt 1978) which thereby "closed" itself and effectively resisted population expansion due to immigration.

Village boundaries were, however, permeable to the movement of people in the other direction. It is possible that mountain populations have long followed a type of gravity model--that is, flows set up by birth rates higher than the replacement level and a comparatively isolated and healthy situation have taken surplus people from the highlands down to the valleys and plains. Törbel citizens have always left their homes permanently as mercenaries, farm laborers, artisans, and clerics, but new opportunities such as the colonizing of the Argentine interior, hotel jobs, or the construction industry in Swiss cities led to increased departures (Netting 1981:101-107). Without this safety valve, local population would have outstripped the potential of the territory to provide viable household subsistence holdings. The late nineteenth century demographic growth would have resulted in both rapid impoverishment and even higher than observed rates of celibacy if everyone had been forced to remain within the village. As it was, Törbel's export of people was one means by which the creation of a landless proletariat was avoided.

Does the conception of a community rejecting settlement by outsiders and allowing its own excess bodies to migrate mean that Törbel remained a homogeneous isolate in biological terms? Again, the self-contained ecosystem proves a poor analogy. Analysis of 917 marriages since 1703 shows that 14% of these were with women from other villages who took up residence in Törbel (Hagaman, Elias, and Netting 1978). Though this represents a very low rate of exogamy (cf. Levine 1977:39), the comparison of genetic contributions shows that in-migrants accounted for nearly 38% of the 1970 gene pool, while the more numerous Törbel ancestors were responsible for only 62% of the living population's genetic constitution. It appears that, for reasons not entirely clear, the fertility of children of in-migrants was significantly higher than that of Törbel natives. A high endogamy rate evidently does not create the conditions of a genetic isolate. The social and economic constraints on marriage between Törbel men and women from other villages have not prevented a substantial flow of "foreign" genetic materials into the local population.

The Ghost of Environmental Determinisms Past: Comparing Alpine Ecosystems

One of the attractions of fitting a structural-functional model of a social system into a more inclusive ecological system is that the significant aspects of the physical environment can be held constant. Where altitude, slope, water availability, and seasonal climate seemed to dictate the possible types and extent of agro-pastoral production under traditional technology in the European Alps, there has been a tendency to conceive of a similarly constrained set of social institutions and demographic patterns. Though I would have heatedly denied the charge that my explanation of Törbel's economic self-sufficiency and the adaptive advantages of its inheritance, marriage, and common property institutions smacked of environmental determinism, it is clear in retrospect that I sought to emphasize the neatness of fit between fixed natural conditions and more malleable human means of organization. If I had some measure of success in this endeavor, it was in part because my picture of an autarchic, closed corporate community with low fertility, emigration as a safety valve, and resource ceiling set by the marginality of mountain agriculture tallied well with the "canonical image of the upland community" in the growing anthropological literature (Viazzo 1989:11-14). In claiming some more than idiosyncratic significance for their work, anthropologists at least implicitly put their village studies forward as representative of an entire region. Without ever having set foot in a French, Italian, or Austrian Alpine village, I may have done just that.

The great corrective to the tacit over-generalization of any local model of ecosystem functioning is, of course, comparison. What might be termed the new, ecologically-informed ethnology examines spatially distinct populations through time in similar alpine environments, using the methods of historical demography and political economy to measure and analyze variation in a series of controlled comparisons (Eggan 1954). Pier Paolo Viazzo (1989) has rigorously applied these methods, demonstrating that the Italian community of Alagna, sharing the same habitat, material culture, legal tradition, language, and ethnicity with Törbel was in many respects strikingly different. Rather than being relatively closed, isolated, and economically marginal, this village displayed an enormous amount of migration, with men up to the 1930's going out every summer as skilled plasterers to France, and outsiders entering the village as miners, beginning in the 1530's. These activities were heavily dependent on international economic conditions and on national trade and emigration policies. Women and hired laborers maintained the agro-pastoral subsistence economy, but there was no communal pasture, no irrigation, and food grains were imported (Viazzo 1989:110-115). Mining fluctuated with both the external market and the stock of exploitable ore resources, and the immigrant wage-labor population had differing interests from those of the local land-owning peasantry.

Low marital fertility was combined with late age of marriage, considerable celibacy, and partible inheritance as in the Törbel case, but there is in Alagna evidence of conscious family limitation. A low pressure demographic regime with restraints on local population growth could evidently be maintained by checks on nuptiality as in Törbel, by emigration as in the Western Alps of France and Italy, or by birth control (Viazzo 1989:219). A system of elevated ages of marriage and high celibacy such as I described for Törbel does appear to act as a crucial homeostatic mechanism in population growth, but levels of nuptiality can vary regionally and over time, and other factors may produce similar effects. The picture of a single pattern of traditional Alpine nuptiality rests on the postulates "that similar ecological imperatives must invariably produce similar responses and that recent changes must have been preceded by a long period of static equilibrium" (Viazzo 1989:220). I would agree that formulations stressing the close functional integration of environmental and demographic factors in a stable ecosystem are too rigid and in the last analysis untenable, but I have taken pains to avoid suggesting the applicability of the specific ecological analysis of Törbel to other alpine cross-cultural situations.

Complex multiple and extended family households in northern Italy contrasted with the Valaisan pattern. Though proportions of household types show some adaptive variation through time (Netting

1979; Viazzo 1989:96-98), it is possible that the stem family in
Alagna was well suited to the periodic absence of younger migrant
males (Viazzo 1989:245-250) as it was in French Savoy households with
early marriage and low fertility rates (Siddle and Jones 1983; Viazzo
1989:202). There are, however, regional similarities in the
prevalence of stem family organization that occur in northern
Portugal and Spain, southern France, northern Italy, and Austria in a
variety of farming systems and landscapes. Peter Laslett (1984)
refers to this as the Mediterranean pattern of domestic organization
and suggests that it may have considerable antiquity in this
extensive area. There are also political factors that favor the
maintenance of stem family households and impartible inheritance in
the Austrian Alps. The nobles who colonized this mountainous region
wanted to ensure the persistence of landed holdings capable of
producing a surplus to be marketed in the lowlands (Viazzo
1989:264). The Hapsburg state also wished to promote a stable tax
base and prevent fragmentation of peasant farms. These policies
contributed to pronounced inequalities between the heir to a
household farmstead and his landless siblings (Khera 1972). The
Törbel corporate community of relatively egalitarian citizen
smallholders (McGuire and Netting 1972) resident in independent
nuclear family households contrasts sharply with the hierarchical
class structure of upland Austrian society.

Even the degree of economic self-sufficiency or "autarchy" in
Törbel, and by extension in Valais, Grisons, and Ticino cantons that
supported a certain peasant autonomy from the market did not
characterize other northern Swiss communities. By the mid-eighteenth
century, villages of the *Hirtenland* or pastoral region had
given up grain cultivation to specialize in summer pasture for
commercial livestock or dairy production. Richer farmers leased the
Alp and bought out the meadows and gardens of the poorer citizens
(Viazzo 1989:184-185, 272-273). In some agricultural villages, a
slow growth of population could continue with the new subsistence
opportunities offered by the potato, while in other areas, the
proliferation of cottage industry among the landless allowed a rapid
demographic increase. In both areas, greater dependence on the
market and the control of land by a large-farmer elite introduced
greater stratification and pauperized a section of the population
(Viazzo 1989:274). The closed corporate community that gave rights
in the commons to all its citizens, allowing a rational exploitation
of the diffuse summer grazing and forest resources, and at the same

time protecting them from overuse and degradation (Netting 1976), may have been an institutional bastion against such differentiation. But even this historic and egalitarian system of local government is not a constant in Alpine social organization (Viazzo 1989:280). The intricate and interlocking factors in even the most aesthetically compelling ecosystem model are not inevitable. Environment and history show regularities and causal relationships, but they are never unitary or necessary.

Tilting the Environment/Population Balance

The variety of regional and national economic and political arrangements that have influenced local Alpine production and social organization historically suggest that a diversity of viable ecosystems may exist. Some measure of stability may be reached at different levels of population, under contrasting fertility and nuptiality regimes, and with variable modes of land tenure and wealth distribution. It is possible, for instance, that a single private owner could stock and manage an alpine summer pasture with more efficiency and in a more optimal manner than does a corporate community with common property rights. Non-communal Alps do exist, and about one-fifth of Swiss Alp territory belongs to a Kanton or to a private owner or group of owners (Picht 1987). It is possible that corporate groups may put less than the optimal or maximally sustainable number of cows on the Alp. Milk yields from common property Alps fall below those of private property Alps (though production and transaction costs may be higher), and grazing pressure on the Swiss commons is *lower* than on private land (Stevenson 1990, cited in Ostrom 1990).

In the early years of this century, surveys of the management of all the Alpine pastures in upper Valais recommended that Törbel reduce the number of cow-rights to prevent over stocking and that a series of improvements be made. Specifically, the tasks of cleaning the Alp of loose stones, removing brush such as raspberries and wild roses (a job estimated at a total of 800 man days of work), draining a swampy area, and constructing a 100 m protective wall were suggested (Schweizerische Alpstatistik 1900, 1909). The inspectors of the Swiss Alp Economy Association appeared to feel that conservative communities like Törbel were unwilling to invest the labor that could intensify the husbandry of local resources and substantially raise production (Schweizerische Alpstatistik 1909:4).

It is also difficult to reach an objective scientific judgment of the forestry practices on communal lands. A rising demand for timber in the market led to significant deforestation of many Swiss mountainous areas in the nineteenth century (Picht, personal communication). Törbel itself had to provide firewood by the 1870's for twice as many people as it had a hundred years before, and one authority expressed dismay about the scraping of the forest floor for livestock stall bedding material (Stebler 1922). Despite rigid controls on cutting live trees and the public auction of windfalls and dead standing timber by the forest steward (*Waldvogt*) after Sunday mass (Netting 1981:68), major house construction in Törbel for more than a century has been possible only with timber purchased outside the community. But communally managed forest cutting and the accompanying protection against avalanche and slope erosion damage seem to have remained an optimal solution to threats of environmental deterioration. Where Swiss commonly-owned forests were divided among villagers to become individually owned woodlots, "the lots were generally too small for effective management and degenerated until [government] intervention occurred in the nineteenth century" (Ostrom 1990; citing Ciriacy and Bishop 1975). Admirable as the institution arrangements may be for allocating rights to scarce resources, the question of whether they maintain optimal, sustainable yields remains a matter of empirical determination. It is also difficult to reconstruct past patterns of resource use, and the contemporary appearance of equilibrium may conceal historic episodes of overuse and environmental degradation.

Cultural Conception, Folk Model, and Ecosystem

In order to adequately conceptualize the ecological relationships of human groups, it may be necessary to treat them *as if* they were parts of a functionally integrated, persisting, homeostatic, isolatable ecosystem. Since Geertz (1963:9) cogently recommended the biologists' term ecosystem, ecological anthropologists have used it with characteristic alacrity and looseness. The danger was not so much in the flexible inclusiveness by which ecosystem could embrace a wide range of cultural, biological, and physical variables as in the tendency for heuristically-drawn boundaries to harden into the familiar shapes of geographical regions or self-conscious social groups. The tendency is strong for the anthropologist to accept the members of a peasant closed corporate community at their word and emphasize the historic identity, economic self-sufficiency, population continuity, and socio-political autonomy that they claim for their village. The direct reality of participant observation combines with the villagers' own behavioral spheres of kinship, neighborliness, farm labor, and religious participation to emphasize everything that is bounded, familiar, and parochial. Living in Törbel, the very mountains and streams and serpentine road to the valley became for me identified with 600-year-old log houses and parchment documents and peasant genealogies vanishing into medieval mists (Reader 1988: Fig. 29). The marvel that any people had lived so long and so well on these alpine slopes led me to see and describe Törbel as an ecosystem. But for all the intricate adaptive mechanisms and the balance between human needs and environmental potentials, the village was never encapsulated or cut off. Subtly variable flows of goods and money and people tied it to a wider world. The significance of these movements of salt, iron, cattle, soldiers, coins, New World migrants, and in-marrying wives was often hidden from me and from my Törbel friends. But without these surges and trickles of energy in both directions, the local system could never have survived. The concept of a human ecosystem, like the idea of a niche or a lineage or a community, does not help us to create an airtight case but to model a useful, well-wrought urn of the imagination.

References Cited

Bates, Daniel G. and Susan H. Lees
    1979   The Myth of Population Regulation. *Evolutionary Biology and Human Social Behavior: An Anthropological Perspective*, Edited by N.A. Chagnon and W. Irons, North Scituate, Mass.: Duxbury, pp. 273-289.

Brush, S. B.
    1975   The Concept of Carrying Capacity for Systems of Shifting Cultivation. *American Anthropologist* 77:799-811

Buchbinder, Georgeda
    1978   Nutritional Stress and Post-contact Population Decline among the Maring of New Guinea. *Malnutrition, Behavior, and Social Organization*, Edited by L. S. Greene, New York: Academic.

242   *Robert McC. Netting*

Ciriacy-Wantrup, S. V. and R. C. Bishop
   1975   Common Property as a Concept in Natural Resource
             Policy. *Natural Resources Journal* 15:713-727.
Dubois, A.
   1965   *Die Salzversorgung des Wallis 1500-1610: Wirtschaft
             und Politik.* Winterthur: P. G. Keller.
Eggan, F.
   1954   Social Anthropology and the Method of Controlled
             Comparison. *American Anthropologist* 56:743-763.
Franscini, Stephan
   1848   *Neue Statistik der Schweiz.* Bern: Haller'schen.
Gaunt, David
   1977   Pre-Industrial Economy and Population Structure.
             Scandanavian Journal of History 2:183-210.
Geertz, Clifford
   1963   *Agricultural Involution: The Process of Ecological
             Change in Indonesia.* Berkeley: University of
             California Press.
Hagaman, Roberta M., Walter S. Elias, Robert McC. Netting
   1978   The Genetic and Demographic Impact of In-Migrants in a
             Largely Endogamous Community. *Annals of Human
             Biology* 5:505-515.
Hayden, B.
   1975   The Carrying Capacity Dilemma. *Population Studies
             in Archaeology and Biological Anthropology*, A. C.
             Swedlund, ed. Memoir 30, pp. 11-21. Washington:
             Society for American Archaeology.
Howell, Nancy
   1979   *Demography of the Dobe Area !Kung.* New York:
             Academic.
Jochim, Michael A.
   1981   *Strategies for Survival: Cultural Behavior in an
             Ecological Context.* New York: Academic.
Khera, Sigrid
   1972   An Austrian Peasant Village Under Rural
             Industrialization. *Behavioral Science Notes* 7:
             9-36.

Laslett, Peter
  1984   The Family as a Knot of Individual Interests.
          *Households: Comparative and Historical Studies of
          the Domestic Group*, Edited by R. McC. Netting, R.
          R. Wilk, and E. J. Arnould, Berkeley: University of
          California Press, pp. 353-379.

Levine, David
  1977   *Family   Formation   in   an   Age   of   Nascent
          Capitalism.* New York: Academic Press.

Lewontin, R. C.
  1970   Units of Selection. *Annual Review of Ecology and
          Systematics* 1:1-18.

Macfarlane, Alan
  1978   *The Origins of English Individualism: the Family,
          Property   and   Social   Transition.*   New   York:
          Cambridge University Press.

McGuire, Randall and Robert McC. Netting
  1982   Levelling Peasants? The Demographic Implications of
          Wealth Differences in an Alpine Community.
          American Ethnologist 9:269-290.

Morren, G. E. B.
  1977   From Hunting to Herding: Pigs and the Control of Energy
          in Montane New Guinea. *Subsistence and Survival:
          Rural Ecology in the Pacific*, Edited by T. P.
          Bayliss-Smith and F. G. Feachem, New York: Academic
          Press.

Netting, Robert McC.
  1971   Of Men and Meadows: Strategies of Alpine Land Use.
          Anthropological Quarterly 45:132-144.

  1974   The System Nobody Knows: Village Irrigation in the
          Swiss Alps.   *Irrigation's Impact on Society*,
          Edited by T. E. Downing and M. Gibson, Tucson:
          University of Arizona Press, pp. 67-75.

  1976   What Alpine Peasants Have in Common: Observations on
          Communal Tenure in a Swiss Village. *Human
          Ecology* 4:135-146.

  1979   Household Dynamics in a Nineteenth Century Swiss
          Village. *Journal of Family History* 4:39-58.

  1981   *Balancing on an Alp: Ecological Change and
          Continuity in a Swiss Mountain Community.*
          Cambridge: Cambridge University Press.

Orlove, Benjamin S.
1980   Ecological Anthropology. *Annual Review of Anthropology*, 9:235-273.
Ostrom, Elinor
1990   *Governing the Commons: the Evolution of Institutions for Collective Action.* Cambridge: Cambridge University Press.
Picht, Christine
1987   Common Property Regimes in Swiss Alpine Meadows. Paper presented at the Conference on Advances in Comparative Institutional Analysis at the Inter-University Center of Post Graduate Studies, October 19-23, Dubrovnik, Yugoslavia.
Rappaport, Roy A.
1971a   Nature, Culture, and Ecological Anthropology. *Man, Culture, and Society*, Edited by H. L. Shapiro, London: Oxford University, pp. 237-267.
1971b   The Flow of Energy in an Agricultural Society. *Scientific American* 244:116-123.
Reader, John
1988   *Man on Earth: A Celebration of Mankind.* New York: Harper and Row.
Richerson, P. J.
1977   Ecology and Human Ecology: A Comparison of Theories in the Biological and Social Sciences. *American Ethnologist* 4:1-26.
Schofield, R. S.
1970   Age-Specific Mobility in an Eighteenth Century Rural English Parish. *Annales de Demographie Historique*, pp. 261-274.
Schweizerische Alpstatistik
1900   *Schweizerische Alpstatistik: Die Alpwirtschaft im Ober-Wallis.* Solothurn: Schweizerische alpwirtschaftlichen Verein.
1909   *Bericht über die Alpen-Inspektionen in Kanton Wallis im Jahre 1909.* Vol. 1, Ober-Wallis. Solothurn: Schweizerischen alpwirtschaftlichen Verein.
Siddle, D. J. and A. M. Jones
1983   Family Household Structures and Inheritance in Savoy, 1561-1975. *Liverpool Papers in Human Geography*, No. 11.

Stebler, F. G.
  1922   *Die Vispertaler Sonnenberge. Jahrbuch des Schweizer Alpenclub.* Sechsundfunfzigster Jahrgang. Bern: Schweizer Alpenclub.

Stevenson, G. G.
  1990   *The Swiss Grazing Commons: The Economics of Open Access, Private, and Common Property.* Cambridge: Cambridge University Press.

Street, J.
  1969   An Evaluation of the Concept of Carrying Capacity. *Professional Geographer* 21:104-107.

Vayda, A. P. and B. McCay
  1975   New Directions in Ecology and Ecological Anthropology. *Annual Review of Anthropology* 4:293-306.

Vayda, A. P. and R. A. Rappaport
  1968   Ecology, Cultural and Non-Cultural. *Introduction to Cultural Anthropology*, Edited by J. A. Clifton, Boston: Houghton Mifflin, pp. 476-498.

Viazzo, Pier Paolo
  1989   *Upland Communities: Environment, Population, and Social Structure in the Alps Since the Sixteenth Century.* Cambridge: Cambridge University Press.

Wiegandt, Ellen
  1978   Past and Present in the Swiss Alps. *Hill Lands: Proceedings of an International Symposium*, Edited by J. Luchok, J. D. Cawthon, M. J. Breslin, Morgantown: West Virginia University Books, pp. 203-208.

CHAPTER 9

THE ECOLOGY OF CUMULATIVE CHANGE

Susan H. Lees
Daniel G. Bates

Introduction

In 1968, A.P. Vayda and Roy Rappaport articulated a new approach to the study of human-environment interaction, representing a marked departure from previous anthropological approaches. Vayda and Rappaport's formulation advocated the adoption of "ecosystems" and component "populations" as conceptual analytical units in place of earlier conceptual units, namely "culture" and "environment." The advantages of their approach were immediately apparent to those who had become dissatisfied with the vague and ambiguous notion of culture when used to explicate human-environment interaction, and who were impressed by the achievements of the emerging fields of systems theory, ethology, and population biology. This formulation was particularly attractive in that its treatment of humans as one population among many enabled anthropologists to use theoretical models developed in other fields. Among the ecosystemic models that were soon used to examine questions of human behavior were optimal foraging strategy, population regulation, carrying capacity, and bioenergetics.

However, in the ensuing years critics, both outside and within the subfield of ecological anthropology, discovered a wide variety of shortcomings of "ecosystemic" approaches. Perhaps the most telling anthropological critiques came from A.P. Vayda and Bonnie McCay (1975, 1977). As Vayda and McCay saw it, the main challenge should be to discover and identify the actual problems people face and to delineate their ways of coping with these problems. In their opinion, ecosystemic approaches had fallen short of meeting this challenge. While ecosystemic approaches appeared to offer a

247

practical means of modeling material transfers and flows of energy
among populations, they did not address the questions of *why*
these relationships were organized as they were, or *how*
people respond when these relationships change. In focusing on the
ecosystem as the object of study, the anthropologist was often led to
confuse or misidentify problems that might threaten the defined
parameters of the analytic unit, the ecosystem, with problems
constituting a hazard to the human population itself (see also Bates
and Lees, 1979).

In this paper we will suggest one way of formulating research
procedures that is particularly useful for identifying environmental
problems and people's responses to them. We will review what we
consider to be interesting and successful examples of studies that
have taken this approach, and discuss some of their general
findings. We will contend, too, that there is a link between the
shift in perspective that has resulted in these studies and the
emergence of development and modernization research, and we will
argue that even closer linkages between these two lines of research
will be beneficial to both.

Shortcomings of Ecosystemic Approaches

The notion of ecosystem specifically incorporates the idea of
self-regulation (Odum, 1971). Even though it is often presented with
suitable disclaimers, the ecosystem is almost inevitably envisioned
as a "superorganic" entity. The term refers to a distinctive level
of organization emergent from its organic and non-organic
components. Ecosystem models assume that self-organization exists at
the scale of the observer-delineated system. The literature is
replete with organic analogies such as "the death," "maturity,"
"senescence," etc., of an ecosystem (Golley, 1984). The
inappropriateness of the organic analogy has been widely commented
upon in the modern ecological literature, and the meager theoretical
results of the ecosystem concept are often noted (see, for instance,
Colinvaux, 1973:229; Smith, 1984).

To describe a human ecosystem is to describe the roles that
humans play in the maintenance or mutual regulation of relationships
between themselves, other living species, and non-organic elements
with which they interact. Ecosystemic models involving humans

organize description around an equilibrium or goal state defined as "carrying capacity" and interpret the structure of energy flows and the working of regulatory mechanisms in terms of dampening oscillations around this hypothetical point (see Bates and Lees, 1979, for a discussion). The habitat, in this view, is seen as the source of limiting factors: food items, water, disease, etc. An objective of the analysis, then, is to see how a human population deals with these limitations. This has also resulted in the concept of carrying capacity being anthropocentrically redefined in a way that places it beyond empirical measurement: it is defined as the level of human activities that can be sustained indefinitely without "damage" to the system.

It is possible, however, to describe human-environment relationships systemically without assuming the self-regulating properties inherent in the ecosystem concept. We might look for evidence of "self-regulation" but not find it. At times, what appears to be "self-regulation" is simply an artifact of the period of observation or of the boundaries delineated for the ecosystem (Ellen, 1984 and this volume). That is, while there is "organization" in the sense of systemic interrelationships, there is not necessarily a restoration of relationships built into the system itself. In human systems, relationships are often not restored in any meaningful sense but undergo continuous change. This probably has something to do with the character of human responses themselves; they are often cumulative because of our ability to organize ourselves in increasingly more comprehensive groups, to rapidly establish social and political hierarchies, and to subvert the material interests of many to those of a few. Humans have a unique ability to extend response devices through technological means and through effective communications.

Thus human over-use of an essential environmental resource, such as locally available protein, water, or topsoil, may not result in some mitigating response, such as the reduction of the local human population, but rather intensification of efforts to acquire the resource from alternative sources (Bennett, 1984). Similarly, the effect of a local catastrophe, such as a drought or a flood, may not be the reduction of the local human population such that sufficient resources will be available over the long run to provide for those remaining. Rather, an increase in human population may result from increased demand for workers to supply labor for the technical means of averting the ill-effects of the environmental hazard. Indeed, it appears that a common human response to the oft-perceived problem of "population pressure" is to increase population to provide labor for intensified resource extraction (see Boserup, 1965). This can be described "systemically" but does not entail self-regulation.

Closely related to the problem of assuming self-regulation and treating what is essentially an analytic unit as a distinctive entity in nature, is the over-reliance on explanations using "latent function." Explanations that rely on reference to latent function often implicitly assume a particular goal, state, or condition toward which change is directed. Often particular patterns of human behavior are explained not in terms of their ostensible or demonstrable objectives and effects, but in terms of how such behaviors may maintain an ecosystem in some given state. An often cited example of this form of explanation is Rappaport's interpretation of ritual among the Tsembaga as serving to regulate a wide range of environmental relations (1968). A more recent example is to be found in analyses of cultural behavior by Marvin Harris (1974, 1977, 1985). Harris explains such diverse practices as food taboos, male supremacy, cannibalism, and aspects of religion in terms of presumed environmental limitations such as protein availability. For a critique of Harris' approach see, for example, Vayda (1987: 493-510) Vayda's main objection is that this approach ignores the mechanisms by which such behavioral practices become established (see Harris, 1987: 511-517 for Harris's response). Very few ecologists have gone as far as anthropologists in utilizing this form of explanation, although it is found in the earlier literature on biomes and succession. Explanations based on "latent function" usually involve ascribing teleological attributes to the analytic unit.

A final shortcoming of "ecosystemic" approaches is an over-emphasis on constraints to the neglect of innovation on the part of organisms in general, and humans in particular. While it is valuable to recognize constraints affecting interactive systems, this recognition must be derived from observing their effects. Individual actors are not simply subject to constraints; they also deal with them.

Changing Perspectives in the 1970s and the 1980s

Even more than an awareness of the theoretical shortcomings of ecosystem approaches, research on economic development has influenced anthropologists' perception of what constitutes an environmental problem and how such problems should be studied. Economic development studies focus attention on issues that are not usually emphasized, or even readily accommodated, in ecosystemic research (Barlett, 1980a). These include: (1) the critical roles of external factors, including political ones; (2) the historical background to contemporary environmental circumstances; (3) the changes taking place in local systems; (4) the diversity and differentiation within local groups or populations; and (5) the various options that people have in their efforts to adjust to change. These issues have come to the fore in human ecological research in part as a consequence of growing awareness of the immediacy and acuteness of problems faced by local populations that are caught up in the political and economic transformations underway throughout the world (Little, Horowitz, and Nyerges, 1987).

Major geopolitical events of the 1970s and 1980s fostered an atmosphere in which ecological models developed earlier, emphasizing stability in closed systems, were found wanting. A number of events have focused the attention of ecologists and others on the interrelationships between climatic and social factors. Notable among these are the Sahelian drought of 1968-1973, the devastation of the Amazon, the "acid rain" phenomenon, the so-called "greenhouse effect," and the destruction of the Sea of Aral. We will take up a couple of these in more detail.

In the context of various emergency measures to relieve widespread famine in the Sahel, scientists had the opportunity to document and to participate in responses to the pressures of change. It became clear that progressive alteration in human patterns of exploitation had interfered with the capacity of the affected human populations to cope successfully with drought conditions. The survival of many people had come to depend on outside intervention.

These events triggered intense concern and interest in the processes of desertification elsewhere. It has been suggested that some 9.1 million square kilometers of desert are the direct result of human activity, and that desertification is accelerating; an additional 30 million square kilometers in over 100 countries, particularly in Africa, Asia, and Latin America, are threatened with desertification today (Babayev and Gerasimov, 1980:137). These studies have also encouraged scholars to look into historical precedents for contemporary desertification processes. Sonqqiao Zhao (1980) points out that desertification in the Mo-Usu Sandy Land of Eastern Mongolia, for example, accelerated during certain historical periods. Adams (1965, 1981) similarly reports on long-term processes of desertification in Mesopotamia, which involved periods of land reclamation as well as times of accelerated rates of loss of arable land. Such studies indicate the necessity of incorporating large temporal scale into environmental analyses (see Wobst, 1979, for general discussion of this problem; also Netting, 1984; Adams and Kasakoff, 1984; Ellen, 1984).

Southern Africa, for example, has experienced several periods of above- and below-normal rainfall since the turn of the century (Tyson, 1980). Major demographic and cultural changes have occurred in this region during the same period, but we have yet to learn about their interrelationships. Carmel Schrire (1980) has suggested that our understanding of the !Kung San population of this region would be considerably enhanced were we to take their history and changing environment into account. Rather than viewing the San as "pristine relics of the Paleolithic," she argues that they shift from hunting/gathering to sedentary herders/cultivators and back, depending on environmental conditions favoring one or another life style. The nature of San interactions with their immediate environment would be understood differently, Schrire argues, if two points were taken into account: (1) they sometimes leave it, and (2) their opportunity to leave or stay depends at least in part on the immediate environmental circumstances of other groups, today white farmers, administrators and Bantu herders. More recently Edwin Wilmsen has published a social history of the Kalahari that explicitly addresses these issues, placing the San-speakers of the region within a changing political and economic context rather than viewing them as self-sufficient and relatively isolated foragers (Wilmsen, 1989).

Obviously, environmental events need not simply arise from "acts of nature" such as abnormal rainfall. Even technologically sophisticated systems are vulnerable to the unanticipated negative effects of human activities, very often arising from policy decisions and centralized planning. Such phenomonena are increasingly attracting the interest of geographers, anthropologists and sociologists. For example, the Soviet Union is today facing an ecological disaster of almost unparalleled magnitude. The Aral Sea, in 1960 the fourth largest lake in the world, is drying up so rapidly that by the end of the century it will be nothing more than a vast, briny swamp (Micklin, 1988: 1170-1173). The waters which formerly fed the Aral have been diverted to complex and distant irrigation schemes, including some over 1300 kilometers away. The water that does flow into the Aral from surrounding agricultural schemes is contaminated with chemical fertilizers and pesticides. Gone is the rich fishing industry, and communities once strung around the lake shore are being stranded on the receding shoreline. The rapidly diminishing surface area of the lake is already resulting in hotter, drier summer temperatures and lower winter temperatures for the surrounding regions; dropping water tables have disrupted oasis farming in the region and are accelerating desertification.

In order to understand the causes of this disaster, one has to bear in mind that those who are making the decisions and carrying them out are myriad individuals united in a bureaucratic hierarchy, each concerned with such mundane matters as career advancement, job security, and their day-to-day livelihood. Keeping to "productivity" goals, limiting one's liability and responsibility for mistakes, demonstrating bureaucratic achievement in extending the scope of one's authority are critical to the success of individuals in bureaucracies. In this context, local information and early warning signs that might signal impending environmental or social problems are easily ignored by those interested in maintaining short-term production levels. "Rational" decision-making on the part of individuals may not result in environmentally sound organizational behavior (see Clarke, 1989, for a discussion of this problem).

As a consequence of their involvement in such issues, human ecologists have come to view environmental change as common, often inevitable, and interesting for its own sake. The questions that current human ecological researchers are most concerned with have to do with the impact of linkages between local and extralocal factors in human-environment interaction. The theoretical approach that predominates is characteristically drawn from one or another version of dependency theory, which emphasizes the loss of local autonomy and its detrimental outcomes as a result of capitalist exploitation of rural resources (see Oxaal, Barnett and Booth, 1975). The historical process generally observed is one of rural impoverishment and increasingly unequal access to resources (see, for example, Gross and Underwood, 1971; Wisner and Mbithi, 1974; Wisner, 1982).

Studies of human adaptive strategies have increasingly focused on decisions and choices made by individuals faced with new or changing circumstances (Barlett, 1980b; Orlove, 1980; Barth, 1981; Boyd and Richerson, 1985). Moreover, researchers no longer are positing that it is "the group," "population," or "culture" that is the basic adaptive unit (see Irons, 1979; also Bates and Lees, 1979). Many have abandoned the idea of cultural adaptation in favor of behavioral approaches to human adaptive responses, in which group or population characteristics are simply viewed as the outcome of individual action (Vayda and McCay, 1975, 1977; Orlove, 1980; Barth, 1981). In fact, there has been a wholesale shift away from approaches in which the ecosystem is treated as a discrete level of integration and toward particularistic studies of response patterns of individual actors (Orlove, 1980). This move is in keeping with the prevalent dissatisfaction in both biology and anthropology with the use of superorganic concepts.

Looking at Environmental "Events"

One important outcome of the new directions in ecological study has been an increased focus on human adaptability (Moran, 1982). Researchers have come to document the diverse and sometimes unexpected or unpredictable innovations devised by individuals acting alone or in groups to cope with new or different situations (Moran, 1981).

Environmental change is too inclusive and vague a term for the purpose of defining a research strategy. Vayda and McCay (1975) urged human ecologists to focus on life-threatening hazards that are actually experienced, not just potentially faced, by the group under study. Such hazards might be, for example, recurrent frost-induced crop failure, drought, or a disease for which there is no local resistance. They argue that much effort in ecosystemic research has been misdirected to the identification of non-problems—for instance, the question of protein availability in populations who are apparently well nourished. This has occurred to the neglect of problems that people regularly experience and that do require some adjustment in their behavior. There are, in this view, no universal problems, not even those as seemingly basic as energy constraints, but rather different populations are seen as experiencing different hazards or constraints that can be only empirically discovered.

We find the "hazard approach" a useful one which reflects what in practice many researchers are doing as they conduct field research. But it still leaves the question of how in practice one differentiates among the many problems or hazards a group may face, or how one evaluates the relative adequacy of coping strategies. We suggest that a useful alternative to efforts to evaluate a generalized problem or hazard is to look instead at the impact of particular events. That is, rather than attempting to show how the Kalahari San are adapted to desert life with scarce water resources, one should look at what they did during a drought, or, instead of discussing the ways that New Guinea highland agriculture conserves soil fertility, examine instead what the highlanders did during the frost of 1972.

Research focusing on events can also look at organizational responses and processes using decision theory, as did Lee Clarke in examining the consequences of one important hazardous accident: the 1981 toxic chemical contamination of an eighteen story building in Binghamton, New York (1989). In analyzing this case, Lee Clarke sees two distinct phases in risk management that might have wider applicability. Phase One he calls "the inter-organizational garbage can": a multitude of competing lines of authority, no clear definition of the problem, and a lack of political accountability on the part of decision makers (p. 168). Decisions occur in a chaotic swirl of competing organizational interests. Phase Two (which in Binghamton followed Phase One after nine months) he terms "the action set": all but one or two governmental organizations pull out or are forced out of the problem area, and key individuals belonging to groups of semi-independent non-governmental organizations establish a division of labor representing victims (for example, unions, health workers, merchants, environmental activisits). The head of one organization takes official responsibility for directing operations and announcing formal risk assessments (1989: 171).

Regardless of the level at which responses are studied, an event-focused approach to ecological study has numerous methodological advantages: it offers a convenient "point of entry" for the description of complex and changing relationships without an overburden of difficult assumptions. It leads the investigator to place the data collected in a diachronic perspective. It allows her/him to establish the spatial scope or range of the study according to the behavioral responses observed and to change the scope of the study as research progresses. Vayda (1983) calls this strategy "progressive contextualization."

An event-focused approach has significance for the development of theory as well. Just as Vayda and McCay (1975, 1977) noted, looking at specific events and their consequences can facilitate generalization about response hierarchies and the ordering of human behavior in terms of costs and risks. More important, in our view, is the value of an event-oriented approach in framing and testing hypotheses about environmental change and human behavior. Events provide the circumstances for testing hypotheses having to do with immediate or proximate causality and thus take us one step beyond simple correlation or association.

In the examples that follow, we have provided illustrations of "event-focused" approaches applied to extreme types of events--catastrophes or disasters. It should be emphasized that these cases represent one end of a continuum and are used here because they illustrate our argument most clearly.

Raymond Firth pointed out in 1959 that it is curious how little attention has been paid by anthropologists to natural disasters, such as hurricanes. He could cite only one exception, a brief article about hurricanes on Yap by David Schneider (1957). Firth's own interest in such events stemmed from a devastating hurricane that had struck the Pacific Island of Tikopia a month or so before his return, destroying houses, facilities, and gardens; subsequently, there was a drought. During his stay in 1952, he witnessed the ensuing famine and period of recovery. Firth took pains to describe the course of these events, in sequence, and to analyze factors which contributed to that particular chain of events.

Firth described a wide variety of responses to food shortage. Household strategies involved changing diet, reducing and then eliminating hospitality, reducing ceremonies, using unripe crops, increasing labor to restore agricultural production, collective planning, and theft. Fallow was shortened, planting and collecting rights were restricted, and land boundaries were more clearly demarcated. Kinship obligations were reduced to a narrower sphere, and their content was altered. Labor was pooled but food was not. In Firth's view, the success of the Tikopia in coping with famine rested in large part on social mechanisms for control of manpower, specifically the organization and control of movement and activity by the Tikopia chiefs. It was the chiefs who directed facility repairs, reduced opportunities for theft, enforced labor in planting rather than fishing, and recruited laborers to be sent abroad for wage work. Thus, while life in Tikopia became more "privatized" in the context of famine, the hierarchical social structure was sustained.

Although one could describe the presence of the chief-dominated hierarchy as a mechanism for "maintaining" or "restoring" equilibrium, this would tell one very little about how it actually worked, when it would be activated in a crisis, and through what processes. Firth' analysis, focused as it is on diverse individual strategies, rewards, and constraints, makes it clear how the chiefs exercised control, what they gained, and why people submitted to chiefly control.

Firth was also concerned to discover whether, and to what extent, the circumstances he witnessed were "abnormal." He found that such hurricanes were not unknown, but apparently occurred on the average of once every 20 years, or about once a generation, and were often accompanied by famine. However, Firth suggests that because the population increased over time the consequences may have become progressively worse. It was clear that hurricanes and famines did not reduce the population to pre-disaster levels: there was no mechanism to establish a population level at any point with regard to the available resources in times of stress.

For Firth the study of the events connected with hazard and disaster had critical value for understanding the "strengths and weaknesses" of a social system, in that one was able to see a test of the extent to which the system could "withstand the strain of competing demands upon [its] agents" (1959:51). Indeed, Firth might have gone further, and asked to what extent were patterns of social behavior and environmental interaction actually shaped by previous and anticipated hazardous events.

Eric Waddell (1975) asked these questions in his model study of how the New Guinea Highland Enga cope with frosts. Beginning with a specific event, the frost of 1972, he discussed previous adjustments to frosts of varying intensity and frequency among Enga groups in different locations. Their adjustments entailed varying levels of behavioral organization—individual, local, and regional, depending on the intensity and duration of the problem. The effectiveness of their strategies depended on their ability to take recourse to appropriate measures. In the 1972 instance, external intervention to relieve the local populations interfered with "normal" processes; in Waddell's view, this intervention was misguided and inappropriate. He believed, for example, that the provision of government relief would undermine the motivation of kin to help one another in future crises.

Many of the societies studied by human ecologists are situated in environments described as "harsh." Other environments, such as Tikopia and the New Guinea Highlands, are also subject to particular conditions that pose serious problems for the health or well-being of populations. In fact, for most populations such problems arise with considerable frequency. While sometimes mentioned in the anthropological literature, stressful events are usually treated as abnormal or aberrant conditions. Yet they have considerable evolutionary significance as students of animal behavior have long noted. In the study of the characteristics of the members of non-human populations, not only the occurrence, frequency, and intensity of environmental events but their *order of occurrence* is of critical importance (Winterhalder, 1980).

Two brief examples may serve to illustrate this. C.S. Holling, in describing the collapse of the Lake Michigan trout population, noted that it was not simply heavy human predation that suddenly precipitated the population crash, but rather the fact that a year of heavy harvesting followed hard upon the effects of another predator, the lamprey eel (1973). A second case concerns the arid interior of northeastern Brazil. In the 1960s numerous small farmers adopted a new crop, sisal, in response (among other things) to a series of droughts. In the absence of the series of droughts, Gross and Underwood suggest, small farmers would not have undertaken this risky and ultimately deleterious shift in cropping strategy (1971). In both these examples, the sequence or ordering of events is important in understanding the consequences.

As we have noted, some geographers and anthropologists, like Firth and Waddell, have paid close attention to disasters and their behavioral consequences. Kates *et al.* (1973), for example, studied human response to the Managua earthquake of 1972, describing how people used social networks to avert the worst consequences of that disaster and the various measures taken at different levels for recovery (1973). Anthony Oliver-Smith (1977) described the responses of urban residents in the city of Yunguay, North Central Peru, to the earthquake disaster of 1970, pointing out the rationality of their reluctance to relocate. The negative consequences of disaster relief are the topic of a study by William Torry (1978). He suggests that disaster relief nurtures long-term risks through short-term remedies by weakening local support structures and increasing dependence on "remote, unpredictable, and poorly devised bureaucratic solutions to disaster management" (p. 302).

But significant environmental events are not necessarily catastrophic. Thomas Rudel (1980) provides an example using a politically-induced resource shortage, the U.S. gasoline crisis of 1973-74, which resulted in changing response patterns as the intensity and duration of the problem increased. The order and range of responses, from individual to community to more inclusive units, parallels those of the Fringe Enga of New Guinea in response to frost, described by Waddell (1975).

The events under study need not be as brief as an earthquake or cyclic season variations in climate. Thomas McGovern and other archaeologists specializing in Norse or Viking history have attempted to unravel the checkered history of Norse settlement in the North Atlantic (1980, 1988). Numerous islands were settled between 790 AD and 1000 AD including the Shetland and Faroe Islands, Iceland, Greenland and very likely the east coast of North America as well. The colonies had rather different histories, with the western-most ones, Iceland and Greenland, marking the outer limits of significant Viking settlement. The once thriving settlements of Greenland, which ultimately failed by 1500, and Iceland, which suffered a significant decline in population several centuries after being founded, offer some insights into processes of long-term adaptation.

The Viking settlers brought with them an established food procurement system based on raising cattle and sheep, fishing, and growing, where feasible, wheat or barley. They also came equipped with a social and political hierarchy encompassing quite rigid ranks separating freemen from slaves, and recognizing distinctions among servants, tenants, landowners, and chiefs. The entire colony was run by an elite comprising chiefs and, in the later periods, the Norwegian king's appointees and church dignitaries. The early colonial period was quite successful; most of the settlers were free, and established independent holdings on which to raise sheep and cattle wherever pasturage was sufficient. The settlers were quick to incorporate the rich marine life into their diets which greatly supplemented the products of animal husbandry. The population of Greenland's settlements grew, as did that of Iceland's, reaching perhaps over 60,000 when taken together. With success, however, came the gradual transformation of the Viking colonies. In the beginning each settlement was relatively autonomous, and the predominant form of homestead was that of a free, landowner family working pastures and lands that they owned and on which they paid taxes. Slaves, although brought over initially, were not further imported, and such labor as was performed by non-family members was provided by servants and tenants. As time passed, settlements grew, churches were erected, and homesteads spread to the outer limits of pasturage. Socially and politically there were gradual changes towards stratification and political centralization that had profound effects.

In 1262-64 Greenland and Iceland came under the direct control of the Norwegian state, and most of the land came to be controlled by church and crown. The church sent bishops to rule, and encouraged the building of monumental structures quite disproportionate to the size and resources of the colonies. Taxes and tithes were collected by an administration increasingly in the hands of foreign-born appointees. The colonies were closely integrated into a growing North European economic system, but this integration did not bring prosperity and, ultimately, had dire consequences. Environmental degradation through soil erosion and depletion of marine resources caused hardship. Much of the pasturage became barren rock, and the colony on Greenland became extinct by 1500—unable to cope with the demands of its top-heavy administration. With a depleted resource base and with the ever-harsh climate, Greenland was once more left to its North American inhabitants—the Eskimos, whose time-proven adaptation the Vikings had chosen to ignore or dismiss. By 1600, 94% of Iceland's people were reduced to tenant farming, and the population had declined sharply.

The extinction of the Norse colony could be described in ecosystemic terms by detailing the constriction of vital food energy flows available to the human population as a result of failure of various regulatory mechanisms as embodied, for example, in elite status-marking activities. What McGovern's analysis offers, and what is not usually inherent in the ecosystemic approach, is the sequence of events to which people had to respond and how their responses were shaped by previous experiences with similar problems. In this context, he pays close attention to the linkages joining the colony to other non-local economic systems.

A number of human ecologists have attempted to formulate generalizations with respect to human adaptations to hazards and other environmental events, drawn largely from theories formulated by L. Slobodkin, C.S. Holling, and G. Bateson (see McCay, 1978; and Winterhalder, 1980). These deal primarily with the "economics of flexibility," namely the adaptive advantages to an organism of having a structured sequence of responses available. Minimally costly responses are tried first so as not to over-commit the organism before it is necessary to do so. The direct application of this perspective to a study of the hazards of development is well illustrated by George Morren's study of the drought of 1975-76 in Britain. Morren (1980) shows how the operation of lower-order response mechanisms was gradually superceded through time as responses to earlier hazards made permanent a condition that would otherwise have had limited duration. Population growth, urbanization, and industrialization resulted in increasing water shortages in Great Britain, and control over water sources had become increasingly centralized. By the time of the drought of 1975-76, rural water users had few recourses to individual or local means of adjustment to water shortage, and were forced to submit to the effects of shortages felt elsewhere from which they had previously been buffered.

Natural events and events caused by human activity combine in what Louise Lennihan (1984) calls "critical conjunctures" that set particular developmental trajectories. Her case in point was the emergence of rural wage labor in Nigeria, related to a local drought, labor strikes in England, and pressures by cotton exporters to expand production. A more recent case of such "conjunctures" comes from Israel (Lees, 1988). A local drought, combined with a new irrigation installation resulted in a farming catastrophe with complex consequences arising from experimentation with a novel source of water. In 1986, a new reservoir in the Jezreel Valley, containing treated urban effluents from Haifa, was combined with one containing fresh water. Drought conditions that year, however, had lowered the water supply in the fresh-water reservoir. This resulted in a higher mixture of mud from that reservoir's base with fresh water and treated effluent, which in turn provided a hospitable environment for algae. The algae accumulated in the filters of the irrigation pipes all along the system, plugging up the pipes so that water could not flow to the sprinklers and drippers in the farmers' fields. This occurred at the height of the rainless summer growing season, causing considerable crop losses.

Our final illustration is from a human ecological study specifically intended to inform and guide policy-makers in future development planning. Vayda *et al.* (1980) describe their study of the interaction of people with forests in East Kalimantan (Indonesia) with reference to the rapid economic growth of the forestry sector after 1967. Their focus is on continuing human "responsiveness to changing conditions" and human "capability for situational adjustment." As such, their study emphasizes the kinds of continuous alterations that occur as different individuals and population sectors meet changing demands, constraints, and opportunities in their exploitation of forest resources. Among their various adjustments are: changing norms of labor exchange, population movements, and a variety of procedures to circumvent forestry laws and regulations. The scope of the study encompasses threads of influence beyond the spatial confines of East Kalimantan and a time dimension that traces historical precedents and causes of migration into the area.

In each of the case studies we have mentioned, research organized about the description and analysis of human responses to critical environmental events has led to a better understanding of how features of social and economic organization bear on environmental relations. Each study was able to describe the systems in question in terms of process and change, and to show the repercussions of human behavior for the immediate, and sometimes longer-term, future of the local human populations under investigation. For these reasons, we believe that the approach they exemplify has considerably more descriptive and explanatory power than steady-state or equilibrium assumptions inherent in earlier ecosystemic approaches.

Environmental Events, Ecology, and Evolution

An important theoretical use of ecological theory is to analyze behavioral and other responses to events and to shed light on evolutionary processes. While this kind of study begins with the documentation of historical episodes, the ultimate goal is to discover the ways in which organisms respond to changes in the material conditions that affect their lives and the consequences of these responses.

Basic to evolutionary theory is the observation that members of populations are not uniform in their behavior but rather diverse in their means of coping with environmental exigencies. Thus, a specific environmental event or a change in material circumstances (such as a change is rainfall pattern or the opening of a new migration route), will elicit different responses from, and have different repercussions for, individual members of the affected group. In many respects it is best to regard even a major change in environmental circumstances from a neutral perspective since each event can be construed as a "problem" or an "opportunity," depending on who are involved and what happens to them. This is, perhaps, preferable to looking explicitly for hazards or problems, for to do so involves the assumption of group- or population-level threats and tends to obscure the diversity of actual costs and benefits incurred by affected individuals.

Significant material changes usually result in some members of the group becoming materially more stressed, others less so. The researcher can study why and how this happens and identify the most and the least vulnerable members of the group with respect to a particular event. In short, the event-focused investigation can indicate who coped successfully and who failed to do so, and why.

Evolutionary theory calls for some explicit measure of success or adequacy in coping. The Darwinian measure of reproductive fitness, while basic to evolutionary theory and indeed to theory in ecology, has limited immediate utility for most ecological anthropologists, whose interests lie in relatively short-term behavioral change (see Chagnon and Irons, 1979). More general (and readily verifiable) indicators, such as nutritional adequacy, health, and material well being, are ways to measure successful responses or coping strategies (see Moran, 1982). They are directly or indirectly related to measures of Darwinian fitness in most instances, and they are easier to observe in the short run.

Responses to material changes occur at various levels of organization. While most of the studies we have mentioned in this paper discuss diversity of response at the individual level, they are more concerned with describing the mobilization of group-level responses. Concern with diversity at the individual-actor level is more evident in economic anthropology of the "formalist" variety and in some of the "natural hazards" literature of geography (see White, 1974). While studies in the early 1970s tended to emphasize formal models of decision-making procedure, sometimes without considering the ethnographic context and the limitations of choice (see Johnson, 1980), increasing numbers of researchers have begun to empirically document variation in decision-making within populations. For example, researchers have examined differential adoption of "green revolution" technology (Farmer, 1977) and cash-cropping (Chibnik, 1980) in terms of the immediate costs and benefits of available choices. The implications of individual decisions and strategies for the study of ecological change are obvious. Looking at the environmental repercussions of behavioral continuity or change will tell us what happened but not why. Ecological analysis, which takes into account people's perception of their varying interests, allows for explanation of why they continue to act in a certain way or why they change their behavior.

Mobilized response at local group and more inclusive levels is the focus of a good deal of explanatory and descriptive generalization. Some of this work tries to look at the implications of higher-level response for flexibility and autonomy of response at lower levels (Morren, 1980; Flannery, 1972; Lees, 1974a). Often, analysts conclude that loss of local-level autonomy (increasing dependence) is detrimental to local-level groups, in that they lose what control they once had over access to local resources and the means to use them for their own benefit. However, evidence that loss of local autonomy is detrimental to groups in terms of material well-being remains inconclusive. Studies of the nutritional consequences of local involvement in market economies suggest that rural groups often experience a decline in nutritional status once they become dependent on a larger economic system (Fleuret and Fleuret, 1980; Nietschman, 1972), but demographic measures, particularly of infant mortality, seem to suggest that health care and conditions are often improved.

Development Research and Ecological Anthropology

We began this paper with a suggestion that certain changes in ecological anthropology were influenced by development research--research on economic "modernization" and change. In particular, we pointed out the influence on ecological research of a growing awareness of the inadequacy of ecosystemic models for dealing with the circumstances of certain major "environmental events" of the 1970s, such as the Sahelian drought, which were the product of cumulative social, technological, and climatic change. These events brought home to many ecological anthropologists the importance of history, of viewing environmental problems in the context of very large-scale economic systems, and of considering the sources of disequilibrium and human material vulnerability even in the study of remote and small-scale societies. We suggest that the result of this awareness was a new focus on the problems or hazards people face in interacting with their environments, and that studying specific environmental events in their local settings was a useful way of organizing research on such problems.

In the mid-1970s, for example, the policy of the U.S. Agency for International Development (U.S.A.I.D.) was adjusted to accommodate a Congressional mandate that specifies, by law, that assistance to the poor shall take priority, and that emphasis shall be placed on helping the poor to help themselves by expanding access to the national economy through various means. The law specifies that U.S.A.I.D. agricultural research shall place a priority on the determination of the special needs of small (poor) farmers (Hoben, 1980; Horowitz, 1988). In addition to a moral concern for the poor, there has also been a practical concern among all the development agencies in accountability. That is, agencies and their sponsors are requiring some sort of demonstration that their programs are having the desired effect. For this reason, U.S.A.I.D., the World Bank, and other agencies have launched major evaluation efforts to see what, if any, effects past programs have had, and to see how these evaluations can be put to use to improve future program design and implementation. "Environmental impact" as well as "social impact" studies sponsored by international agencies are becoming increasingly common, as they have been with federally-funded and some state-funded projects within the U.S., thanks to the National Environmental Policy Act. While we might take issue with the actual implementation of such studies, the fact that many ecological researchers have become involved in them and that their existence is known to many others cannot but have influenced the course of ecological research in recent years. The work done by the associates of the Institute for Development Anthropology, published in a monograph series (Westview Press Monographs in Development Anthropology) illustrates this trend.

In a very general sense, the "problem" an ecological anthropologist might study is defined by the agency itself: a local population is identified as an appropriate target for "assistance" with respect to an area of intervention, such as transportation, water, or public health. In what respects are the target population doing well, and in what respects are they doing poorly, and how might they be helped to do better? Who needs this kind of help? In the context of the kind of study of hazards and how people cope with them advocated by Vayda and McCay (1975, 1977), these sorts of questions would be addressed as a matter of course.

However, much anthropological research sponsored by development agencies is instigated after some problem in an assistance program has already occurred; often the investigator is called upon to evaluate and to determine what has "gone wrong." The literature of applied anthropology is replete with such cases. One classical example suffices to illustrate: a small Peruvian town's water supply was polluted by upstream wastes. A health-care worker was assigned the job of convincing the women of the town to boil their families' drinking water. After two years of her concerted effort, only 11 of the 200 households regularly boiled their drinking water. Why? Among the various impediments discovered by the anthropologist-investigator, Edward Wellin, was the difficulty of obtaining sufficient fuel. It was all the women could do, among their other chores, to gather enough fuel to cook meals for the day; the extra fuel for boiling water was too costly in effort (or cash), hence beyond their means. Was fuel shortage their "problem"? Not really. But the limited availability of fuel emerged as a "limitation" specifically in the context of a real problem, water pollution, and an effort to solve it by convincing women to boil their water (Wellin, 1955).

The cases of "environmental events" we described above are mostly instances of disasters or hazards or clear-cut negative pressures on human populations. Development research adds another dimension to the study of environmental events: events generated by the human (social, economic, or political) environment that have environmental components and/or repercussions. Such events might include a change in the market price of a cash crop, or a land-reform law which alters land tenure, or relatively large-scale migration into or out of a locality, or establishment of a quota on certain types of livestock (for a case involving imposition of restrictions on fishing, see McCay, 1981), or a campaign to get women to boil water to improve health. As with events like droughts and hurricanes, explaining what people do in response, and why, requires looking into past events, the existing differentiation among the local population, and interactions between people and other environmental factors. Such events elicit changing response processes through time and sometimes call for shifts from individual to local or more inclusive responses. As in the case of "natural hazards," it is in the context of such events that real problems, limitations, and often strengths of local systems emerge. Horowitz and Salem-Murdock (1987), for example, relate desertification processes in White Nile Province, Sudan, to rural social differentiation and partial proletarianization of rural producers, which were outcomes of the completion of a dam and agricultural intensification through irrigation.

An understanding of the tested strengths of a local system, it has been argued, is particularly important but largely neglected by development agencies (Wisner and Mbithi, 1974; Waddell, 1975; Lees, 1980). Local groups that have experienced environmental hazards before sometimes develop regular and effective means of coping with them. External intervention by development and relief agencies often interferes with the utilization of these means, and reduces their effectiveness. A major contribution that ecological anthropologists have to make to practical development research is to explain how local groups do cope with hazards and to help development planners either to enhance these coping devices, or at least avoid interfering with them where possible (Wisner and Mbithi, 1974; Waddell, 1975; Lees, 1980). J. Terrence McCabe (1987) illustrates this point in a study of the impact of drought on Turkana herding in Kenya. He suggests that "underestimating the effects of drought on the livestock population of pastoral nomads lends support to the argument that pastoralists keep too many animals..", which in turn leads to a misunderstanding of ecological balance in the rangeland (McCabe 1987:387). Development efforts to decrease livestock mortality or to alter rangeland management will have to take both droughts and current management practices into account.

We have pointed out that an important aspect of development research has been recognition and documentation of human adaptability. Development project evaluations repeatedly report that beneficiaries and other participants in such projects frequently do not behave as the planner, sponsors, and administrators expected. At times, the adaptiveness of this unexpected behavior is obscured, neglected, or misunderstood, but it surely deserves further study.

For example, Tony Barnett's study of the Gezira Scheme (1977), a very large irrigation project in Sudan, documents a variety of ways in which local irrigators both dealt with legal constraints of the Scheme and tried to achieve ends of their own, which were not defined or recognized by the Scheme. All Scheme tenants were required to raise the cash-crop, cotton, on some land and to plant sorghum and other subsistence crops in rotation with cotton, all of which are irrigated by the Scheme. Tenants sold their cotton to the Scheme and shared the profits, but their subsistence crops were theirs alone. Because of this and because they found it more difficult to use cotton as a source of credit, tenants contrived with local field inspectors to divert irrigation water from cotton to sorghum whenever water was scarce. Both the reasons for this activity and the means by which it was carried out were embedded in a more complex adaptive system that applied directly to the problem of obtaining labor to work individual holdings. And both the labor problem and its resolution were specific to the way in which the Scheme was operated, that is, its constraints and opportunities. It seems clear that similar kinds of processes of adjustment occur in virtually every development project.

Conclusions: An Event-Focused Approach
to Human Environmental Interactions

Many ecological anthropologists became disillusioned with the notion of ecosystem during the 1970s for a variety of different reasons. Other papers in this volume deal with problems of boundary definition, scale, and time depth, which limit the utility of ecosystemic modeling for describing human-environment interaction. Here we have been concerned with the problem of cumulative change, often described as "development" and the "development of underdevelopment." With Vayda, McCay, and others, we have suggested that for those of us interested in the study of human adaptation our focus must be on human responses to environmental "problems." We have argued that a useful way of organizing research on human responses is to focus inquiry on specific environmental events. Studies of such responses suggest that their effects are usually cumulative.

An article by Margoh Maruyama (1963) brought to the attention of ecological anthropologists the notion of "deviation amplification" as a complementary systemic state to that of equilibrium. Some scholars who sought to apply this notion to human ecological systems saw deviation amplification as a form of systemic pathology (Rappaport, 1969; Flannery, 1972), while equilibrium was seen as a form of health. However, "deviation amplification," another term for cumulative change, seems to be so regular and common a state of affairs for humans that the systems-based term "pathological" appears to us inappropriate. To paraphrase Vayda and McCay (1975, 1977), people (and other organisms) are sometimes obliged to abandon the "system" for the sake of their own survival and well being. While systemic terms used by Rappaport and Flannery, such as "linearization" (the by-passing of lower-order controls by higher-order controls) may be useful for conceptualizing a process of cumulative systemic change, they do not focus attention on the relative utility of the new arrangements for people coping with problems at hand.

McCay (1978) has argued for the replacement of "systems ecology" by "people ecology" in order to accomplish just this. Many of the terms widely in use among ecological anthropologists, such as "perturbation," assume the normative operation of a system which, in reality, may simply not occur because of continuous change. Looking at "events" rather than "perturbations" is a somewhat more neutral stance and retains a focus on situations occasioning material change.

If the people we study exist in a continuously changing context or even if they are relatively isolated and apparently stable, how do we select significant "events" about which to organize research? As in any type of research, this depends on our particular interests and, of course, opportunities. As ecological anthropologists, we are obviously interested in events that have environmental components and/or repercussions involving some sort of change in material circumstances. Most of us are also interested in particular forms of resource extraction or production, such as fishing, hunting, irrigated agriculture, forestry, or livestock production; the "events" that interest us will most likely have some material bearing on these activities. Furthermore, we tend to do our research in specific localities whose environments are typified in part by characteristic "events," such as drought, floods, hurricanes, or frosts, of varying severity and duration. But the events in question need not be catastrophes, nor need their genesis be, strictly speaking, "environmental." The events studied by Bernard Nietschman (1972) among the Miskito turtle hunters were generated by the appearance of an external market for turtle meat in the form of a new processing factory which increased the intensity of exploitation, and soon led to the decreased availability of turtles.

By evaluating the impact of events and people's varying responses to them, we begin to relate our scientific interest to the needs of those whom we study and, perhaps not incidentally, of those who sponsor our research. The practical utility of ecological research, while not the only consideration, is a matter that cannot be ignored. Indeed, some of the scientific benefits of these externally felt needs have already been demonstrated. Human ecological research has been brought into closer communication with other lines of anthropological inquiry, particularly areas related to history, political organization, and economics (Orlove, 1980). The result has been, we believe, greater refinement and specificity in development theory and research, and better understanding of human adaptation and ecological change.

References Cited

Adams, J.W. and A.B. Kasakoff
  1984   Ecosystems over Time: The Study of Migration in "Long
          Run" Perspective. *The Ecosystem Concept in
          Anthropology,* AAAS Selected Symposium 92. Edited
          by Emilio F. Moran, Boulder, Colorado: Westview
          Press. Pp. 205-23.
Adams, Robert McC.
  1965   *Land Behind Baghdad: History of Settlement on the
          Diyala Plains.* Chicago: University of Chicago
          Press.
  1981   *Heartland of Cities: Surveys of Ancient Settlement
          and Land Use on the Central Flood Plains of the
          Euphrates.* Chicago: University of Chicago Press.
Babayev, A.G. and I.P. Gerasimov
  1980   The International UNEP-USSR Project on Combatting
          Desertification through Integrated Development.
          The Threatened Drylands. Edited by J.A.
          Mabbutt and S.M. Berkowicz, New South Wales: University
          of New South Wales.
Barlett, P.
  1980a   Adaptive Strategies in Peasant Agricultural
          Production. *Annual Review of Anthropology*
          9:545-73.
Barlett, P. (ed.)
  1980b   *Agricultural Decision-Making: Anthropological
          Contributions to Rural-Development.* New York:
          Academic Press.
Barnett, Tony
  1977   *The Gezira Scheme: An Illusion of Development.*
          London: Frank Cass & Co.
Barth, Fredrick
  1981   *Process and Form in Social Life.* London:
          Routledge & Kegan Paul.
Bates, D.G. and S.H. Lees
  1979   The Myth of Population Regulation. *Humans in
          Evolutionary Perspective.* Edited by N. Chagnon and
          W. Irons, North Scituate, Mass.: Duxbury.

Bennett, J.W.
1984   Ecosystems, Environmentalism, Resource Conservation, and Anthropological Research. *The Ecosystem Concept in Anthropology.* Edited by Emilio F. Moran, AAAS Selected Symposium 92. Boulder, Colorado: Westview Press. Pp. 289-310.

Boserup, E.
1965   *The Conditions of Agricultural Growth.* Chicago: Aldine.

Boyd, R., and P. J. Richerson
1985   *Culture and the Evolutionary Process.* Chicago: University of Chicago Press.

Chagnon, N.A. and W. Irons eds.
1979   *Evolutionary Biology and Human Social Behavior: An Anthropological Perspective.* North Scituate, Massachusetts: Duxbury.

Chibnik, M.
1980   The Statistical Behavior Approach: The Choice Between Wage Labor and Cash Cropping in Rural Belize. *Agricultural Decision-Making: Anthropological Contributions to Rural Development.* Edited by P. Barlett, New York: Academic Press. Pp. 87-114.

Clarke, L.
1989   *Acceptable Risk? Making Decisions in a Toxic Environment.* Berkeley: University of California Press.

Colinvaux, P.
1973   *Introduction to Ecology.* New York: John Wiley & Sons.

Ellen, R.F.
1984   Trade, Environment, and the Reproduction of Local Systems in the Moluccas. *The Ecosystem Concept in Anthropology.* Edited by Emilio F. Moran, AAAS Selected Symposium 92. Boulder, Colorado: Westview Press. Pp. 163-204.

Farmer, B.H. (ed.)
1977   *Green Revolution? Technology and Change in Rice-Growing Areas of Tamil Nadu and Sri Lanka.* Boulder, Colorado.: Westview Press.

Flannery, K.V.
1972   The Cultural Evolution of Civilizations. *Annual Review of Ecology and Systematics* 3:399-426.

Fleuret, P. and A. Fleuret
    1980   Nutritional Implications of Staple Food Crop
            Successions in Usambara, Tanzania. *Human
            Ecology* 8:311-327.
Firth, Raymond
    1959   *Social Change in Tikopia.* London: George Allen
            & Irwin Ltd.
Golley, F.B.
    1984   Historical Origins of the Ecosystem Concept in Biology.
            The Ecosystem Concept in Anthropology. Edited
            by Emilio F. Moran, AAAS Selected Symposium 92.
            Boulder, Colorado: Westview Press. Pp. 33-49.
Gross, Daniel and B. Underwood
    1971   Technological Change in Caloric Costs: Sisal
            Agriculture in Northeastern Brazil. *American
            Anthropologist* 73(3):725-740.
Harris, Marvin
    1974   Cows, Pigs, Wars, and Witches: The Riddles of
            Culture. New York: Vantage.
    1977   *Cannibals and Kings.* New York: Random House.
    1985   *Good to Eat: Riddles of Food and Culture.* New
            York: Simon & Schuster.
    1987   Comment on Vayda'a Review of Good to Eat: Riddles of
            Food and Culture. *Human Ecology.* 15(4):511-518.
Hoben, Allen
    1980   Agricultural Decision-Making in Foreign Assistance: An
            Anthropological Analysis. *Agricultural Decision-
            Making: Anthropological Contributions to Rural
            Development.* Edited by P. Barlett, New York:
            Academic Press. Pp. 337-369.
Holling, C.S.
    1973   Resilience and Stability of Ecological Systems.
            Annual Review of Ecology and Systematics
            4:1-23.
Horowitz, Michael
    1988   Anthropology and the New Development Agenda. Bulletin
            of the Institute for Development Anthropology.
            Development Anthropology Network 6:1-4.

Horowitz, Michael, and Muneera Salem-Murdock
1987   The Political Economy of Desertification in White Nile
        Province, Sudan. Little, Horowitz, and Nyerges eds.,
        *Lands at Risk in the Third World: Local Level
        Perspectives.* Edited by P. Little, M. Horowitz and
        R. Nyerges. Boulder: Westview. pp. 95-14.
Irons, William
1979   Natural Selection, Adaptation, and Human Social
        Behavior. *Evolutionary Biology and Human Social
        Behavior.* Edited by N. Chagnon and W. Irons, North
        Scituate, Mass.: Duxbury Press.
Johnson, A.
1980   Limits of Formalism in Agricultural Decision Research
        *Agricultural Decision-Making: Anthropological
        Contributions to Rural Development.* Edited by P.
        Barlett, New York: Academic Press.
Kates, Robert W., J. Eugene Haas, Daniel J. Amaral, Robert A. Olson,
Reyes Ramos, and Richard Olson
1973   Human Impact of the Managua Earthquake.
        Science 182:981-990.
Lees, S.H.
1974a   Hydraulic Development as a Process of Response.
        Human Ecology 2:159-75.
1980   The "Hazards" Approach to Development Research:
        Recommendations for Latin American Drylands. *Human
        Organization* 69:372-376.
1988   Algae: A Minor Disaster in the Jezreel Valley, Israel.
        Studies in Third World Societies. 38:155-176.
Lennihan, L.
1984   Critical Conjuncturures in the Emergence of
        Agricultural Wage Labor in Northern Nigeria. *Human
        Ecology.* 12(4):465-480.
Little, Peter D., Michael M. Horowitz, and A. Endre Nyerges, eds.
1987   *Lands at Risk in The Third World: Local-Level
        Perspectives.* Boulder, Colorado: Westview.
Maruyama, M.
1963   The Second Cybernetics: Deviation-Amplifying Mutual
        Causal Processes. *American Scientist* 1:164-179.
McCay, Bonnie
1978   Systems Ecology, People Ecology, and the Anthropology
        of Fishing Communities. *Human Ecology*
        6:397-422.

1981   Optimal Foragers or Political Actors? Ecological
          Analysis of a New Jersey Fishery. *American
          Ethnologist* 8:356-382.

McCabe, J. Terrence
1987   Drought and Recovery: Livestock Dynamics Among the
          Ngisonyoka Turkana of Kenya. *Human Ecology*
          15(4):371-390.

McGovern, Thomas
1980   Cows, Harp Seals, and Church Bells: Adaptation and
          Extinction in Norse Greenland. *Human Ecology*
          8:247-75.

McGovern, T., G. Bigelow, T. Amorosi, and D. Russell
1988   Northern Islands, Human Error, and Environmental
          Degradation. *Human Ecology* 18:225-270.

Micklin, P. P.
1988   Desiccation of the Aral Sea: A Water Management
          Disaster in the Soviet Union. *Science*
          241:1170-1175.

Moran, Emilio
1981   *Developing the Amazon.* Bloomington: Indiana
          University Press.
1982   *Human Adaptability.* Boulder, Colorado:
          Westview Press. (Originally published in 1979 by
          Duxbury Press).

Morren, George
1980   The Rural Ecology of the British Drought 1975-1976.
          Human Ecology 8:33-63.

Netting, R. McC.
1984   Reflections on an Alpine Village as Ecosystem. *The
          Ecosystem Concept in Anthropology.* Edited by
          Emilio F. Moran, AAAS Selected Symposium 92. Boulder,
          Colorado: Westview Press. Pp. 225-235.

Nietschman, B.
1972   Hunting and Fishing Focus Among the Miskito Indians,
          Eastern Nicaragua. *Human Ecology* 1:41-67.

Odum, E.P.
1971   *Fundamentals of Ecology.* Third Edition.
          Philadelphia: Saunders.

Orlove, Benjamin
1980   Ecological Anthropology. *Annual Review of
          Anthropology* 9:235-273.

Oxaal, I., A.S. Barnett, and D. Booth eds.
1975 *Beyond the Sociology of Development.* London: Routledge & Kegan Paul.

Oliver-Smith, Anthony
1977 Traditional Agriculture, Central Places, and Post-Disaster Urban Relocation in Peru. *American Ethnologist* 4:102-116.

Rappaport, R.A.
1968 *Pigs for the Ancestors: Ritual in the Ecology of New Guinea People.* New Haven: Yale University Press.

1969 *Sanctity and Adaptation in the Moral and Esthetic Structure of Human Adaptation.* New York: Wenner-Gren Foundation.

Rudel, Thomas K.
1980 Social Responses to Commodity Shortages: The 1973-1974 Gasoline Crisis. *Human Ecology* 8:193-212.

Schneider, D.
1957 Typhoons on Yap. *Human Organization* 16:10-15.

Schrire, Carmel
1980 An Inquiry into the Evolutionary Status and Apparent Identity of the San Hunters-Gatherers. *Human Ecology* 8:1-32.

Smith, Eric A.
1984 Anthropology, Evolutionary Ecology, and the Explanatory Limitations of the Ecosystem Concept. *The Ecosystem Concept in Anthropology.* Edited by Emilio F. Moran, AAAS Selected Symposium 92. Boulder, Colorado: Westview Press. Pp. 51-85.

Torry, William
1978 Bureaucracy, Community, and National Disasters. *Human Organization* 37:302-8.

Tyson, P.D.
1980 Climate and Desertification in Southern Africa. *The Threatened Dryland.* Edited by J.A. Mabbut and S.M. Berkowicz, New South Wales: University of New South Wales.

Vayda, A.P. and R.A. Rappaport
1968 Ecology, Cultural and Non-Cultural. *Introduction to Cultural Anthropology.* Edited by J.A. Clifton, Boston: Houghton Mifflin.

Vayda, A.P. and B. McCay
   1975   New Directions in Ecology and Ecological Anthropology.
           Annual Review of Anthropology 4:293-306.
   1977   Problems in the Identification of Environment
           Problems. *Subsistence and Survival: Rural Ecology
           in the Pacific.* Edited by T.P. Bayliss-Smith and
           R.G.A. Feachem, New York/London: Academic.
Vayda, A.P.
   1983   Progressive Contextualization: Methods for Research in
           Human Ecology. *Human Ecology* 11(3):265-282.
   1987   Explaining What People Eat: A Review Article.
           Human Ecology 15(4):493-510.
Vayda, A.P., C. Colfer, J. Pierce, and M. Brotokusumo
   1980   Interactions Between People and Forest in East
           Kalimantan. *Impact of Science on Society* Vol.
           30, No. 3, UNESCO:179-190.
Waddell, Eric
   1975   How the Enga Cope with Frost: Responses to Climatic
           Perturbations in the Central Highlands of New Guinea.
           Human Ecology 3:249-273.
Wellin, Edward
   1955   Water Boiling in a Peruvian Town. *Health, Culture
           and Community.* Edited by Benjamin D. Paul, New
           York: The Russell Sage Foundation.
White, G.F. (ed)
   1974   *Natural Hazards: Local, National and Global.*
           New York: Oxford University Press.
Wilmsen, Edwin N.
   1989   *Land Filled with Flies: A Political Economy of the
           Kalahari.* Chicago: University of Chicago Press.
Winterhalder, Bruce
   1980   Environmental Analysis in Human Evolution and
           Adaptation Research. *Human Ecology* 8:135-170.
Wisner, B.
   1982   MWEA Irrigation Scheme, Kenya: A Success Story for
           Whom? Boston: ARC Newsletter.
Wisner, B. and P.M. Mbithi
   1974   Drought in Eastern Kenya: Nutritional Status and
           Farmer Activity. *Natural Hazards: Local, National
           and Global.* Edited by G.F. White, New York:
           Oxford. Pp. 87-97.

Wobst, Martin H.
  1979   The Archaeo-ethnology of Hunters-Gatherers, or the
          Tyranny of the Ethnographic Record in Archaeology.
          American Antiquity 43(2):303-309.
Zhao, Sonqqiao
  1980   Desertification and de-desertification in China.
          The Threatened Drylands. Edited by J.A.
          Mabbutt and S.M. Berkowicz, pp. 80-88. Kensington:
          School of Geography, University of New South Wales,
          Australia.

CHAPTER 10

LEVELS OF ANALYSIS AND ANALYTICAL LEVEL SHIFTING: EXAMPLES FROM AMAZONIAN ECOSYSTEM RESEARCH

Emilio F. Moran

The Problem of Level Definition

The ecosystem approach, because of its focus on the hierarchical level of organization above the population and the community, tends to emphasize macro-level processes such as trophic exchanges, nutrient cycling, and system maintenance. As a consequence, past studies using an ecosystem perspective seemed to have a strong functionalist bent that overlooked historical and evolutionary considerations. This should not be surprising given the dominance of equilibrium approaches in the 1950's and 1960's coming out of the work of Lotka and Volterra. It is only in the 1970's that non-equilibrium approaches began to demonstrate their power in explaining observed data in larger ecological systems. The more dynamic basis of non-equilibrium models promises to take us closer to a more realistic understanding of human ecosystems (cf. Foin and Davis 1987). Ecosystem analysis is at it's best when it takes on a large and important question, when it is willing to use whatever research procedures work, when it includes human management, when it suggests testable hypotheses and when it is concerned with global change and long-term dynamics (cf. Pomeroy and Alberts 1988).

The shift to the use of the ecosystem as a *unit of analysis* may be responsible for obscuring the multiple hierarchical levels within ecosystems. It has not been sufficiently recognized that each level's structural and functional relations obscure relationships observable at other levels, particularly as one moves from micro to macro levels. The greater the scope of the level, the less visible details of group and individual behavior and ideology appear.

One of the major works in shaping the ecosystem view was Howard T. Odum's *Environment, Power and Society* (1971). In that book he invited scientists to use the detail eliminator or "macroscope" to overcome the attention to detail that had kept us from seeing the workings of ecological systems (Odum 1971:10). It is important to note that Odum glossed over the differences between parts and the whole of large, inclusive ecosystems while suggesting that there are enough similarities between the workings of compartments and the whole to permit describing the former at the macro-level rather than as having distinct properties (1971:60). Nowhere in that fundamental work on energetics is there any effort to caution readers as to the problem of level shifting and its analytical consequences. These problems, however, are evident in current Amazonian research and policy-making.

Recent Amazonian studies are characterized by a number of recurrent dilemmas: is the Amazon ecosystem fragile or not? (Farnworth and Golley 1974). Are its soils fertile or sterile? (Alvim 1978; Meggers 1971). Can state-level societies in the region persist without destroying the habitat? (Caneiro 1957, 1961; Meggers 1954, 1971). I would contend that some of the heated debates about Amazonian ecosystems reflect a tendency to generalize about processes at one level of analysis (usually Amazonia as a whole) from data and research carried out at another level (usually specific sites unrelated to other sites by criteria such as climate, biomass, vegetation type or linguistic classification). The problem is traceable to how the processes at one level obscure relationships at other levels. In this paper I explore the implications of level shifting for understanding human ecological interactions generally, and in the Amazon Basin in particular.

The articulation between micro- and macro-levels of analysis is still undergoing conceptual and methodological development. The disciplinary confines of most investigators (see Bennett, this volume) are responsible for this current state and for a tendency to work within a given level, to the exclusion of others. This sort of problem has recently surfaced, for instance, in debates between molecular biologists and evolutionary biologists. While there is little doubt that data on differences in nucleotide sequence found in the DNAs of related species will provide more accurate reconstruction of phylogenies, "the changes in the molecules do not necessarily directly reflect the evolutionary history of the group from which they were taken" (Lewin 1982:1091). Thus, on the one hand, molecular biologists seek to provide us with a neucleotide sequence to understand evolutionary processes whereas evolutionary biologists attempt to provide a time-based model emphasizing species interactions and environmental change. As Lewin has aptly stated, "evolution is a hierarchical process operating at several levels, each important in its own right. Nevertheless, it is prudent to ensure that analytical tools are applied only at appropriate levels. The trick is to agree which levels are accessible to which tools" (1982:1091).

The importance of dealing with issues of scale and hierarchy, if anything, has grown in the late 1980's. Ecosystem studies have moved away from traditional local studies to looking at landscapes and global processes such as global carbon cycling, climatic change, species extinction and nuclear winter. These problems "demand that we accelerate our ability to translate small-scale ecological principles to higher levels" (O'Neill 1987:140). New tools are now available to aid study of higher levels, such as the combination of geographic information systems (GIS) and remote sensing technologies (see Conant, this volume). Relating these global processes to local ones remains problematic despite these advances (cf. C. Smith 1984). The recognition that any ecological research question must be approached at the appropriate scale (temporal, spatial and organizational) is becoming commonplace -- but less often is it reflected in research design and analysis.

Level-specific Data

The relevance of this issue for social and ecological scientists lies in how a given level of analysis may influence one's interpretation of problems such as low agricultural productivity and its social structural implications. Amazonian research, whether anthropological, agronomic or ecological, has been mostly site-specific. Much of the research has taken place in sites about to be affected by "development" and there has been little time for systematic baseline coverage of habitat types, for assessing the impact of various technologies per habitat type, and for collecting representative aggregate data of major social, ecological and economic indicators. Perhaps the most systematic sampling carried out in the Amazon has been by agronomists (Sombroek 1966; Falesi 1972; IPEAN 1974; Nicholaides *et al.* 1983) and, more recently by climatologists (Molion 1987; Salati & Vose 1984; Dickinson 1987; Salati 1987; 1985; Salati *et al.* 1979. Even with adequate sampling at one's chosen level of study, hierarchically more complex levels of organization must be sampled as well if "systemic" conclusions are to be drawn. This issue has grown into significance with the development of concern with global environmental change. The National Science Foundation now has a special program on the Human Dimensions of Global Environmental Change; the Social Science Research Council has a Committee for Research on Global Environmental Change; and the National Research Council has a Committee on the Human Dimensions of Global Change.

The question before us is, can site-specific studies (micro-level) be the appropriate basis of "region-wide" statements and analyses (macro-level)? A moment's reflection will tell us that such extrapolations seldom work. Phenomena at a given level may have analogs at other levels but they are not identical (Gould 1982:386). Sliding between levels, by making statements about individuals from aggregate data, has been termed the "ecological fallacy" (Robinson 1950). It is generally understood that micro- and macro-levels of analysis have distinct systems of relationships and answer different questions. For example, paraphrasing Gould, populations contain so many individuals that small biases in mutation rate can rarely establish a feature but the analog of mutation pressure at the species level, directed speciation, may be a powerful agent of evolutionary trends. Directed speciation (i.e., directional bias toward certain phenotypes in derived species) can be effective because its effects are not easily swamped by differential extinction due to the restricted number of species in a clade and because biases in the production of species may be more prevalent than biases in the genesis of mutations (Gould 1982:386). Macro-level studies rely on aggregate data *from a broad and representative sample of the universe in question.* Micro-level studies rely on careful observation of individuals in a population in order to understand the internal dynamics of that population. A macro-study is not only bigger than the sum of the micro-studies, it is *structurally different.* Economics long ago distinguished between macro- and micro-economics. Demography while still dominated by the macro-approach has in recent years spawned micro-demographics of small communities—even households (cf. Kosinski and Webb 1976; Dewalt and Pelto 1985).

Geographers have been particularly aware of the scale problem in reference to trying to comprehend a large region while studying small areas within it. McCarthy *et al.* (1956:16) noted that "every change in scale will bring about the statement of a new problem and there is no basis for assuming that associations existing at one scale will also exist at another." The caveats of the 1950's have given way to calls for integration of the macro- and micro-levels of analysis but bringing together what are different processes remains a challenge (Beer 1968; Dogan and Hokkam 1969). Hierarchy theory has been proposed as a way of connecting these levels of analysis (Allen and Starr 1982; Eldredge 1985; O'Neill *et al.* 1986; Salthe 1985). So far, however, notions of hierarchy have had little influence in ecological anthropology. It is important to be aware that for hierarchical principles to be helpful it is necessary to first identify such levels and then to test what rates or magnitudes make them distinct (O'Neill 1987:147). Application of time-series analysis has proven useful in detecting distinct levels in ecological systems.

Choice of Level and Analytical Implications

The analytical implication of levels chosen in research is part of the larger question faced by scientists when they delimit their scope of study to "feasible" questions. What are the implications of setting those particular bounds to the relevance or completeness of analysis? What is less frequently noted is that the conclusions at each of these levels are distinct and, yet at the same time, each one is relevant to a complete understanding of human behavior. The results of a community study, for example, may not be generalized to a whole society, but the internal structure of a community is relevant to understanding how a community is affected by larger external forces (Epstein 1964:102; Sheridan 1988; C. Smith 1984). If one started with the "state" or national level it might appear that these larger external forces shape the life of local communities in relatively similar ways (Watson 1964:155; Blok 1974; Schneider and Schneider 1976). The focus on the community, on the other hand, shows individuals responding actively to actually subvert or alter these external forces, not passively accepting them (Bennett 1967; Eder 1981; S. Smith 1984; Sheridan 1988; Moran 1981, 1988a; Balee 1989; Hecht and Posey 1989).

It has not been sufficiently recognized that each level's scope obscures relationships observable at other levels, particularly as one moves from local research to regional or national levels (Devons and Gluckman 1964:211). The greater the scope, the less details of group and individual behavior and ideology are analytically recognized. There are, for example, significant differences in the mean demographic behavior of small and large populations because of greater variation in the former. Since population distribution is usually uneven, small areal units have a wider range and variance of population distribution than larger units such as countries or large regions like the Amazon. Likewise, since mobility mainly occurs over short distances, such patterns are critical to micro-demographics but they appear as insignificant factors when aggregated as compared to natural changes in population (J. Clarke 1976). Thus the *relative* significance of migration versus natural change depends more on the size of the area studied (i.e. the level of analysis) than on *real* demographic differences. This "scale-linkage" problem remains incompletely resolved (cf. Haggett 1965).

What is clear, is that future contributions to ecological studies will require the integration of multiple models, focused at different levels of detail. Local or site-specific models will be able to take into account the role of particular resources, topographic and orographic features and the contingencies that influence individual actors. Meso- or regional models can better deal with longer time analysis, structural contingencies, patterns of exchange, etc. Global models are becoming increasingly important tools in dealing with contemporary environmental problems (Levin 1987:243) and present a new challenge to anthropologists concerned with the human dimension of environmental change.

The ecosystem is a flexible unit defined by the needs of the researcher. Its use seldom, if ever, facilitates replicability nor is it obvious on reading the conclusions of a study what the scope of the study had been, given the tendency to shift levels between field data and theoretical discussion. Reference to the ecosystem as one's unit of analysis has produced results that address important human ecological relations *within a single level*--they need not apply to the whole of the human adjustments. A given ecosystem study is a model of horizontal structural and functional relations and is confined to a given level. Its "holism" is level-specific.

Geertz (1963) was the first anthropologist to argue for the usefulness of the ecosystem as a unit of analysis in social/cultural anthropology. In *Agricultural Involution* (1963) he tested Steward's emphasis on subsistence and found it wanting. He showed by a broad use of historical records that Indonesia's agricultural patterns, for example, could be understood in terms of the economic restrictions of the Dutch colonial authorities. In fact, Geertz used the *region* of Indonesia as his ecosystem level. He identified two contrasting agricultural systems within the broader ecosystem and discovered the explanation for their differing development in the varying historical pressures of Indonesia's colonial economy.

In another use of the concept, Rappaport's study of ritual and ecology in the New Guinea Highlands (1968) defined the ecosystem unit in terms of the material exchanges of a *local population*. How could one compare the Geertz and Rappaport studies which deal with ecosystemic interactions at different levels? In the last chapter of *Pigs for the Ancestors* (1968) Rappaport acknowledged that a local population engages in material and nonmaterial exchanges with *other local populations* which, in the aggregate can be called "regional populations." These, he suggests, are likely to be more appropriate units of analysis for long-range evolutionary studies given the ephemeral quality of local populations (1968:226). Unfortunately, other anthropologists have overlooked this insight regarding the differences between relatively synchronic micro-level studies and diachronic, macro-level approaches like Geertz's.

From this contrast of the varied uses of the ecosystem unit by Rappaport and Geertz, and from the examination of debates surrounding the human occupation of the Amazon Basin (Lathrap 1970; Meggers 1971; Carneiro 1957; and more recently Hames and Vickers 1983; and Posey and Balee 1989; Moran 1990), it is possible to examine the possible association between differing levels of analysis and major points of disagreement in human ecological and Amazonian studies. Anthropologists have long used local communities as their fundamental units of study wherein a cultural or ethnographic method could be applied (Steward 1950:21). Most scholars have been quite aware that individual communities are part of larger wholes but such functional interdependencies have seldom been a part of the analysis. It has been common in anthropology to study local communities to quantify certain variables or to study populations before the full impact of the modern world reaches them (Rappaport 1968; Nietschmann 1973; Waddell 1972; Baker and Little 1976; to name but a few). While all these researchers recognize the value of addressing larger populational and ecological units, they chose to limit the scope of the investigation for the sake of precise and efficient data gathering. In such studies, the community is seen as a "closed" system for the purposes of analysis.

Localized studies provide insight into family structure, subsistence strategies, labor inputs, health and nutritional status, flow of energy, socialization, and cultural institutions. Studies at this level, however, cannot address issues of social evolution, explain changes in the economic structure of society, patterns of economic development, or political economy. These issues can be addressed only by a different type of research method emphasizing historical, geographical, economic and political change over time.

Regional analyses add a very different and much needed insight into the processes of human adaptation. A regional study emphasizes *historical and economic* factors and considers many local-level phenomena as secondary to the historical forces at play (cf. Braudel 1973; Smith 1959; Bloch 1966). One may note that both Geertz (1963) and Bennett (1969) predominantly use historical factors in explanation in their regional analyses. Bennett defined the region of the North American Northern Plains in terms of its historical unity (1969:26). He was able to explain the adaptive strategies of four distinct ethnic groups in terms of differential access to resources, differential access to power loci, and social/cultural differences. Thus, while he was able to flesh out the social/cultural details of the population by local interviews and study of ethnic interactions, a full understanding of the operative forces required aggregate data from social and economic history.

The choice of proper level can come only from a recognition of the appropriate level at which one's questions can be addressed. Julian Steward's difficulties in achieving his goals in his ambitious Puerto Rico research project (Steward 1956) can be traced to a failure to shift from his micro-level analysis of a group's "culture core" to the necessary macro-level analysis of the Puerto Rican political economy. Steward's earlier micro-level studies had successfully generated sophisticated analyses of the internal structure of patrilineal bands and their articulation with selected habitat features (Steward 1955). The multilineal evolution goals of the Puerto Rico study needed to move beyond the study of specific human/habitat interactions towards a dynamic model of structural transformations. The articulation of Puerto Rican communities needed to be related to external social systems through which many community features could be understood through time.

Questions about levels of analysis and the scale of sampling appropriate to given research interests helps shed light on some of the major debates about human occupation of the Amazon. An examination of the levels at which generalizations have been made will serve both to suggest the articulation between levels and to identify some of the major gaps in our current knowledge about the area.

Even though one of the first steps in describing an ecosystem is the construction of energy and material flows, such description is insufficient to advance ecological theory (E.A. Smith 1984). There is a notable absence of general theory dealing with how energy and matter are regulated and how diversity affects the structure and function of ecosystems. The lack of a general nonequilibrium theory makes it very difficult to extrapolate findings from one ecosystem to another and even to predict how non-observed changes might affect a particular ecosystem (Levin 1987:247).

Levels of Analysis and Amazonian Soils

The problem of level shifting emerges at the outset as one of the fundamental problems in the ability to differentiate between Amazonian soils. Most maps available are at a scale of 1:100,000 to 1:500,000. These macro-scale maps show the soils of the Amazon to be primarily oxisols (latosols) with a small area of inceptisols (alluvial soils) along the floodplain (National Academy of Science 1972). If the connection is attempted between these soils and human use of them, as is often the case, discussions will emphasize that these soils are poor and their utilization is restricted to long fallow swidden agriculture with shifting of fields every two to three years due to rapid declines in fertility caused by the loss of the limited nutrients made available after the burn (McNeil 1964; Gourou 1966; Meggers 1971; Goodland and Irwin 1975). A great deal of the pessimism about the potential of the Amazon is based on this level of analysis.

But how representative is the available data? Does it provide an adequate enough representative sample of the whole Basin to permit macro-level generalizations such as those above? Reliance, until recently, on the simple dichotomy between the floodplain and the uplands, comprising 2% and 98% of the area respectively, is at a level of generality not likely to generate systematic scaling of the Amazonian regional system and implies that each of the two areas are more homogenous than is the case. This probably results from a now surpassed Clementsian view of ecological communities (by way of Lotka and Volterra) in which an area is treated as an integral unit - basically homogeneous and in equilibrium (Levin 1987:247) -- when, in fact, they are heterogeneous in space and time. Studies done elsewhere have noted that variability increases with movement towards the more micro-scale units in the sampling process. For the Amazon there is no available soil mapping at a scale that permits observation of specific soils except for a few isolated localities (Furley 1980; Ranzani 1978). Is the absence of such detailed micro-scaled maps critical?

When one changes level from the Amazon as a whole to specific sub-regions, the homogeneity suggested at the regional level rapidly yields to extreme local variability. Instead of two soil types, three to five are noted. Not only is there increased detail in visible soil types but even the areal extent of soil types may be misjudged (Ranzani 1978). A technologically sophisticated aerial survey of the Amazon using sideways-looking radar (RADAM 1974) at a scale of 1:100,000 observed that the dominant soil type in the subregion of Maraba were the *ultisols*. However, a localized study by Ranzani in Maraba (1978), at a scale of 1:10,000, concluded that oxisols constituted 65%, entisols 22%, and ultisols *only* 13% of the soils in the area in question. Scale is important when such variability is present (see Figure 10.1).

Whereas maps at a scale of 1:100,000 to 1:500,000 may be useful in addressing questions about geologic history, geomorphology, and general questions about the relationship between soils and biotic productivity, speciation, and climate, they are of little use in addressing questions about human use of resources, the social organization and structure of human communities, and their adaptive strategies. Maps in the order of 1:20,000 and up are not useful for land management. A planner or a researcher using such a map will assume that all the soils labeled with a particular name will have the same characteristics as does a "typical profile." Such a macro-scale map might suggest that farmers can move from place to place with a uniform land management approach and expect similar results everywhere. This has been, in fact, a dominant viewpoint in anthropological writings about the Amazon tropical rain forest peoples (cf. Wissler 1926; Meggers 1971). Given the lack of micro-scale studies in a sufficient number of areas by systematic sampling, it has been easy for investigators to dismiss variations as

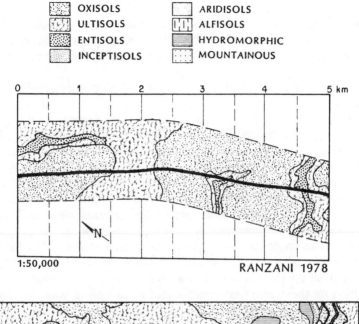

1:50,000   RANZANI 1978

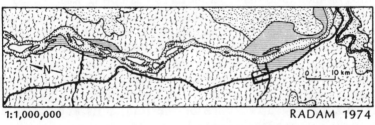

1:1,000,000   RADAM 1974

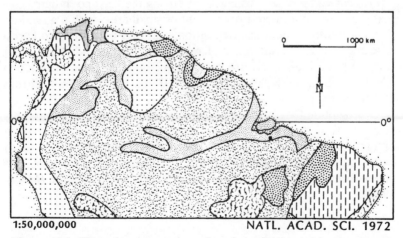

1:50,000,000   NATL. ACAD. SCI. 1972

Figure 10.1 Changing perception of major soils found in the Amazon region as a function of scale.

"non-representative" and to accept the macro-scale as more accurate. Such a decision is incorrect from the point of view of geographical sampling and its analytical implications (Duncan, Cuzzort and Duncan 1961).

From the point of view of regional policy, reliance on macro-scale maps had serious consequences. The decision to focus government directed colonization along Brazil's Transamazon Highway in the Altamira region of the Xingu River was based on political and economic priorities based on the identification of medium to high fertility *alfisols* which appeared to dominate the region (IPEAN 1967). This decision was based on the extrapolation of a few soil samples to the region as a whole. *As a result, colonists were placed on all available lots as they arrived since soil quality was thought to be homogeneous.* Also, a uniform set of crops was required in order for colonists to obtain bank credit. Most farmers who followed the directives of the bank obtained low yields and defaulted on their loans. It was not until the colonists were all settled on their land that micro-level soil sampling was carried out by Moran (1975), Smith (1976) and Fearnside (1978) in the Altamira area and by Ranzani (1978) and Smith (1976) in the Maraba region. These investigators discovered that the soils of the area are a patchwork, with radical differences in nearly every kilometer and even from one neighbor's plot to another. Thus, the soils of Altamira were highly variable with the medium to high fertility alfisols making up *only 8%* of the total soils and scattered in small patches (see Figure 10.1).

Recent studies of soil management among the Kayapo Indians show that they carry out micro-management of small areas within a swidden that takes into account ash deposition, shading, proximity to forest edge and nutrient requirements of each plant (Hecht and Posey 1989). This is not unlike the diverse patterns documented for the Hanunoo (Conklin 1957), the Kekchi (Carter 1969) and the Shipibo (Behrens 1989).

Recognition that land use planning and farmer decisions can only be accurate at the micro-level could have led to a different pattern of land occupation, a less homogeneous effort at agricultural extension, reduced likelihood of loan defaults, and fewer problems that affected the performance of farmers in the Transamazon (Moran 1981).

A literature in ecology has developed that permits treatment of heterogeneity and patchiness. Those approaches have culminated in the emergent field of landscape ecology (Levin 1976, 1987:248). Advances are taking place in dealing with fragmented landscapes and with the ever fluctuating spatial and temporal mosaics that are characteristic of ecosystems (Harris 1984; Forman and Godron 1986).

Levels of Analysis and the Provision of Agricultural Inputs

Macro-level analyses often have noted that in the Amazon the lack of sufficient credit was a major obstacle to increases in food production. In the Amazon Basin, analysts have added that the archaic system of *aviamento*, wherein riverine traders controlled the supply of goods to the Amazon interior, and long-term credit was extended at exorbitant rates, was a fundamental cause of the region's perpetual state of underdevelopment (Wagley 1953; Santos 1968). The macro-level solution to this situation was to provide credit at favorable rates through the normal channels of the Bank of Brazil. In making such a policy decision, planners failed to take into consideration micro-level constraints to the use of capital resources such as (a) the structure of local social relations, (b) the costs of monitoring the credit worthiness of a rural population in a rain forest region, (c) the differential experience of farmers with bank credit, and (d) the traditional forms of allocating cash inputs. Nor did the planning process allow for the imperfections in the local-level administration of resources by government agencies. Credit institutions were unable to release funds in accordance with the agricultural schedule of the specific areas in question. Extension agents were unable to gain access to fields in order to monitor the progress of farm work yet continued to require elaborate procedures designed to monitor credit worthiness (Moran 1981).

The aggregated data showed a high rate of credit default. However, the reasons for such defaults could only come from farm management surveys--micro-level studies at the level of the individual farm, which were not part of the monitoring process. Micro-level analysis of credit has shown its costs to be unreasonably high. Loans cost not the 7 percent per annum "face" cost but 50 percent, due to lost labor time in obtaining release of funds (Moran 1975). Moreover, lack of previous experience in using agricultural credit led to misallocations.[1] Farmers with little managerial background tended to consume their loans rather than apply them to the intensive use of limited areas (Moran 1981).

Credit institutions were also not attuned to the micro-level agronomic constraints to cereal production in some parts of the Amazon Basin. While cereals can be grown in the Amazon Basin, they are more susceptible to pests and diseases and require soils of higher initial fertility than do root crops (Moran 1975, 1981; Smith 1977, 1982). Though cereals could be grown well in alfisols, they do not fare so well on oxisols and ultisols. They also usually require substantial fertilizer inputs after the first or second year of cultivation, but levels of fertilizer input for tropical soils have as yet to be worked out per crop and per soil type (Tropical Soils Research Program 1976:137). The banks, however, gave credit for cereal crops only, despite the patchy availability of alfisols and the higher cost of inputs that they required.

Not only was planning affected by an over-specific, yet macro-perspective to land planning, but so was the evaluation of farmer performance. Whereas some farmers familiar with the Amazon refused to go along with the practices promoted by the government and obtained good yields from their diversified agricultural operations, the use of aggregate production data, rather than individual farm management surveys hid the differential performance of farmers and led to a reduction of government support to small farmers. Elsewhere I have shown (Moran 1979b, 1981) that the Amazonian *caboclo* population[2] had precise knowledge of forest resources, soil types, and had better results than farmers following practices promoted by the planners. The caboclos' use of the region's resources (Moran 1974) was more complete, more rational and more efficient than that of outsiders. Instead of identifying farm strategies that worked, the aggregate analysis provided no details on what management practices worked but only that the output did not meet national expectations set before the project began.[3] At the evaluation stage the agencies involved used inappropriate quantitative tools to measure farmer's productivity, to identify limiting factors and to correct actions. All that the aggregate analysis could do was balance the output of the sector against the total inputs provided. The analysis did not show, *and could not show*, that the inputs were not timed to the needs of farmers, that the institutional performance was a constraint in itself, and that the technological inputs were in part responsible for the low yields in two out of the three years measured.[4]

The aggregated data from the banks and the government agencies showed that low amounts of rice, corn and beans had been marketed by Transamazon farmers. These low production levels were attributed to the "low level" of technology in use and the "lack of entrepreneurial spirit" among the farmers. It was a case, as Wood and Schmink (1979) have reminded us, of "blaming the victim." The negative evaluation of farmer performance comes as no surprise, though, when one considers the inappropriateness of the data. Input/output data for a sector created a mere three years before could have hardly yielded results capable of explaining poor performance. Only micro-level farm management investigations could have revealed those factors.[5]

Discussion

Human adaptation and social differentiation do not occur in a vacuum. This process of social reproduction reflects the adaptation of the population to local habitat, to the economic and structural relations within nation-states, and to the ability to function within the social field provided by an incipient socio-economic setting in a rain forest environment.

One proposed solution to the limitations found in past ecological analyses has been to move the field towards the adoption of microeconomic models (see E. A. Smith, 1984; Winterhalder 1984). However, this is one of the most problematic of all solutions for ecological anthropological research given the crucial difference between ecological and economic theory (see Rappaport, this volume). Both economic and ecological theory recognize that individual decisions, in the context of hierarchies of group organization, structure the system. Economics describes the process as one of individual firms seeking to survive and maximize their utility within the constraints present--making them analogous to ecological systems. However, ecologists tend to argue that the behavior of individuals at *all* levels of group organization is subject to natural selection on the basis of relative fitness. Economists, by contrast, make a distinction between the decision-making process at the individual level and the level of the firm. Whereas firms are assumed to act to ensure survival, individuals are assumed to aim at maximizing consumption. By making it an assumption that individuals are more concerned with immediate consumption that they are with survival and reproduction, classical and neoclassical economists have created a profound theoretical problem for the study of human ecosystems.

Human ecosystems are inextricably bound to systems of *values*, however culturally-defined (see Bennett, this volume). Thus, it is of fundamental importance to understand how our values, and economic systems relate to ecological systems. Bernstein (1981:326) believes that the absence of feedback about the effects of macro-decisions on the local environment in larger economic and political systems is responsible for the failure of such systems to respond adequately to perturbations. Whereas it is relatively easy to see the connection between crop rotation and yield in a village plot, it is difficult to establish the causes and effects of acid rain on several states downwind from coal-fired power plants. Ecological and economic scientists need to develop a coherent theory of decision-making that avoids the inconsistent assumptions between levels of organization currently present. While it may not be erroneous to assume that individuals seek to maximize their utility, neither is it far-fetched to note that they also seek their survival and reproduction and that the latter is more fundamental to individuals than is maximization of consumption. However, in human ecosystems issues of production and reproduction, whether biological or social, are inextricably bound in units above the level of the individual. In anthropological economics this has taken the form of "household economics" (Wilk 1989; Netting, Arnould, Wilk 1984; Collins 1988; Maclachlan 1987; Guyer 1981) which seems to fit better with observed economic behavior than either an individual, a firm, a community, a culture, or a nation-state unit of analysis. Ecological anthropology shows signs of moving in this direction as well (e.g. Orlove 1980; Netting *et al.* 1990; Moran 1990; Wilk, this volume) thereby more effectively dealing with household variability and environmental patchiness.

The social and ecological future of the Amazon and its people are constrained by a centralized macro-level approach to planning and implementation which did not begin with the military takeover in 1964 (*contra* Hecht and Cockburn 1989) but has deep historical roots (Roett 1972). The priorities set by the State for the development of the Amazon were inspired by Vargas in the 1940's and did not concern itself "with details." Its aims were national: to improve Brazil's foreign exchange balance, to promote national integration, and to reduce social tensions in the Northeast. In none of the planning documents is there mention of how the macro-economic objectives of the project would articulate with the complex micro-economic and micro-ecological processes at the level of individual farms.

The choice to use macro-economic planning followed long-standing patterns built into the structure of the Brazilian bureaucracy with its preference for aggregate inputs in formulating policies. Hirschman has pointed out that planners tend to be biased against programs that involve technological uncertainties and prefer to provide support to large corporations rather than many local small holders (1967:39-44). Thus, *the problem is general to all complex bureaucratic structures.* In Brazil's case that structure is also remarkably centralized (i.e. authoritarian), which makes the structure of decision-making even less amenable to inputs from micro-level studies. The more centralized the structure of decision-making, the less able it is to process complex information incorporating the variability present in any areally extensive system. As a result, decision-making is insensitive to micro-level variability and tends toward homogenization of both environmental and social variables. Economists have noted that economic policies in Brazil since 1964 have increased the gap between income groups to a more evident distinction between haves and have-nots. While economists agree that such a long-term process is destabilizing, it has not led to a reconsideration of the basic assumptions responsible for this process. Efforts at reform have faltered because of the resistance of bureaucratic organizational priorities over time (cf. Adams 1988). Ecologists have also noted a tendency to treat the Amazon as a forest, of which any part can be cleared with equivalent results. The results have been destructive of forest and of the store of value they represented. The question which is central to the future of Amazonian ecosystems is whether or not the structure of the Brazilian bureaucracy is capable of adjusting its policies to include inputs from specific sites to balance productivity and conservation per site. The implications of such a change are explored elsewhere (Moran 1990).

There is increased recognition that the Amazon is very heterogeneous and probably varies a great deal in fragility and/or resiliency from place to place. We already know that white sand/black-water river watersheds are particularly susceptible to "desertification" (Herrera et al. 1978; Uhl 1983; Moran, n.d.) The soils are extremely variable throughout the Basin and demand site-specific strategies of utilization. Clearly, areas with low initial soil fertility should be protected from predatory forms of exploitation and reserves created to prevent a breakdown in the closed nutrient cycle of the forest. Macro-level approaches are too insensitive to variation and to feedback from local environmental and social impacts to permit the development of the necessary site-specific strategies of resource use in heterogeneous ecosystems. On the other hand, recent advances in geographic information systems and in the resolution of images from satellites like LANDSAT, SPOT and NOAA permit growing sophistication in studying climate change, rates of deforestation, and global flows of energy and matter (Dickinson 1987). What is needed is a "nested" approach to resource use that builds up systematic sampling of subregions, districts and localities so that information feedback flows from specific sites through each level of the hierarchy in order to permit adjustment to variation.

Conclusion

For the purposes of field research, it is seldom practical to try to investigate more than one level. But as the above suggests, to shift levels between data and analysis is analytically inappropriate and obscures the complex processes being studied. Levels are hierarchically structured and exhibit both vertical and horizontal interactions. Ecosystem research normally focuses upon horizontal interactions within a given level. Vertical hierarchical organization has received much less attention in ecological anthropological research and remains problematic both theoretically and methodologically (cf. Pattee 1973; Gould 1982; Lewin 1982; E. A. Smith 1984; Winterhalder 1984; Roughgarden, May and Levin 1987).

Minimally, the first requirement in overcoming the current dilemma is the recognition of the distinctiveness of levels of analysis. Once they are recognized, the differences between levels pose little difficulty for specific empirical studies with limited and clear objectives (cf. Gross, this volume). A study of a sample of groups in a rural community makes it possible to generalize about group structure, although not about the structure of, say, "peasantries." To arrive at the latter type of generalization requires systematic comparisons of a representative sample of rural groups and of their interactions with the larger society. In short, the research design must be adapted to the level of organization to be explained and explanations confined to that level. Each explanation "nests" within the other level and operates within the general constraints set by the other level. Thus, while each hypothesis is restricted to a particular level of analysis, structural and functional aspects of ecosystems are affected by processes at other levels. Each hierarchical level adds a layer to our understanding of the total human adaptational situation.

Ellen (this volume) stresses the need to remain open and flexible with regards to the number and types of units in a particular study. His notion of "graded" ecosystems, whose boundaries expand and contract in spatial and temporal scales, seems both inductive and operational in research practice. In turn, these graded systems are nested in hierarchies of varying complexity defined for analytical purposes.

While we should all aspire to the integration between macro- and micro-level explanations, this integration cannot be achieved by mixing levels between the data-gathering and the interpretational stages. Synthesis can only result from preliminary separation of micro- and macro-analyses within nested hierarchies along gradients of both space and time. Only after such initial separation can we hope to understand how the various levels are linked. This is of more than trivial significance. Growing concern with the destruction of entire ecosystems—and the folk and native populations that live in those areas—means that anthropology must engage these local and global crises with the appropriate tools. For dealing with some of these ecological problems, the ecosystem approach has proven value if stripped of equilibrium assumptions and enriched with attention to history, boundaries, scale, hierarchy and the decision-making processes at various levels that influence the structure and function of ecosystems.

Acknowledgements

The research on which this paper is based is the result of six fieldtrips over the past two decades. Research trips were made possible by funds from the Social Science Research Council, the National Institutes of Mental Health, the Council for the International Exchange of Scholars, the National Science Foundation, and the Tinker Foundation. None of these organizations are responsible for the views espoused herein. I with to thank Dennis Conway, William Denevan, Jim Eder, Roy Ellen, and Ivan Karp for their critical comments on an earlier version of this paper. Revisions for this new edition were undertaken while the author was a Fellow at the Institute for Advanced Study, Indiana University.

Notes

[1]   Fewer than 21% of farmers had had previous experience in dealing with banks for financing agricultural work (Moran 1979c).

[2]   Caboclo refers to the racially mixed rural population of the Amazon. As a sub-culture it adopted aboriginal subsistence techniques and Portuguese social forms (cf. Wagley 1952, Moran 1974).

[3]   The projected yields were unrealistic in the extreme given the uncertainties of farming and the lack of baseline surveys of soils, climate and input prices. In fact, farmers did reach those yields–but only on the fifth year after settlement. The projected yields have persisted well into the 10th year of the project (Moran 1988b) showing that farmers adjusted and learned over time how to manage their new land. See also forthcoming Stanford University history honors thesis by Douglas Stewart.

[4]   The 1972 rice harvest was reduced in no small part due to the promotion of a seed type inappropriate for the humid tropics. (i.e. developed for the semi-arid Northeast with a short growing season). The 1973 season was hampered by unusually high rainfall, flooding and fungus infestations resulting from the high moisture. Unfinished roads made it impossible for farmers to market the produce they obtained (Moran 1981).

[5]   While it can be argued that the evaluation process was political rather than economic, it can be argued that *even if* political aspects could be left out, the choice of level of analysis could not have yielded any other results.

References Cited

Adams, R.N.
    1988   *The Eighth Day.* Austin: University of Texas
            Press.
Alvim, P. de T.
    1978   Perspectivas de Produção Agrícola na Região Amazônica.
            Interciencia 3(4):243-249.
Allen, T.F.H. and T. B. Starr
    1982   *Hierarchy: Perspectives   for   Ecological
            Complexity.* Chicago: University of Chicago Press.
Baker, Paul T. and M. Little eds.
    1976   *Man in the Andes.* Stroudsburg, PA: Dowden,
            Hutchinson and Ross. US/IBP Synthesis Series, No. 1.

Balee, W.
    1989   The Culture of Amazonian Forests. *Advances in Economic Botany* 7:1-21.
BASA (Banco da Amazônia, S.A.)
    1971   *Programa Especial de Crédito Rural.* Belém, Pará: BASA.
Beer, Stafford
    1968   *Management Science.* Garden City: Doubleday.
Behrens, Clifford
    1989   The Scientific Basis for Shipibo Soil Classification and Land Use: Changes in Soil-Plant Associations with Cash Cropping. *American Anthropologist* 91(1): 83-100.
Bennett, John
    1967   Microcosm-Macrocosm Relationships in North American Agrarian Society. *American Anthropologist* 69:441-54.
    1969   *Northern Plainsmen.* Chicago: Aldine.
Bernstein, B.B.
    1981   Ecology and Economics: Complex Systems in Changing Environments. *Annual Review of Ecology and Systematics* 12:309-30.
Bloch, M.
    1966   *French Rural History.* Berkeley: University of California Press. Originally published in 1931.
Blok, Anton
    1974   *The Mafia of a Sicilian Village, 1860-1960: A Study of Violent Peasant Entrepreneurs.* New York: Harper and Row.
Braudel, Fernand
    1973   *The Mediterranean and the Mediterranean World in the Age of Philip II.* Two Volumes. New York: Harper and Row.
Carneiro, Robert L.
    1957   Subsistence and Social Structure: An Ecological Study of the Kuikuru. Ph.D. Dissertation, University of Michigan.
    1961   Slash-Burn Agriculture: A Closer Look at Its Implications for Settlement Patterns. *Man and Culture.* Edited by Anthony F. Wallace. Fifth International Congress of Anthropological and Ethnological Sciences.

Carter, William
    1969   *New Lands and Old Traditions.* Gainesville:
           University of Florida Press.
Clarke, J.
    1976   Population and Scale. *Population at*
           Microscale. Edited by L. Kosinski and J. Webb.
           New Zealand: Commission on Population Geography.
Collins, Jane
    1988   The Household and Relations of Production in S. Peru.
           Comparative Studies in Society and History
           28:651-671.
Conklin, H.C.
    1957   *Hanunóo Agriculture.* Rome: FAO.
Devons, E. and M. Gluckman
    1964   Conclusion: Modes and Consequences of Limiting a Field
           of Study. *Closed Systems and Open Minds.*
           Edited by M. Gluckman. Chicago: Aldine.
Dewalt, B. and P. Pelto, eds.
    1985   *Micro and Macro Levels of Analysis in*
           Anthropology: Issues in Theory and Research.
           Boulder: Westview Press.
Dickinson, R.E. ed.
    1981   *The Geophysiology of Amazonia.* New York: Wiley.
Dogan, M. and S. Rokkan eds.
    1969   *Social Ecology.* Cambridge, MA: MIT Press.
Duncan, O., R.P. Cuzzort, and B. Duncan
    1961   *Statistical Geography: Problems in Analyzing Areal*
           Data. Glencoe, IL: Free Press.
Eder, Jim
    1981   *Who Shall Succeed?* New York: Cambridge
           University Press.
Eldredge, N.
    1985   *Unfinished Synthesis: Biological Hierarchies and*
           Modern Evolutionary Thought. New York: Oxford
           University Press.
Ellen, Roy
    1978   Problems and Progress in the Ethnographic Analysis of
           Small-scale Human Ecosystems. *Man*
           13(2):290-303.
    1979   Introduction: Anthropology, the Environment and
           Ecological Systems. *Social and Ecological*

Systems. Edited by P. Burnham and R. Ellen. London: Academic Press.

Epstein, A.L.
1964 Urban Communities in Africa. *Closed Systems and Open Minds.* Edited by M. Gluckman. Chicago: Aldine.

Falesi, Ítalo Claudio
1972 *Solos da Rodovía Transamazônica.* Belém, Pará IPEAN. Boletim técnico No. 55.

Farnworth, Edward and Frank Golley eds.
1974 *Fragile Ecosystems: Evaluation of Research and Applications in the Neotropics.* New York: Springer-Verlag.

Fearnside, Philip
1978 Estimation of Carrying Capacity for Human Populations in a Part of the Transamazon Highway Colonization Area of Brazil. Ph.D. Dissertation: University of Michigan, Department of Biological Sciences.

Foin, T.C. and W.G. Davis
1987 Equilibrium and Non-equilibrium Models in Ecological Anthropology. *American Anthropologist* 89(1): 9-31.

Forman, R.T.T. and M. Godron
1986 *Landscape Ecology.* New York: Wiley.

Furley, Peter
1980 Development Planning in Rondonia based on Naturally Renewable Resource Surveys. *Land, People and Planning in Contemporary Amazonia.* Edited by F. Scazzocchio. Cambridge: Cambridge University, Centre for Latin American Studies.

Geertz, Clifford
1963 *Agricultural Involution.* Berkeley: University of California Press.

Gluckman, Max ed.
1964 *Closed Systems and Open Minds: The Limits of Naivety in Social Anthropology.* Chicago: Aldine.

Goldscheider, C.
1971 *Population, Modernization and Social Structure.* Boston: Little Brown.

Goodland, R.J. and H.S. Irwin
1975 *Amazon Jungle: Green Hell to Red Desert?* Amsterdam: Elsevier.

Gould, S.J.
1982   Darwinism and the Expansion of Evolutionary Theory. *Science* 216:380-7.
Gourou, Pierre
1966   *The Tropical World.* 4th edition. New York: Wiley.
Guyer, Jane
1981   Household and Community in African Studies. *African Studies Review* 24:87-137.
Haggett, P.
1965   Scale Components in Geographical Problems. *Frontiers in Geographical Teaching.* Edited by R.J. Chorley and P. Haggett. London: Methuen.
Harris, Larry
1984   *The Fragmented Forest: Island Biogeography Theory and the Preservation of Biotic Diversity.* Chicago: University of Chicago Press.
Hecht, S. and A. Cockburn
1989   *The Fate of the Forest.* New York: Verso
Hecht, S. and D. Posey
1989   Preliminary Results on Soil Management Techniques of the Kayapo Indians. *Advances in Economic Botany* 7:174-188.
Herrera, R., C. Jordan, H. Klinge, and E. Medina
1978   Amazon Ecosystems: Their Structure and Functioning with Particular Emphasis on Nutrients. *Interciencia* 3(4):223-231.
Hirschman, A.O.
1967   *Development Projects Observed.* Washington, D.C.: Brookings Institution.
IPEAN (Instituto de Pesquisa e Experimentação Agropecuária do Norte).
1967   *Contribuição ao Estudo dos solos de Altamira.* Belém, Pará: IPEAN. Circular No. 10.
1974   *Solos da Rodovía Transamazônica: Trecho Itaituba-Rio Branco.* Belém, Pará: IPEAN.
Katzman, M.T.
1976   Paradoxes of Amazonian Development in a "Resource-starved" World. *Journal of Developing Areas* 10(4):445-460.
Kosinski, L. and J. Webb eds.
1976   *Population at Microscale.* New Zealand: Commission on Population Geography.

Lathrap, D.
1970   *The Upper Amazon.* London: Thames and Hudson.

Levin, S.A.
1976   Population Dynamic Models in Heterogeneous
        Environments. *Annual Review of Ecology and
        Systematics* 7:287-311.
1987   Challenges in the Development of a Theory of Community
        and Ecosystem Structure and Function. *Perspectives
        in Ecological Theory.* Edited by J. Roughgarden, R.
        May and S.A. Levin. Princeton: Princeton University
        Press. pp. 247-255.

Lewin, R.
1982   Molecules come to Darwin's Aid. *Science*
        216-1091-2.

Maclachlan, Morgan ed.
1987   *Household Economies and their Transformations.*
        Washington D.C.: University Press of America. Society
        for Economic Anthropology Monograph Series No.3.

McCarthy, H.H., J.C. Cook, and D.S. Knos
1956   The Measurement of Association in Industrial
        Geography. Department of Geography, University of
        Iowa.

McMaugh, T.H.
1982   Adaptation can be a Problem for Evolutionists.
        Science 216-1212-14.

Meggers, Betty
1954   Environmental Limitations on the Development of
        Culture. *American Anthropologist* 56:801-824.
1971   *Amazonia: Man and Culture in a Counterfeit
        Paradise.* Chicago: Aldine.

Ministério de Agricultura
1972a   *Altamira 1.* Brasília D.F.: INCRA.
1972b   Amazônia: Uma Alternativa para os Problemas Agrários
        Brasileiros. Mimeographed Manuscript.

Molion, L.C.B.
1987   Micrometeorology of an Amazonian Rain Forest. *The
        Geophysiology of Amazonia.* Edited by R. Dickinson.
        New York: Wiley pp. 255-270.

Moran, Emilio F.
1974   The Adaptive System of the Amazonian Caboclo. *Man
        in the Amazon.* Edited by C. Wagley. Gainesville:
        University of Florida Press.

1975   Pioneer Farmers of the Transamazon Highway: Adaptation
          and Agriculture Production in the Lowland Tropics.
          Ph.D. Dissertation, University of Florida.
1977   Estrategias de Sobrevivencia: O Uso de Recursos ao
          Longo da rodovia Transamazonica. *Acta Amazonica*
          7(3):363-379.
1979a   *Human Adaptability: An Introduction to Ecological
          Anthropology.* N. Scituate: Duxbury Press.
          Published in 1982 by Westview Press.
1979b   Strategies for Survival: Resource Use along the
          Transamazon Highway. *Studies in Third World
          Societies* 7:49-75.
1979c   Criteria for Choosing Homesteaders in Brazil.
          Research in Economic Anthropology 2:339-359.
1981   *Developing the Amazon.* Bloomington: Indiana
          University Press.
1988a   Social Reproduction in Agricultural Frontiers. Edited
          by J. W. Bennett and J. Bowen. *Production and
          Autonomy.* Washington D.C.: University Press of
          America pp. 199-212. Society for Economic
          Anthropology, Monograph Series No. 5.
1988b   Following the Amazon Highways. *People of the
          Tropical Rain Forest.* Edited by J.S. Denslow and
          C. Padoch. Berkeley: University of California Press.
          pp. 155-162.
1990   *A Ecologia Humana das Populacoes da Amazonia.*
          Petropolis, R.J. (Brasil): Editora Vozes
n.d.   Rivers of Hunger: Adaptive Strategies in Amazonian
          Blackwater Ecosystems. *American Anthropologist*
          . Under Review.
Moran, Emilio F. ed.
1983   *The Dilemma of Amazonian Development.* Boulder,
          Colorado: Westview Press.
National Academy of Science (NAS)
1972   *Soils of the Humid Tropics.* Washington, D.C.:
          National Academy of Sciences.
Nelson, Michael
1973   *The Development of Tropical Lands: Policy Issues
          in Latin America.* Baltimore, MD: The Johns
          Hopkins University Press.

Netting, R., E. Arnould, R. Wilk eds.
1984   *Households: Comparative and Historical Studies of the Domestic Group.* Berkeley: University of California Press.

Nicholaides, J., et al.
1983   Crop Production Systems in the Amazon Basin. *The Dilemma of Amazonian Development.* Edited by E.F. Moran. Boulder, Colorado: Westview Press.

Nietschmann, Bernard
1972   *Between Land and Water.* New York: Seminar Press.

Nugent, Stephen
1981   Amazonia: Ecosystem and Social System. *Man* 16(1):62-74.

O'Neill, R.V.
1987   Perspectives in Hierarchy and Scale. *Perspectives in Ecological Theory.* Edited by J. Roughgarden, R.M. May and S.A. Levin. Princeton: Princeton University Press, pp. 140-156.

O'Neill, R.V. *et al.*
1986   *A Hierarchical Concept of Ecosystems.* Princeton: Princeton University Press.

Orlove, Benjamin
1980   Ecological Anthropology. *Annual Review of Anthropology,* 9:235-273.

Pattee, H.H. ed.
1973   *Hierarchy Theory.*New York: Brazilier.

Pomeroy, L.R. and J.J. Alberts eds
1988   *Concepts of Ecosystem Ecology.* New York: Springer-Verlag. Ecological Studies Monog. Series No. 67.

Posey, D. and W. Balee eds.
1989   *Resource Management in Amazonia.* New York: New York. Botanical Garden. Adv. in Econ. Bot. Monog. Series, no. 7.

RADAM (Radar da Amazônia)
1974   *Levantamento de Recursos Naturais.* Vol. V, Rio de Janeiro: Ministério de Minas e Energía.

Ranzani, G.
1978   Alguns Solos da Transamazônica na Região de Marabá. *Acta Amazônica* 8(3):333-355.

Rappaport, Roy
1968   *Pigs for the Ancestors.* New Haven, CT: Yale
        University Press.
Robinson, W.S.
1950   Ecological Correlations and the Behavior of
        Individuals. *American Sociological Review*
        15:351-57.
Roughgarden, J., R.M. May, and S.A. Levin eds.
1987   *Perspectives in Ecological Theory.* Princeton:
        Princeton University Press.
Salati, E.
1987   The Forest and the Hydrological Cycle. *The
        Geophysiology of Amazonia.* Edited by R.E.
        Dickinson. New York: Wiley. pp 273-296.
1985   The Climatology and Hydrology of Amazonia. *Key
        Environments: Amazonia.* Edited by G.T. Prance and
        T. Lovejoy. Oxford: Pergamon
Salati, E. and P.B. Vose
1984   Amazon Basin: A System in Equilibrium. *Science*
        225:129-138.
Salati, E. et al.
1979   Recycling of Water in the Amazon Basin: an Isotopic
        Study. *Water Resources Research* 15:1250-1258.
Salthe, S.N.
1985   *Evolving Hierarchical Systems: Their Structure and
        Representation.* New York: Columbia University
        Press.
Sanchez, Pedro and S.W. Buol
1975   Soils of the Tropics and the World Food Crisis.
        Science 188:598-603.
Sanchez, Pedro, et al.
1972   *A Review of Soils Research in Tropical Latin
        America.* Raleigh: North Carolina Agricultural
        Experiment Station Technical Bulletin 219.
Santos, Roberto
1968   O Equilíbrio da Firma Aviadora e a Significação
        Económica. *Pará Desenvolvimento* 3:7-30.
Schneider, J. and R. Schneider
1976   *Culture and Political Economy in Western
        Sicily.* New York: Academic Press.

Schuh, G. Edward
1970   *The Agricultural Development of Brazil.* New York: Praeger.

Sheridan, Thomas
1988   *Where the Dove Calls: The Political Ecology of a Peasant Corporate Community in Northwestern Mexico.* Tucson: University of Arizona Press.

Smith, Carol
1984   Local History in Global Context. *Comparative Studies in Society and History* 26:193-228.

Smith, E. A.
1984   Anthropology, Evolutionary Ecology and the Explanatory Limits of the Ecosystem Concept. *The Ecosystem Concept in Anthropology.* Edited by E. F. Moran. Washington D.C: American Association Adv. of Science, pp. 51-86.

Stone, G., R. Netting, P. Stone
1990   Seasonality, Labor Scheduling and Agricultural Intensification in the Nigerian Savanna. *American Anthropologist* 92 (1):7-23

Smith, Nigel
1976a   *Transamazon Highway: A Cultural Ecological Analysis of Settlement in the Lowland Tropics.* Ph.D. Dissertation, University of California, Berkeley, Department of Geography.
1977   Influéncias Culturais e Ecológicas na Produtividade Agrícola ao longo da Transamazônica. *Acta Amazônica* 7:23-28.
1982   *Rainforest Corridors.* Berkeley: University of California Press.

Smith, Thomas
1959   *The Agrarian Origins of Modern Japan.* Stanford: Stanford University Press.

Sombroek, W.
1966   *Amazon Soils.* Wageningen: Centre for Agricultural Publ. and Documentation.

Steward, Julian
1950   *Area Research: Theory and Practice.* New York: Social Science Research Council.
1955   *Theory of Culture Change.* Urbana, Illinois: University of Illinois Press.

1956   *The Peoples of Puerto Rico.*   Urbana, Illinois:
        University of Illinois Press.
Tropical Soils Research Program
    1976   Annual Report 1975. Raleigh, North Carolina: Soil
            Science Department, North Carolina State University.
Vickers, William
    1979   Native Amazonian Subsistence in Diverse Habitats: The
            Siona-Secoya of Ecuador. *Studies in Third World
            Societies* 7:6-36.
Waddell, Eric
    1972   *The   Mound-Builders.*   Seattle,   Washington:
            University of Washington Press.
Wagley, Charles
    1952   *The Folk Culture of the Brazilian Amazon.*
            Proceedings of the XXIX Congress of Americanists.
            Chicago: University of Chicago Press.
    1953   *Amazon Town.* New York: Macmillan.
Wagley, Charles ed.
    1974   *Man in the Amazon.*   Gainesville, Florida:
            University of Florida Press.
Watson, W.
    1964   Social Mobility and Social Class in industrial
            Communities. *Closed Systems and Open Minds.*
            Edited by M. Gluckman. Chicago: Aldine.
Wilk, R. R. ed.
    1989   *The Household Economy.*   Boulder, Colorado:
            Westview Press.
Winterhalder, Bruce
    1984   Reconsidering the Ecosystem Concept. *Reviews in
            Anthropology.* 11(4):301-307.
Wissler, Clark
    1926   *The Relation of Nature to Man in Aboriginal
            America.* New York: Oxford University Press.
Wood, C. and M. Schmink
    1979   Blaming the Victim: Small Farmer Production in an
            Amazon Colonization Project. *Studies in Third
            World Societies* 7:77-93.

CHAPTER 11

ECOSYSTEMS AND METHODOLOGICAL
PROBLEMS
IN ECOLOGICAL ANTHROPOLOGY

Daniel R. Gross

General ecology is hierarchically subordinate to evolutionary biology. It deals with the structure and function of living systems and provides insight into the mechanisms of microevolutionary change, particularly that of selection. The ecosystem concept is a leading tool of ecology, but it is understood in different ways by different authorities. For some, it refers to any delimitable area of nature; for others it refers to specific models of energy flow or nutrient cycling. Some definitions attach notions concerning change and stability in ecosystems. All the definitions include the notion of a system whose variables interact in definite ways, including elements of the living and non-living environment. Other papers in the volume have critiqued some uses of the ecosystem concept, e.g., to assume homeostasis without specific evidence, to reify the ecosystem as if it were an organism itself, to confuse different levels of analysis and others. These are weighty criticisms and should be taken into account in any use of the ecosystem approach.

One utility of the ecosystem approach in anthropology has been to extend the ecological approach first proposed by Steward to cover aspects of human behavior as belonging to a more general class of biobehavioral phenomena and not to a presumably unique class of cultural phenomena. Within this general paradigm, a number of original and useful studies emerged such as Rappaport's (1968), Vayda's (1961), Netting's (1968), and Thomas' (1973). These investigators asked some serious questions about the problems posed by human ecological studies although none so serious, in my opinion, as to require that the ecological approach be discarded.

One of the most impressive gains of the ecosystems approach was that it led investigators to make more and better measurements in the field than had previously been attempted and then to investigate the interrelations of variables. Not only could environmental and behavioral variables be precisely measured, but also quantification made possible mathematical modeling and testing of hypotheses. Precision itself, as critics pointed out, was not sufficient to establish the validity of some of the claims that were made. The present state of affairs appears to be one in which even more and better data are required in order to validate ecologically-oriented hypotheses. This has both advantages and dangers. The advantage is that our studies will become more reliable and replicable as the scope and precision of measurement increases. There is a danger of becoming obsessed with measurement, hyper-specialized and chronically unable to complete studies. A further danger, on which I shall comment later, is that ecological studies may become so micro-focused that their relevance for broader phenomena of widespread occurrence may be seriously reduced.

The ecosystem approach, while it may have led some investigators astray, provided a guide to the interactions between people and their environments. For any particular problem, reference to the ecosystem may suggest the kinds of measurements to be made in the field, and it provides models of how these variables may be related. Upon first confronting the ecosystem approach, one might gain the impression that the investigator must measure virtually everything in order to understand the interrelationships in nature. In practice, this is seldom the case. Ecological anthropologists are not generally simple-minded inductivists who relate everything to everything else. They usually examine the relationship between a limited number of variables in accordance with a model of ecosystem functioning, e.g., a model of predator-prey relationships, a model of an optimal diet, or a model of nutrient cycling in a particular biome. In any given case, a large number of potentially influential variables must be ignored in order to examine the specific question at hand. Thus, paradoxically, while the ecosystem approach proposes that everything is systematically related to everything else, investigators in the field restrict the number of variables to a selected few. The use of an ecosystem model, in biology or in anthropology, cannot substitute for theory that is coherent and that can yield testable hypotheses.

Hypothesis-testing--or what many anthropologists call "problem-oriented research"--has become dominant in contemporary academic anthropology. Very often the hypothesis is only implicit, or it is stated in the form of a critique of a previous theory. For example, an ethnographer goes to the field to try to show the inadequacy of a model of peasant cognition as a way of understanding peasant agricultural decisions. Ecological theory, and the ecosystem approach, may be useful in orienting the formulation of hypotheses and framing of a research design. In my opinion, we cannot frequently draw hypotheses directly from ecological theory. Rather, the concept of the ecosystem serves as a framework within which we can formulate and test hypotheses. Thus ecological theory in anthropology takes a hybrid form, combining with ideas generated in other contexts or even under different paradigms. At a minimal level, the investigator may choose a biobehavioral variable as the indirect measure of something he or she wants to examine, e.g., relative well-being. At another level, an investigator may examine the relative success of two populations in adapting to an environment.

Ecological research is compatible with materialist and political economic paradigms in anthropology. This is because the starting point of their analysis is the way in which people interact with nature in order to produce and reproduce. Some critics attack the ecological position for allegedly treating religion and other aspects of culture as epiphenomena (e.g., Godelier 1972). In spite of the apparent differences, there is still much on which ecologically-oriented anthropologists and political economists ought to agree. There is general agreement that a particular social formation is the outcome of an historical process and that populations and cultures cannot be investigated independently of their context. Both approaches stress the influence of production and exchange. Finally, both approaches are nomothetic, seeking to establish valid generalizations about sociocultural development.

I should like to devote most of the remaining space to a consideration of some basic methodological issues faced by ecological anthropologists, and others with similar objectives. The basic question may be stated as follows: how can anthropologists with their particular research techniques stressing intimate, contextualized knowledge of living human communities, and their recognized expertise in conducting case studies, achieve valid generalizations about social and ecological process? It may be

helpful to examine the methodological <u>forms</u> which many of our studies
take in order to clarify the benefits and pitfalls of each.

Let us take as a point of departure, a study designed to explain
the consequences of a particular kind of change in a particular area
or region, e.g., the introduction of a new cash crop in northeastern
Brazil, the development of European slaving in West Africa, or
perhaps the response of the Plains Indians to the horse. There are
many other kinds of questions to be asked; I have merely picked a
very common type. There are at least three basic ways of approaching
such questions; all are aimed at the same basic objectives, i.e.,
understanding the basic processes of stability and change in human
society. The different modes may be summarized under the following
headings: comparative analysis, cross-sectional analysis, and
longitudinal analysis. Each is a variety of the experimental method
in a context in which the variables cannot be manipulated as in a
laboratory. The principal obstacle which these approaches attempt to
overcome is that social processes unfold within time frames which
often make them inaccessible to direct observation. The three modes
are different ways of conducting studies which take time into
consideration but which are more controlled than conventional
historical research. The three modes are compatible with each other
and often are employed simultaneously. The choice of mode depends on
the constraints of time and resources on the investigation.

Comparative analysis involves selecting homologous social
units widely separated in space which are believed to be similar in
certain aspects of structure and function. Often such comparisons
are conducted through the literature or comparison of the community
studied. Techniques are rarely standardized, and the investigator
must proceed on the assumption that significant differences are
not due to differences in field technique. Comparative
analysis can yield generalizations about causal or developmental
processes, but it is most revealing when correlations can be detected
between variables in systems which differ in one or more key
respects. Because rigorous controls cannot usually be applied,
fine-grained statistical analysis is usually impossible. Reliance on
the literature for cases inevitably skews the sample in favor of the
biases of the available literature.

There are many examples of comparative analysis. One which may
be a kind of "classic" is Murphy and Steward's "Tappers and Trappers"
(1956) which compared two primarily village-organized societies, the
Mundurucu who occupy a tropical environment in Central Brazil and the

Montagnais-Naskapi who occupy the sub-arctic boreal forests of Labrador. The comparison deals with the introduction of trader-based commodity exchange involving the gathering of a dispersed resource (natural latex for the Mundurucu, animal pelts for the Montagnais-Naskapi). The study suggests that both societies, while different in many respects, converged on a family-level of organization due primarily to the organizational demands of their productive system. The novelty of the study lies in the fact that the habitats and earlier cultural styles of the two groups are highly contrastive. The proposed mechanism accounting for the convergence is very compelling, and nothing reported in the two studies suggests that any other factor might be responsible. Nevertheless, it is possible that some other variable, unnoticed by the investigators, was in fact the cause of the convergence. Additional case studies would help to clarify this issue, especially if they were chosen in such a way as to avoid excluding cases where the connection does not obtain.

Cross-sectional analysis is usually conducted in a field setting within a single, relatively homogeneous area. The purpose of cross-sectional analysis is to detect differences in otherwise similar segments of cultural or social units that have been exposed to known influences which may have produced changes. Cross-sectional analysis also aims at discovering regularities in processes occurring over time spans longer than the time allotted for the investigation. It accomplishes this by treating different segments of a unit as if they represented different points in a time series. For this reason, cross-sectional analysis may be particularly important in applied studies where recommendations must be forthcoming within a short period.

The basic idea of this mode is to examine different communities or segments of a larger population which have been exposed to known influences at different points in time. The key assumption in cross-sectional analysis is that the various units of analysis differ only in regard to a specific, known influence. Where the influence is a recent disturbance, the assumption is that all units were once very similar to each other. In other words, cross-sectional analysis attempts to find comparable units in order to hold all variables but one constant. Measurement techniques can often be standardized in cross-sectional analysis and samples of adequate size can be drawn.

The assumption that two communities are comparable may be difficult to demonstrate. In fact this is one of the thorniest issues of contemporary ethnography. The absence of conventions and criteria for establishing the comparability of two human groups is a reflection of the underdevelopment of anthropological theory. In fact, one of the most frequent objections raised to ecological research in anthropology is that the investigator has not considered some additional factor which was not a variable in the original research. A highly particularist focus is still dominant in many areas of ethnology. Multiple events may impinge in different ways on otherwise comparable communities. The investigator in the field may have difficulty in ascertaining exactly what perturbations took place and just when they occurred. Increasing the sample size increases the confidence that exotic or unknown influences have not affected the results. But unless resources are unlimited, it may be impossible to build a large sample precisely because of the level of detail at which so much anthropological research is carried out.

An example of this mode is a study done by Gross et al. (1979) of acculturation among four native groups of Central Brazil. The four groups live in similar (but not identical) habitats, speak related languages and have similar histories. Through time-allocation studies, it was determined that two of the groups were far more involved in market exchange than the others. The history of each village population was reconstructed using documents and native informants and approximate dates were reconstructed using an "event calendar." The study revealed the effects of long-term environmental circumscription on shifting cultivators. As the availability of land diminished and fallow periods perforce reduced, the productivity in food per unit of labor went down, and once abundant nutrients became scarce. These factors account for the variation in commitment to market activities. Other ecological studies using cross-sectional analysis include Moran's comparison (1981) of different settlements along the Trans-Amazon Highway in Brazil in which he identified colonist experience with tropical forests as a key factor in explaining variation in cropping and other farm management strategies.

Cross-sectional analysis is more powerful than comparative analysis because greater standardization in measurement can be achieved and because the investigator is more assured of the comparability of the units. The trade-off between ethnographic detail and sample size still obtains, however, and the possibility that some unknown factor has affected the dependent variable cannot be definitely eliminated.

Longitudinal analysis is possible when there is sufficient time to observe directly the changes taking place in a community at various points in time. Because of the rates at which sociocultural change occurs, longitudinal analysis may require tens of years to yield results. But, there is no more reliable means for understanding change over time in human populations. In longitudinal analysis, processes unfolding in time may be directly observed together with the factors thought to be responsible for them. One assumption which longitudinal analysis makes is that the particular time span over which observations are made is "typical" in terms of time-dependent variables such as rainfall and "historical" factors such as warfare, etc. In longitudinal analysis, there usually is no question of comparability of units since specific sites may be revisited repeatedly. Longitudinal analysis is perhaps the "ultimate" in terms of controls and accuracy in naturalistic observation. Its greatest drawback lies in the time-span required to conduct longitudinal studies. Planners can rarely wait long enough to obtain results from longitudinal studies before implementing programs. Some familiar examples of longitudinal studies in anthropology include Mervyn Meggitt's long term study of the Mae Enga (1977) and the long-term field research conducted by Richard Lee (1980), Henry Harpending (1970), Silberbauer (1981), Yellen (1977), and others among the !Kung San of Botswana. Nevertheless, even when sites can be revisited over intervals as long as several decades, it may be difficult to observe processes of change which unfold very slowly. The modern world with vastly accelerated communication and transportation presents yet another problem. The very rapidity of change means that changes often occur in bundles combining several factors which we would prefer to maintain analytically separate. Thus, even the most painstaking longitudinal study of a human community may not be adequate to permit discrimination of the particular factors at work.

Archeologists, and ethnohistorians, of course, have the luxury of analyzing change over very long periods of time. The main drawback for archeologists is that they cannot observe the changes themselves taking place but must rely on the physical residues of human activity as evidence. Ethnohistorians must rely on the sometimes haphazardly recorded evidence which has accumulated. An example of longitudinal analysis in ethnohistory is provided by Adams and Kasakoff's novel study of marriage patterns in New England communities (1984). Even though they cannot directly verify the accuracy of the data they are using, they can have some confidence that the demographic variables they employ were measured with relatively high accuracy and validity.

As stated above, the three modes of analysis are not mutually exclusive and they may be interwoven in such a way as to provide important support for each other. For many research problems, an optimal field technique might be a hybrid, "micro-macro" approach. First, micro-focused field studies should be carried out in one or more communities, or other social units, in order to gain a first-hand account of the actual social and ecological dynamics. If possible, the selection of units of analysis should be "stratified" in accordance with known variation within the inclusive social unit. Micro-focused field studies lasting for periods of months or years are still the most accurate way of obtaining valid and reliable data on social and ecological structure and function. This is because the ethnographer gains intimate knowledge of local matters, witnesses more aspects of life, and may be permitted to observe much more than any casual visitor. Still, there is no way to verify how representative the local level study was unless the investigator casts a wider net. This is done by systematically investigating selected parameters of the phenomenon under investigation by means of a random sample at the regional level (however this may be defined). In other words, the ethnographer conducts a survey to verify the extent to which processes observed at a local level are replicated globally.

One example of this approach may be found in Gross and Underwood's study of undernutrition induced by work and related caloric deficits among the children of sisal workers in Northeastern Brazil. The process was first identified by detailed dietary intake monitoring at the level of the commensal unit (usually a co-resident household). One class of worker households was judged especially liable to caloric deficits because of extraordinarily high energy requirements of tasks performed by the principal wage earners. The extent of this phenomenon was verified by a survey of 100 households which showed that children in certain workers' households were more likely to show nutritional deficits than others (Gross and Underwood 1971). The next step, unfortunately not feasible, would be to compare survey results for different communities within the region which showed different levels of involvement with the sisal industry. This would constitute a cross-sectional analysis whose results would theoretically show a greater tendency toward undernutrition in communities where the dependence on sisal was greater. Finally, a longitudinal study may be undertaken, in which the same individual children and communities surveyed in 1968 would be surveyed again, say, in 1988, to determine how the influence of sisal agriculture had actually been played out in the long run. One paradox of the longitudinal method is that other conditions might have changed, making the points along the time series non-comparable. Thus, under some conditions, cross-sectional analysis may be more accurate.

There has been an unfortunate tendency, in recent years, to cast doubt on the validity of an ecological model. While critics have pointed to real shortcomings, the result has been that many anthropologists have either prematurely abandoned the ecological approach or engaged in highly particularistic data-collecting sprees. The early proponents of the ecological approach (e.g., Steward, Sahlins, Service) urged that it be comparative and they inaugurated a new era of studies which were very influential, e.g., Sahlins' *Social Stratification in Polynesia* (1958), and Steward's *People of Puerto Rico* (1956). As later investigators urged the adoption of the ecosystem as the principal unit of analysis, the objections of critics became more harsh. But the reaction has been out of all proportion with what was proposed. Pejorative phrases like "vulgar materialism" have been heaped on the ecologists because they sought to understand cultural behavior in its environmental context.

One line of response has been to suggest that the only context within which to understand culture is within the meaningful aspects of culture itself (e.g., Sahlins 1976). Another tack has been to proclaim that the social organization of production has causal priority over adaptive behavior (e.g., Godelier 1972). The consequence of these approaches is to return anthropology to 1915, to Kroeber's proclamations that anthropology is a purely historical science, that it is folly to speak of causality and that "culture comes from culture." The ecological approach suggests that culture grows out of the concrete problems which people must solve in their everyday life. It does not purport to explain every aspect of culture, but it does suggest a way of understanding the fundamental outlines of social action in any society. The ecosystem approach is one useful item in the ecologist's toolkit because it provides a framework within which to test hypotheses. While the fundamental questions of anthropology will continue to grow out of general social and historical issues, the ecosystem approach may yet serve us well as a guide to the investigation of those questions.

References Cited

Adams, J. and A. Kasakoff
   1984   Ecosystems over Time: The Study of Migration in "Long-Run" Perspective. *The Ecosystem Concept in Anthropology.* Edited by E. F. Moran. Washington DC: AAAS. pp. 205-223.
Godelier, Marice
   1972   *Rationality and Irrationality in Economics.* London: NLB.
Gross, Daniel R. and Barbara A. Underwood
   1971   Technological Change and Caloric Costs: Sisal Agriculture in Northeastern Brazil. *American Anthropologist* 73(3): 725-740.
Gross, Daniel R., George Eiten, Nancy M. Flowers, M. Francisca Leoi, Madeline Lattman Ritter, and Dennis W. Werner
   1979   Ecology and Acculturation among Native Peoples of Central Brazil. *Science* 206:1043-1050.
Harpending, Henry
   1976   Regional Variation in !Kung Populations. R.B. Lee and I. DeVore, eds. *Kalahari Hunter-Gatherers.* Cambridge, Mass.: Harvard University Press.
Kroeber, Alfred Louis
   1915   The Eighteen Professions. *American Anthropologist* 17(2):283-288.
Lee, Richard
   1980   *The !Kung San.* New York: Cambridge University Press.
Meggitt, Mervyn
   1977   *Blood is their Argument.* Palo Alto: Mayfield Publishing Company.
Moran, Emilio
   1981   *Developing the Amazon.* Bloomington: Indiana University Press.
Murphy, Robert F. and Julian Steward
   1956   Tappers and Trappers: Parallel Processes in Acculturation. *Economic Development and Culture Change* 4:335-355.

Netting, Robert McC.
   1968   *Hill Farmers of Nigeria: Cultural Ecology of the Kofyar of the Jos Plateau.* Seattle: University of Washington Press.
Rappaport, Roy A.
   1968   *Pigs for the Ancestors: Ritual in the Ecology of a New Guinea People.* New Haven, Connecticut: Yale University Press.
Sahlins, Marshall
   1958   *Social Stratification in Polynesia.* Seattle: University of Washington Press.
   1976   *Culture and Practical Reason.* Chicago: University of Chicago Press.
Silberbauer, George
   1981   *Hunter and Habitat in the Central Kalahari Desert.* Cambridge, UK: Cambridge University Press.
Steward, Julian H. ed.
   1956   *The People of Puerto Rico.* Urbana, Illinois: University of Illinois Press.
Thomas, R. Brooke
   1973   *Human Adaptation to a High Andean Energy Flow System.* University Park, PA: Department of Anthropology, The Pennsylvania State University. Occasional Papers in Anthropology, No. 1.
Vayda, Andrew P.
   1976   *War in Ecological Perspective.* New York: Plenum Press.
Yellen, John
   1977   *Archaeological Approaches to the Present.* New York: Academic Press.

PART IV
TOWARDS INTERDISCIPLINARITY
IN ECOSYSTEM RESEARCH

CHAPTER 12

HOUSEHOLD ECOLOGY:
DECISION MAKING AND RESOURCE FLOWS

Richard R. Wilk

Introduction

This paper grows out of dissatisfaction with anthropological treatment of the household as an economic unit. Households are a problem for anthropologists for a number of reasons, historical, sociological and intellectual. In avoiding some of these problems we have left some significant gaps in ethnographic and ethnological knowledge. Our most glaring failure is the comparative study of household budgets, including the flows of resources and decisions about how to allocate and consume them.

The more substantive goal of this paper is to suggest some tools and standards of comparison for household economies and budgets. Anthropology has a rich vocabulary of terms for describing and comparing kinship systems and behavior, but very few that can be applied to household systems. I will put some concepts borrowed from ecology to use in suggesting how this particular gap can be filled.

The problem of household budgeting and decision making was forced on me during ethnographic research on the adaptations of Kekchi Maya swidden farmers in southern Belize to economic change (Wilk 1981, 1984, 1986, 1989). Some households were taking advantage of new economic opportunities, growing new crops using agrochemicals, expanding their cash crop production, setting up small retail shops, trading in hogs, and buying trucks for hauling freight. Other households with the same family composition and the same access to labor and basic resources, spent most of their cash income on consumer goods and foods. Each household made different decisions about mixing subsistence production and participation in the cash economy, but some were actively accumulating capital while others did not.

I could find no differences in 'entrepreneurial spirit' between the members of the two kinds of households, nor were there sharp differences in desires for cash income or consumer goods. Rather, what enabled some households to invest their cash income in new economic enterprises, appeared to be the way they *managed* their household economies. Simply put, some households pooled their labor and money and channeled it into productive investments, while other households did not. Within the same culture, with the same set of kinship roles and the same normative values of how household members should behave, there was still variation in household behavior. I could not reduce this variation to individual personalities, to differential exposure to external norms, or to different modes of articulation with capitalist forms of production.

A number of authors have focused attention on the structural problems that households face in combining subsistence production with market-oriented farming or wage labor (e.g. Collins 1986, Meillasoux 1981, Adams 1988, Painter 1984). World-systems theorists have recently focused on the household as the crucial social unit, where pooling of different forms of income from household and non-household production reproduces labor (see papers in Smith et al. 1984, Young et al. 1981). Recent marxist and feminist literature gives a great deal of attention to the division of labor within the household, and to the ways that power and production roles change during proletarianization (e.g. Hartmann 1981, Minge 1986, Creighton 1980). Yet others have looked at the changing economic basis of power and inequality in the household (Young et al. 1981, Curtis 1986, Bould 1982)[1].

This work makes important points about what households do, how they fit into larger-scale processes, and how their economic and social functions change over time. But the details of what goes on *inside* households, of how they manage and combine their production, exchange, investment, inheritance, sharing, minding, pooling, preparing, and consuming, are rarely the central focus (see Van Esterik 1985:79). Anthropology has tended to treat each of these activities and functions separately, leaving their conjunction untouched. Households end up being treated as things instead of activities and relationships (Wilk and Netting 1984).

Why is the Household a "Black Box"?

The different behavior of Kekchi households can only be explained by what takes place inside them, in the intimate space of 'householding.' But we cannot break open the "black box" of the household simply by refining existing typologies. Households that may look the same, with the same number of members and the same kinship structure, at the same stage of the developmental cycle, can have very different economic structures. They may organize their work, and apportion costs and benefits quite differently.

Two theoretical issues block an analytical approach to intra-household processes. First is the question of whether or not households function as corporate social units. Do the members of households act in their own self-interest, or are they behaving in an altruistic way, acting to further the interests of the group at the expense of their own (see Laslett 1984, Creighton 1980, Anderson 1971, Peters 1986)? The first option suggests that the individual should be the unit of study, and the second implies that we can treat the household as an economic unity, a proxy individual with what economists call a "joint utility function" (Becker 1981, see Folbre 1984 for discussion).

The "new home economics" based on neoclassical approaches to individual behavior, has tried to show that household behavior, and the gender-based division of labor within the household, can be generated from individual rational goal seeking (e.g. Blau and Ferber 1986). While this mode of analysis has led to some convincing arguments about, for example, the costs and benefits of raising children (Schultz 1981, Caldwell 1981), values and utilities are assumed to be exogenous. The question of where the values come from is left out of the analysis completely, a crippling problem when studying social change.

The anthropological view has tended to stress altruism, rather than self-interest, as the basis of household behavior. Sahlins' 'domestic mode of production' defines domestic behavior as inherently unselfish, founded on generalized reciprocity and uncalculated pooling (1972:196). Individuals use a different economic logic, founded on morality rather than self-interest, in dealing with their close kin.

For anthropologists, therefore, raising the issue of intra-household resource distribution leads into uncomfortable territory. It requires that we pry into the details of altruism, generosity, kinship, and the moral basis of social life (see Bloch 1973). The degree to which individuals are willing to submerge their own interests in that of a collectivity is often a matter of principle instead of the object of empirical research. Medick and Sabean label the two poles of opinion as interest and emotion, and find they are often treated as opposites, when in fact all domestic life is composed of their rich dialectic (1984:10-16).

Marxist analysis has tended to focus on issues of class and increasingly gender, in determining how households respond to economic change. But the world systems and dependency theorists still tend to assume that within the household a moral or pre-capitalist economy prevails. The survival of peasant enterprises and simple commodity production and the continuing subordination and exploitation of the peasantry depend on the household's non-capitalist rationality (see Lehman 1986). In contrasting this literature with traditional neoclassical approaches, however, Folbre observes that:

Whether or not peasants and capitalists, households and firms...respond in similar ways to economic constraints. It is not clear whether they respond as individual households or as members of economic classes, or both. But both theoretical perspectives tend to treat the household in much the same way as traditional Neoclassical economists treat the capitalist firm - as a black box whose internal workings are uninteresting and irrelevant. (1984:23)

This lack of attention becomes especially crippling in trying to understand issues of equity and inequality. Who gets the benefits of increased income? Who will eat and who will go hungry? Van Esterik says "The subject has become everyone's "Black Box" -- the great residual category." (1985:79)

The second obstacle is the problem of setting boundaries around the household unit. How can we talk about what is inside the household if we can't agree on the boundary between inside and outside? What functions or attributes can provide a uniform guide to membership? As Laslett (1984) and Sahlins (1972) point out, there is no society in which households are totally isolated and self-sufficient. Households are always connected to each other, and penetrated by other affiliations through age, kinship, gender and class. In a number of cultures, especially those in west Africa, individuals may belong to a number of domestic units. The actual conjugal unit may be of little economic importance in their lives (eg. Woodford-Berger 1981, Guyer 1981, 1986). The slippery and insubstantial nature of many households leads many to question its corporate nature and its usefulness as a unit of analysis (see Wilk and Netting 1984).

These problems challenge us to refine our concept of the household and re-examine its definition, and in doing this we need to pay much closer attention to what goes on among household members. We need to see the household as social relations and practices that integrate a number of functions and activities, distributing the products of labor, and allocating work and resources. A focus on integrative activities, on the ways that things are shared, and the ways decisions are made, is logically inseparable from the issue of household boundaries.

The inner workings of the household have not been completely ignored by ecological and economic anthropologists. Attention has been paid to the ways that the labor of household members is managed and apportioned to various tasks (e.g. Barlett 1982, Maclachlan 1983), the ways that property and wealth are managed and transmitted between household members (e.g. Goody 1977, Carter 1984), the ways that food is apportioned among members for consumption (e.g. Van Esterik 1985), and the ways that rights to the use of household resources are divided between members (e.g. McMillan 1986).

Nevertheless these separate pieces have not been assembled into a model of the household as a system. In anthropology, the beginnings of such a *household ecology* can be found in studies of consumer/worker ratios in households, Chayanovian balancing of labor and resources, and the juggling of waged and unwaged labor in 'simple commodity production' (all in some way building on Sahlins' discussion of the domestic mode of production [1973]). These studies are implicitly concerned with the balance of resources, labor, and consumption within households, and recognize that the household is not a closed system. But the 'domestic mode of production' itself has not been cracked open, and the conceptual tools of ecological anthropology have not been applied to the contents.

What is peculiar is that anthropology has developed comparative techniques and terminology for almost every aspect of human culture *except* the daily conduct of household relationships and the handling of funds. There is no comparative 'Household Economics' on

a par with comparative studies of systems of production. It seems odd
that the very heart of domestic life, the daily activities and
interactions that are the 'habitus' of the household, is not an
ethnological subject in and of itself[2]. We need a
systematic language to fill this gap. Useful concepts can be found in
the work on household decision making in consumer research and the
ecological study of energy and resource flows within ecosystems.

Approaches to Household Decision-Making

Anthropological interest in household decision-making can be
traced back to studies of post-marital residence choice (e.g. Korn
1975, Goodenough 1955, Barth 1967, Ross 1973). More recently, the
study of economic change and innovation has led many to the study of
choice and decision-making (see Cancian 1972, Barlett 1980, 1982,
Orlove 1980). Formal modeling and graphic methods have been used to
show the crucial factors affecting decisions.

But the range of choices analyzed in this way has been limited,
concentrating on discrete choices among defined alternatives, or on
the allocation of a finite quantity (especially labor and land) among
a number of alternative uses (Nardi 1983). "Strategies" are often
post-hoc deductions based on actual performance, rather than
something discovered through ethnography (Schmink 1984:95). These
unitary strategies conceal dissent, confusion, and the complex
processes by which goals are set and solutions are structured
(Mathews 1987).

While decision theorists in anthropology treat households and
individuals as the crucial players in economic adaptation, group
decision processes have largely been ignored. Often a single
individual in the household is singled out as the decision maker;
between the farm and this lone farmer, the household gets squeezed
out (e.g. Gladwin and Zabawa 1987). The household is depicted as a
single entity, or the kinds of interactions between members are
subsumed under the term of 'generalized reciprocity,' assuming that
everything is shared equally in a primitive Communism, while the term
exchange is reserved for what goes on between households (Hardesty
1977:82)[3].

Consumer researchers have given close and detailed attention to a
particular kind of household decision-making -- purchasing behavior
-- in Euro-American households (O'Connor et al. 1983, Davis 1976).
They have been motivated by the desire to predict purchasing

behavior, especially in response to recent changes in American household composition and organization (see Roberts and Wortzel 1984). In the process they have generated a wealth of concepts and tools that can be used to study cross-cultural household patterns.

At its best, consumer research approaches ethnographic detail in discussing the ways that purchasing decisions are made within households. Topics include the different means of group decision making, including bargaining, consensus-making, and negotiation, and the variety of means which husbands and wives use (from threats to barter) to influence each other (Spiro 1983, Ferber and Lee 1974). The research asks what kinds of household decisions are made by a group ("syncratic" or consensual), and which kinds are made by husband or wife alone ("autocratic" or accommodative) (Ferber 1973, Bonfield et al. 1984, Green et al. 1983, Douglas-Tate et al. 1984). These sophisticated typologies of modes of decision making, conflict, and the different roles played within households often draw on game theory and cognitive studies of choice. A major research question is why some households are characterized by conflict and disagreement, while others handle decisions in a routine fashion, why some households are highly structured and others are informal (Reviewed in Bonfield et al. 1984).

Some of the more interesting studies propose that American households cope with disagreement by specialization, with husbands and wives each taking different roles at different points in the decision process, for example, initiator, shopper, gatekeeper, and information seeker (Sternthal and Craig 1982, Hempel 1974, O'Connor et al. 1983). Park suggests that some important decisions that appear to involve both husband and wife are not made jointly at all; instead the individuals pursue their own decisions, while attempting to minimize conflict, in a recursive, discontinuous process (1982). While husband and wife may end up thinking they made a decision together, they actually pursued a "disjointed, unstructured, and incremental strategy", the main goal of which was to "muddle through", keeping the marriage going whatever the actual decision reached (1982:52). Thus, household decisions are seen as inherently bivocal and recursive; they are simultaneously concerned with their ostensible object, and with maintaining or changing the household itself.

This school of research also has its problems. Though self-reports of decision making are unreliable and spouses rarely agree on what happened, most studies use self-reported survey responses instead of recording actual behavior. The resulting models depict household decision making as a linear process with a beginning and an end, rather than as parts of continuing social relationships.

A more fundamental problem stems from the intellectual roots of this research in social psychology and social exchange theory. The source of variation is always sought in measurable variables like age, occupational status, and income. Strategies of decision making are often simplistically classified as either husband-dominant, wife-dominant, or joint. Then exchange theory (Heer 1963, Spiro 1983) or resource theory (Blood and Wolfe 1960, Davis 1976, O'Connor 1963, Rodman 1972) is used to explain patterns of occurrence. The gist of both theories is that the balance of power and influence belongs to the spouse who brings the most economic and educational resources into the marriage (see Oppong 1970, Scanzoni 1972). These common sense notions (similar to those of the 'new home economists') contradict anthropological studies that find household roles and power firmly entrenched in wider cultural and social concepts like gender and age (e.g. Hartmann 1981, Curtis 1986, Leacock 1986). Those studies that recognize the cultural nature of household roles tend to reduce it to a single unidirectional variable, often called 'traditional role ideology' (Davis 1976, Green et al. 1983, Kenkel 1961).

The lack of a concept of culture or social structure cripples the consumer research approach to household decision making and economic allocation[4]. Lacking a comparative sample of households in other cultures, or a concept of corporate social structure as instituted process, each actor is seen as an isolated economic node, guided only by norms and values. The major mode of analysis is to classify the nodes into empirical 'types', which are then correlated with an arbitrary selection of economic and social constructs and measures, called 'variables' (Ferber 1973, Roberts and Wortzel 1984)[5].

To summarize, consumer research has elaborated a set of concepts and tools for understanding certain kinds of household decision-making in a narrowly defined cultural context. Anthropologists would find many of the goals of consumer research in studying American households to be rather trivial (for example, the question of who holds the balance of power in decisions to buy soap). Nevertheless, the descriptive tools and vocabulary they have developed can form the basis for cross-cultural studies and ethnological analysis of group decision-making, even if consumer researchers have not taken on this task themselves. Anthropologists need to venture more often and deeply into this literature, and should think about pursuing cross-cultural projects in cooperation with colleagues in consumer research and marketing, to take advantage of their expertise. I will move on now to examine what tools and concepts ecological anthropology has to offer.

The Ecology of Households

Ecological anthropology offers concepts and tools that can be used to study the internal operations of households and the ways they change. Most of these ideas come directly from biological ecology, but like most ecological anthropologists I will treat them as heuristic tools and analogies (Moran 1984a:xv).

The boundary problem mentioned above has been persistent in household studies. How can we treat households as corporate budget units when they are so interconnected and their boundaries are permeable? If instead we treat households as *systems* analogous to ecosystems, the problem of the discreteness of the household becomes less pressing, and even expectable. Ecosystems are not naturally bounded units either. We put arbitrary lines around sub-units for analytical purposes (Moran 1984b:19, 1982:9-11). Closure can never be assumed, though degrees of permeability can be defined. An analogy between ecosystems and household systems research suggests that we should place boundaries where we want during our analysis, as long as we remember the boundaries are arbitrary, and specify the flows across them (Golley 1984:44).

Ecology also offers tools for describing discrete systems and the ways they interrelate, without obscuring the dynamic and changeable nature of those relationships. A common method is formal modeling (Smith 1984:53, Moran 1982), using a variety of graphic and statistical methods to simplify and represent the systems under study. While biological ecology has tended to concentrate on models of energy flow, and some ecological anthropologists have also modeled the flow of energy within cultural systems (Ellen 1982:96-109), other kinds of human interactions can be modeled. A major goal has been to make the models less static, and more useful in depicting socially relevant flows and connections (for example Foster's use of network models [1984]). The rest of this paper will draw selectively on consumer research and ecological anthropology to model intra-household processes in several ways. Some of the symbols I use are loosely based on those of Odum (1971 also Moran 1982:82-90).

Figure 12.1 is a model of the flow of resources, and the major decision points in the Anglo-American household, as usually assumed in consumer research and household economics (e.g. Fitzsimmons and Williams 1973, Ferber 1974). Flows of cash and labor from household members enter a single general fund (using Odum's symbol for a 'storage' function). A proportion of this fund is drained off for necessities (determined by 'Engel's Law', which states that the proportion of income spent on necessities goes down as the total income goes up), and the leftover, the overflow, is apportioned to different uses by the household head (Prais and Houthaker 1971).

This model lies behind the research agenda of consumer research, and many problems stem from using it as a model for all households. The crucial node is the decision box (decision by head in Figure 12.1), and most consumer research is concerned with how this allocation function works. Most 'resource theory' in consumer research looks for linkages between who produces the inputs to the general fund, and who gets to allocate those resources (e.g. Davis 1976), or at the ways that household members interact in making those allocation decisions (e.g. Olshavsky and King 1984). Anthropologists and economists have often assumed that household budgets in other cultures also take the form of a general fund or pool. McGuire *et al.* (1986) see household change in Mexico as a product of changes in the kinds of contributions to the pool (i.e. cash vs. labor), as well as in the kinds of kin who are contributing to and drawing from the pool (men, women and children). In other words, they are not concerned with the shape and nature of the pool, only with the kinds of things going into it and who provides them.

This 'general fund' model of household budgets has been challenged by anthropologists working in Africa, who have found the notion of a single household fund or pool, and a single nexus for allocation decisions to be inaccurate and deceptive in understanding household change (see Moock 1986). Their objection has much wider applicability, because some degree of individual ownership and self-provisioning is found in every culture studied by anthropologists (e.g. children often procure food that is not contributed to the household pool). The single household fund is as difficult a concept as the idea of perfect altruism or complete self-interest.

More accurate models for the ways that household funds are differentiated can be found implicitly in the work of several anthropologists. Figure 12.2 is derived from Eric Wolf's (1966) proposal that the peasant household maintains a series of different funds, each obligated for a particular purpose. He implies that they are hierarchically arranged by order of priority, as I have shown here with the notion of overflow. Once the most basic fund is filled, the overflow goes to the next, and so on, leaving discretionary funds for the last. This model implies that for peasants the allocation of funds is largely pre-determined, and that the only place for the exercise of decision making power is with the small 'profit' fund left when the crop is in[6].

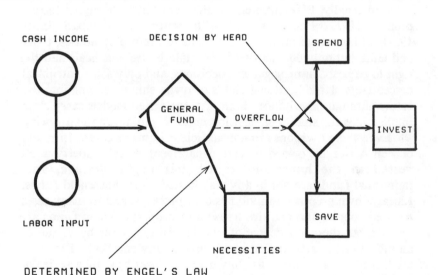

Figure 12.1. The Normative model of the American household, as usually assumed in the social sciences.

However, this model also implies that the rights to household funds are equally shared among all household members. Figure 12.3 shows two ways to model total household resources as a series of nested boxes. On the left we have a system of nested funds, as found in some African households (see McMillan 1986). A pool of land, money, food, and goods belongs to an entire multiple-family household, and then each nuclear family unit within the household has its own pool. Finally, each adult also controls personal property and funds. As Whitehead points out, inequality between the sexes may be manifested when women are expected to donate their earnings to the household pool, while men get to keep their earnings and spend them at their own discretion (1981).

But ownership and control of property are not absolutes. On the right is an adaptation of Goody's idea that large households are often internally differentiated by different kinds of rights to real property (1973). For example, in 17th century rural Japan (Smith 1959), all household members had rights to the use of houses, tools, and land, but only collaterals of the male household head had the right to be heard in management decisions, and only some patrilineal descendants of the head had rights of ownership extending into the future through inheritance. Both of these nested models carry clear implications for studies of decision making, ranking, and authority.

Even these divisions are not sufficient in many cases. In much of west Africa, the ownership and management of real property is not vested in the household at all, but in the *lineage*, and individual funds are not in any way nested inside household funds. Lineages own property, individuals own property, and in many cases married couples also acquire household property. These funds are distinct, but there is a degree of overlap and transfer by virtue of an individual's participation in all three at any one time. Lineages are linked to each other, as shown in the inset box, by a series of individual and household funds. Men contribute to a household fund of capital and permanent goods of value, while women do not (Abu 1983). Women put more into the household maintenance fund than their husbands. Both husband and wife eat out of this fund, which the wife controls. Obviously, modelling this situation is much more complicated when step-children and polygynous marriages are included, and if the contributions of grown children are added to the household budget.

In some societies (for example the Japanese Dozoku, urban Ashanti and Effutu fisherfolk in Ghana), there is no communal household property at all. With duolocal residence, husband and wife do not even share the use of a house. In these cases, conjugal economic relationships are best modeled as exchanges (as suggested in Clark 1986). Among the coastal Effutu in Ghana (Hagan 1983) males live in patrilineal men's' houses, while women live in matrilineal women's houses. Children live with the mother until about age ten, when boys go to live in their father's house. Husbands and wives have two different kinds of exchange relationship. During the fishing season the husband gives his share of the fish he catches (with his agnates) to his wife, who smokes and markets them. At the end of the season she gives him the cash value of the fish (discounted below open market price). This delayed exchange is counted as a debt if it's not

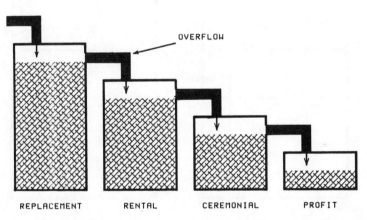

Figure 12.2. Household budgets as a series of funds (Wolf 1966). Overflow from each fund goes into the next lowest in priority.

paid on time (there are many other cultures in which husband and wife make loans to each other). The husband and wife also have a direct exchange relationship in which she provides one meal a day to him and his sons, in exchange for cash. He is also supposed to give her cash payments towards the feeding, schooling, and medical costs of his daughters and his younger children. The most important *productive* economic relations for both men and women are with the members of their lineages, but the marriage relationship forms a crucial link in the processing and marketing of a seasonal and variable resource. Divorce is frequent, and men and women often have single-stranded exchange relationships with partners outside the marriage.

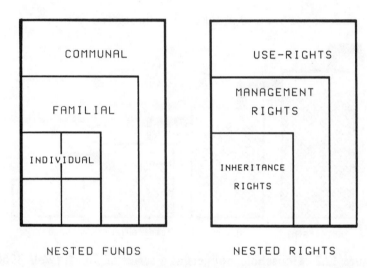

NESTED FUNDS   NESTED RIGHTS

Figure 12.3 Two ways to model household economies as a series of nested boxes. On the left, funds are held individually or pooled at the familial or household level. On the right, the household is differentiated into concentric spheres of rights to household resources.

It may appear that we are dealing here with two basic types of household; one like the Effutu which has no household pool or fund, and one like the Euro-American single-account household. But most of the world's households actually lie in between, with some communal or conjugal funds, and other funds that are individually managed. A combination of mapping funds *and* flows helps distinguish between various kinds of mixed-fund households.

Figure 12.4 shows a working-class English household. The husband splits his income, keeping part for his own personal discretionary fund, and giving the rest to his wife as a housekeeping fund for food and daily expenses. Part of his discretionary income goes for personal items like beer and tobacco, and another part goes for periodic household expenses like large bills, furniture, or educational costs. His personal expenditures are his own business, but the control of the household fund is negotiated between husband and wife. The wife's wage goes into the "housekeeping money" (which presumably pays the costs of substituting services like child care for her own labor). Children's wages go to the mother, who puts some in the housekeeping money, and returns some as allowance (Segalen 1986:266). This model is a much more accurate representation than Figure 1, and it shows that the issue of control is more complex than is often assumed; *there are both group and individual decisions.* The model also serves as a better basis for predicting what will happen if some of the flows change (see Whitehead 1981).

These maps of exchange relationships are especially useful in understanding households in parts of the world where long-term wage migration is common. In the rural Caribbean, for example, men may leave and send cash payments to a number of households where they have children, sisters, or parents, and women may leave their children with a caretaker and do the same (good examples are Rubenstein 1987 and Palacio 1982). Households that may look the same in terms of social *structure* (i.e. the number of people and their relationships to each other) may actually prove to have quite different economic structures, when mapped out in this way.

Conclusion: A Return to the Kekchi

My goal has been to show that households can be seen in other ways than just through lists of activities, members, inputs and outputs. In particular, we need to differentiate cash, labor, and material flows, and to define (and perhaps create symbols for) more kinds of funds and decision processes. It should be possible to distinguish the degree to which the household budget and household processes are *structured*. For example, Segalen (1986:269) finds that lower middle class French households have very organized budgets, while upper middle class households handle every expenditure through a separate decision process. She also points out that models must incorporate aspects of rhythm and timing if they are to be complete.

Another issue that needs further investigation is that of real vs. ideal behavior. Abbott (1976) found that there were consistent and patterned differences between cultural ideals and actual practice in

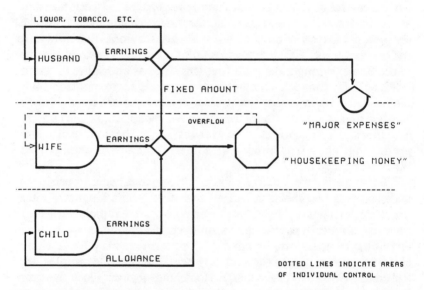

Figure 12.4 A model of funds in a working-class English household, as discussed by Segalen (1986). The dotted lines show the boundaries between areas of individual control.

the gender-based division of authority in Kikuyu households. Yet even her measures of 'real' were self-reported behavior. There are actually three levels of data that need to be taken into account therefore; the stated cultural ideal, the self-reported instance (how people perceive their own actions), and the actual behavior. We should not expect congruence between any of the three levels. The major challenge remains the elicitation and recording of actual decision-making behavior. Discourse analysis may be our most powerful tool in this task (e.g. Mathews 1987, Gaskins and Lucy 1986).

The ultimate goal of this attempt at a "household ecology" is not just a cross-cultural catalog of the various forms that households can assume. As Firth pointed out in his study of Tikopia, households adapt to changing circumstances by changing their internal arrangements, their structures of sharing and decision-making, to serve new purposes (see Lees and Bates 1984). Economic change leads to alterations in boundaries, in the economic bargains and balances between household members, in the allocation of labor and resources to different funds, and in the economic roles taken by different people (McKee 1986 and Jones 1986 for excellent examples). By adding this kind of analysis to existing studies of household decision making, households can be seen to do much more than passively adapt to changing environments.

In the Kekchi case, and perhaps in others (Hyden 1986, Moran 1981; Rudie 1971) the *form* of the household ecology actually restricts and channels the household's ability to adapt to new circumstances, and affects the kinds of strategic choices made by members. To return to my brief example at the beginning of the paper, I should ask once again, why some Kekchi households are able to pool resources and allocate them to productive investment, while other households do not.

One part of the problem is that money, labor, and food are not equivalent within the household economy. As money becomes more important in the household, it provides an uncontrolled and therefore objective standard of value that undercuts existing concepts of equivalence (Wilk 1989, Maher 1981). As the cultural definitions of costs and benefits change, the balances and bargains that underlie the household economy are also changing. Monetary values have certainly penetrated some households more quickly and deeply than others. A crucial event is often when sons who still live in the household take wage-earning jobs. Parents must then decide how to treat those earnings, and reach some accommodation with the son about them.[7] But even among the households where some members earn wages, there is a good deal of variation in how wages, income from crops sales, and subsistence production are pooled and managed. The two most common Kekchi household systems are shown in figures 12.5 and 12.6.

In Figure 12.5 we have a patriarchal system where the male household head controls a single central fund which includes most agricultural products, cash from selling crops, and the cash income earned by all members, including children. He allocates some of the agricultural products and some cash income to an operating fund which is managed by his wife. She transforms the agricultural products into food, and spends the cash on clothing and other consumable goods as needed. The labor she spends in raising pigs does not result in autonomous cash income, for the money from selling pigs goes into the central fund. Similarly, sons' cash earnings go into the central fund, where they are allocated by the male household head.

Decision-making about expenditures from the central fund in these households tends to be authoritarian, and wives and children often resort to overt bargaining to get cash for their own personal use ("I will provide meals so you can participate in the ritual dance, if you will give me money for the childrens' schoolbooks".) The man has a general obligation to meet his family's needs, but these needs are vague and nonspecifically defined, and are often themselves the subject of dispute. Because women and children lose control over the products of their labor when they contribute to the central fund, both try to divert some of their production into their own individual fund; sons keep some of their earnings for themselves, and women sell small amounts of eggs, forest and craft products, and food in order to get their own cash.

In Figure 12.6, the central fund and the operating fund have been merged, and management is not patriarchally controlled by the male household head. All household members contribute to the single fund, and share rights to its use. Each decision is a matter of joint management, conceived as a group decision over group resources, rather than as a process of bargaining between individuals over funds controlled individually. While the operating fund could be analytically separated, because it is not emically separate, goods and money can flow back and forth between them on an *ad hoc* basis. School supplies, for example, are paid from the household general fund, rather than from the wife's operating fund. Paying for school supplies is therefore a household decision, rather than an occasion for the wife to ask the husband for money[8].

In households with budget structures like figure 12.6, all household members have a say in the management of the central fund, but this greater participation is balanced by less autonomy. The male household head is less autonomous in controlling household resources, and women and children do not have their own individual funds to manage as they want.

In the short run, the patriarchal household budget is capable of motivating and concentrating its resources for particular goals, like buying young hogs to feed, or box hives for keeping bees. But these projects run into problems, because while they require the help and labor of all household members, they are not household projects, but

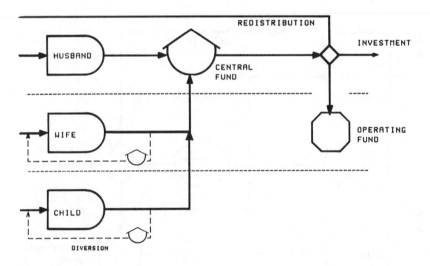

Figure 12.5 Kekchi households with central funds under the control of the male household head.

those of the patriarch. The patriarch must now convince, cajole,
threaten, or bargain to get his wife and children to tend the hogs,
separate the honey, or process the cocoa. What is their incentive?
Only the promise that some of the increased income will improve their
standard of living, but they have only vague 'moral' claims on this
fund. Often the new income just goes back into the business -- his
business. The result is resistance; everyone is reluctant to
contribute their labor, their products, or their money to the
household pool. In the long term, wives and daughters cut back their
labor in household production, and hoard whatever resources they can
keep for themselves. Sons work on their own fields if they can, and
they keep aside their profits or their wages. Each person tries to

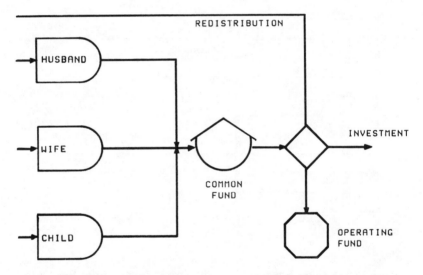

Figure 12.6 Kekchi households with a central fund that is treated as
a common fund under the joint control of all household members.

conceal at least a portion of their income. As soon as sons have a chance, they leave the household to set up their own farm, so they will have some control over the products of their labor. As a business, the household fails in the long run because it provides no incentives to its members.

In the long run the most successful households have adopted the second form of budgeting. In these households there is more overt discussion of goals and expenditures. The household may be slow to take new opportunities or expand its enterprises in response to changes in the market or the labor supply. But when the members of the household agree to pursue a strategy, from buying a horse to transport crops, to planting several acres of cocoa, or opening a small village shop, they act together. The consequence is that each household member is more willing to contribute extra work time, extra effort, extra attention, to the project.

Many Kekchi men begin projects with income-earning in mind, but later abandon them because they did not have time to manage them during the dry season when they have to devote full time to subsistence farming. They were unable to get their wives and children committed to the project. Managing both income-generating projects and subsistence production, in this environment, requires a great deal of coordination and informal task-sharing that must be based on mutual interest. Households that organize their budgets along the centralized lines depicted in Figure 12.6 have an easier time promoting this mutual interest. Household ecology is therefore a crucial variable in the economic performance of households. Articulating two economies, subsistence and cash, is a difficult and complex management task that places very specific demands on the household budgeting process.

My analysis of Kekchi household budgeting turns some of Sahlins' ideas about the dynamics of the domestic mode of production around. Sahlins argues that generalized reciprocity within and between households -- pooling -- reduces the incentive to produce (1973:84-88,94). Households set their production targets in relation to other households, below their optimum capacity. The mechanical solidarity of the household encompasses a simple gender-based division of labor that is linked to low work intensity. Sahlins implies that relations of authority and inequality, of coercion, extraction and competition, provide the incentive for increased production. The Kekchi case shows that in specific circumstances, more equal pooling of household labor and resources may increase production, and authoritarian relationships may lead to a long-term decline.

This mode of analysis points out the need to go beyond the study of personality or of the composition of households in understanding economic and social change.It may be that a great deal of household-centered research has focused on the wrong variables. Perhaps studies of household typology and change would be more productive if they shifted from descriptions of kinship composition, and household size or developmental cycles, to patterns of sharing, exchange and decision making.

Acknowledgments

The paper depends heavily on insights accumulated while working with Robert Netting on other papers, and on his comments on an earlier draft. Ben Orlove, Henry Rutz, and Peggy Barlett have also made extremely helpful comments. Please do not place any burden of responsibility for the ideas and arguments presented here on any of those who have helped.

Notes

[1] Much of the recent literature on the role of households in capitalist penetration has little historical depth, counterpoising the timeless encapsulated self-sufficient household of pre-capitalist times with the households of a fully-monetized proletariat or marginal peasant society (eg. Wallerstein 1984). My historical study of the Kekchi economy found, on the contrary, that their households had been mixing various forms of subsistence farming, cash crop production, and wage labor for at least 400 years (Wilk 1986). Yet Kekchi villages still seem traditional and unacculturated, and each time a new wave of peripheral-capitalist development washes over them, it appears to be the first time they have emerged from the primeval forest.

[2] I have appropriated Bourdieu's (1977) term here with some understanding of its implications. I find his concept fruitful in understanding how seemingly corporate and structured aspects of households exist without being normatively determined.

[3] The issue has some theoretical weight since it is parallel to the controversy about group selection in evolutionary ecology. What is the unit of human adaptation (see Lees and Bates

1984)? Some researchers feel that treating the household (or any group) as an adaptive unit raises the issue of altruism. In rejecting altruism or group selection, they assume that groups themselves do not adapt, except through the cumulative action of individuals or the manipulative actions of powerful individuals (Ellen 1982:246). In this way the issue of what goes on inside households, the most fundamental human social group, questions and threatens some of the core concepts of ecological and economic anthropology, and poses difficult problems about the origin, structure, and function of all social groups.

[4] It is clear that people working on household decision making are aware that something is lacking in their analysis. This is reflected in the somewhat confused response to understanding changes in household decision making that are occurring because of increasing female participation in the workforce (Hill and Klein 1973, O'Connor 1983, Douglas-Tate et al. 1984, Rodman 1982).]

[5] Without serious cross-cultural experience or a concept of culture, consumer researchers have not even asked the question of whether ideology and customary rules, or economic rationality and pragmatic practice are the primary cause of household and family behavior, an issue that anthropologists are deeply concerned with (Yanagisako 1984, Bourdieu 1977). The cross-cultural research done by consumer researchers seems concerned mainly with finding differences in the relative power of husbands and wives; researchers assume that the variables established in North American studies are cross-culturally valid. Sometimes the typology of cultures used to 'explain' differences in decision making is dangerously simplistic and naive (e.g. Green et al. 1983).

[6] This model, as well as the previous one, assumes that there is a sharp dividing line between luxuries and necessities, when in fact the definition of what is necessary is largely itself a cultural matter.

[7] The disposition of cash earned from selling crops is much less problematical for Kekchi households. Crop surplus falls within the traditional definition of the male economic sphere, and male household heads retain rights to them whether they are sold or not.

[8] Both of these budget arrangements are 'traditional.' Neither is a recent product of capitalist penetration, for the Kekchi have been producing for the marketplace for hundreds of years. Instead the two forms of budgeting are perceived as two forms of

marriage, resulting from differences in personality, relative ages of spouses, the different political positions and ambitions of men and women, and such accidents as birth order.

References Cited

Abbott, S.
  1976   Full-Time Farmers and Week-End Wives: An Analysis of Altering Conjugal Roles. *Journal of Marriage and the Family* 38:165-174.
Abu, K.
  1983   The Separateness of Spouses: Conjugal Resources in an Ashanti Town. In *Female and Male in West Africa*. Edited by C. Oppong. pp. 156-168. London: George Allen and Unwin.
Adams, Jane
  1988   The Decoupling of Farm and Household: Differential Consequences of Capitalist Development on Southern Illinois and Third World Family Farms. *Comparative Studies in Society and History* 30:453-482.
Anderson, Michael
  1971   *Family Structure in Nineteenth Century Lancashire*. Cambridge: Cambridge University Press.
Barlett, Peggy
  1980   Introduction: Development Issues and Economic Anthropology. In *Agricultural Decision Making*. Edited by P. Barlett. New York: Academic Press. pp. 1-18.
  1982   *Agricultural Choice and Change*. New Brunswick N.J: Rutgers University Press.
Barth, F.
  1967   On the Study of Social Change. *American Anthropologist* 69:661-669.
Becker, Gary S.
  1981   *A Treatise on the Family*. Cambridge, Mass: Harvard University Press.
Blau, F. and M. Ferber
  1986   *The Economics of Women, Men, and Work*. Englewood Cliffs: Prentice-Hall.
Bloch, Maurice
  1973   The Long Term and the Short Term: The Economic and

Political Significance of the Morality of Kinship. In *The Character of Kinship*. Edited by J. Goody. Cambridge: Cambridge University Press. pp. 75-87.

Blood, R.O. Jr. and D.M. Wolfe
1960 *Husbands and Wives*. The Free Press: New York.

Bonfield, E., Kaufman, C., and S. Hernandez
1984 Household Decisionmaking: Units of Analysis and Decision Processes. In *Marketing to the Changing Household*. Edited by M. Roberts and L. Wortzel. Cambridge, Mass: Ballinger. pp. 231-263.

Bould, Sally
1982 Women and the Family: Theoretical Perspectives on Development Working Paper 13, Women in International Development Program, Michigan State University.

Bourdieu, Pierre
1977 *Outline of a Theory of Practice*. Cambridge: Cambridge University Press.

Caldwell, John
1981 *The Theory of Fertility Decline*. Homewood, Illinois: Irwin Publishers.

Cancian, Frank
1972 *Change and Uncertainty in a Peasant Economy:The Maya Corn Farmers of Zinacantan*. Stanford: Stanford University Press.

Carter, A.
1984 Household Histories. In *Households:Comparative and Historical Studies of the Domestic Group*. Edited by R. Netting, R. Wilk and E. Arnould. Berkeley: University of California Press.

Clark, Gracia
1985 Domestic Work and Trading: Pressures on Asante Wives and Mothers. Paper presented at the annual meeting of the Society for Economic Anthropology, Urbana-Champaign.

Collins, Jane
1988 The Household and Relations of Production in Southern Peru. *Comparative Studies in Society and History* 28(4):651-671.

Creighton, Colin
1980 Family, Property and Relations of Production in Western Europe. *Economy and Society* 9:129-164.

Curtis, Richard
   1986   Household and Family in Theory on Inequality.
             American Sociological Review. 51:168-183.
Davis, Harry
   1976   Decision Making within the Household. *Journal of
             Consumer Research* 2:241-260.
Douglas-Tate, M., J. Peyton, and E. Bowen
   1984   Sharing of Household Maintenance Tasks in
             Married-Couple Households. In *Marketing to the
             Changing Household.* Edited by M. Roberts and L.
             Wortzel. Cambridge, Mass: Ballinger. pp. 205-216.
Ellen, Roy
   1982   *Environment,   Subsistence,   and   System.*
             Cambridge: Cambridge University Press.
Ferber, Robert and Lucy Lee
   1974   Husband-Wife Influence in Family Purchasing Behavior.
             Journal of Consumer Research. 1:43-50.
Ferber, Robert
   1973   Family Decision Making and Economic Behavior: A
             Review. In *Family Economic Behavior: Problems and
             Prospects.* Edited by E. Sheldon, pp. 29-64.
             Philadelphia: Lippincott.
Fitzsimmons, C. and Williams, F.
   1973   *The Family Economy: Nature and Management of
             Resources.* Ann Arbor: Edwards Brothers.
Folbre, Nancy
   1984   Cleaning House: New Perspectives on Households and
             Economic Development. Paper Presented at the XII
             International Congress of the Latin American Studies
             Association Albuquerque, New Mexico.
Foster, B. L.
   1984   Family Structure and the Generation of Thai Social
             Exchange Networks. In *Households: Comparative and
             Historical Studies of the Domestic Group.* Edited
             by R. Netting, R. Wilk and E. Arnould, pp. 84-107.
             Berkeley: University of California Press.
Gaskins, S. and J. Lucy
   1986   Passing the Buck: Responsibility and Blame in the
             Yucatec Maya Household. Paper Presented at the meetings
             of the American Anthropological Association,
             Philadelphia.

Gladwin, C. and R. Zabawa
 1987 Transformation of Full-Time Family Farms in the U.S.:
 Can They Survive? In *Household Economies and Their
 Transformations.* Edited by M. Maclachlan. Lanham:
 University Press of America, pp. 212-227.

Golley, Frank
 1984 Historical Origins of the Ecosystem Concept in Biology.
 In *The Ecosystem Concept in Anthropology.*
 Edited by Emilio Moran, pp. 32-50. Boulder: Westview
 Press.

Goodenough, W.
 1955 A Problem in Malayo-Polynesian Social Organization.
 American Anthropologist 57:71-83.

Goody, J.
 1973 Strategies of Heirship. *Comparative Studies in
 Society and History* 15:3-20.

Green, R., J. Leonardi, J. Chandon, I. Cunningham, B. Verhage, and A.
Strazzieri
 1983 Societal Development and Family Purchasing Roles: A
 Cross-National Study. *Journal of Consumer
 Research* 9:436-442.

Guyer, Jane I.
 1981 Household and Community in African Studies. *African
 Studies Review* 24:87-137.
 1986 Intra-Household Processes and Farming Systems Research:
 Perspectives from Anthropology. In *Understanding
 Africa's Rural Households and Farming Systems.*
 Edited by J. Moock, pp. 92-105, Boulder: Westview.

Hagan, G.
 1983 Marriage, Divorce and Polygyny in Winneba. In
 Female and Male in West Africa. Edited by C.
 Oppong, pp. 192-203. London: George Allen and Unwin.

Hardesty, D.
 1977 *Ecological Anthropology.* New York: John Wiley
 and Sons.

Hartmann, Heidi
 1981 The Family as a Locus of Gender, Class, and Political
 Struggle: The Example of Housework. *Signs*
 6:366-394.

Hareven, Tamara
 1982 *Family Time and Industrial Time.* Cambridge:
 Cambridge University Press.

Heer, David
    1963   The Measurement and Bases of Family Power: An Overview.
            Marriage and Family Living 25:133-139
Hempel, D.J.
    1974   Family Buying Decisions: A Cross Cultural Perspective.
            Journal of Marketing Research 11:295-302.
Hill, R. and D. Klein
    1973   Towards a Research Agenda and Theoretical Synthesis.
            In *Family Economic Behavior: Problems and
            Prospects.* Edited by E. Sheldon, pp. 371-404.
            Philadelphia: Lippincott.
Hyden, Goran
    1986   The Invisible Economy of Smallholder Agriculture in
            Africa. In *Understanding Africa's Rural Households
            and Farming Systems.* Edited by Joyce Moock, pp.
            11-35. Boulder: Westview Press.
Jones, C. W.
    1986   Intra-Household Bargaining in Response to the
            Introduction of New Crops: A Case Study from North
            Cameroon.   In *Understanding Africa's Rural
            Households and Farming Systems.* Edited by J. Moock,
            pp. 105-123. Boulder: Westview Press.
Kenkel, W. F.
            Husband-Wife Interaction in Decision-Making and
            Decision Choices. *The Journal of Social
            Psychology* 54:255-262.
Korn, S. R. D.
    1975   Household Composition in the Tonga Islands: A Question
            of Options and Alternatives. *Journal of
            Anthropological Research* 31(3):235-260.
Laslett, Peter
    1984   The Family as a Knot of Individual Interests. In
            *Households: Comparative and Historical Studies of
            the Domestic Group.* edited by R. Netting, R. Wilk
            and E. Arnould, pp. 353-382. Berkeley: University of
            California Press.
Leacock, E.
    1986   Postscript: Implications for Organization. In
            *Women's Work: Development and the Division of Labor
            by Gender.* Edited by E. Leacock and H. Safa, pp.
            253-265. Massachusetts: Bergin and Garvey.

Lees, Susan and Daniel Bates
1984   Environmental Events and the Ecology of Cumulative Change. In *The Ecosystem Concept in Anthropology.* Edited by Emilio Moran, pp. 133-159. Boulder: Westview Press.

Lehman, David
1986   Two Paths of Agrarian Capitalism, or a Critique of Chayanovian Marxism. *Comparative Studies in Society and History* 28(4):601-627.

Lofgren, Orvar
1984   Family and Household: Images and Realities: Cultural Change in Swedish Society. In *Households: Comparative and Historical Studies of the Domestic Group.* Edited by R. Netting, R. Wilk and E. Arnould. Berkeley: University of California Press.

Machlachlan, Morgan
1983   *Why They Did Not Starve.* Philadelphia: ISHI.

Maher, V.
1981   Work, Consumption and Authority within the Household: A Moroccan Case. In *Of Marriage and Market: Women's Subordination in International Perspective.* Edited by K. Young, C. Wolkowitz and R. McCullagh, pp. 69-87. London: CSE Books.

Mathews, H. F.
1987   Predicting Decision Outcomes: Have We Put the Cart before the Horse in Anthropological Studies of Decision-Making? *Human Organization* 46(1):54-61.

McGuire, R., J. Smith, and W. Martin
1986   Patterns of Household Structures and the World Economy. *Review.* 10(1):75-97.

McKee, K.
1986   Household Analysis as an Aid to Farming Systems Research: Methodological Issues. In *Understanding Africa's Rural Households and Farming Systems.* Edited by J. Moock, pp. 188-198. Boulder: Westview.

McMillan, Della
1986   Distribution of Resources and Products in Mossi Households. In *Food in Sub-Saharan Africa.* Edited by A. Hanson and D. McMillan. Boulder: Lynne Rienner.

Meillassoux, Claude
    1981   *Maidens, Meal and Money.* Cambridge: Cambridge
            University Press.
Minge, Wanda
    1986   The Industrial Revolution and the European Family:
            "Childhood" as a Market for Family Labor. In
            *Women's Work: Development and the Division of Labor
            by Gender.* Edited by E. Leacock and H. Safa, pp.
            13-24. Massachusetts: Bergin and Garvey.
Moock, Joyce
    1986   Introduction. In *Understanding Africa's Rural
            Households and Farming Systems.* Edited by J. Moock,
            pp. 1-10, Boulder: Westview Press.
Moran, Emilio
    1984a   Preface.   In   *The   Ecosystem   Concept   in
            Anthropology.* Edited by Emilio Moran, pp. xiii-xvi.
            Boulder: Westview Press.
    1984b   Limitations and Advances in Ecosystems Research. In
            The Ecosystem Concept in Anthropology. Edited
            by Emilio Moran, pp. 3-32. Boulder: Westview Press.
    1982   *Human Adaptability: An Introduction to Ecological
            Anthropology.* Boulder: Westview Press.
    1981   *Developing the Amazon.* Bloomington:Indiana
            University Press.
Nardi, Bonnie
    1983   Goals in Reproductive Decision Making. *American
            Ethnologist.* 10(4):697-714.
Odum, H. T.
    1971   *Environment, Power, and Society.* New York:
            Wiley-Interscience.
O'Connor, P. J., G. L. Sullivan, D.A. Pogorzelski
    1983   Cross Cultural Family Purchasing Decisions: A
            Literature Review. *Advances in Consumer
            Research* 10:59-64.
Olshavsky, R. and M. King
    1984   The Role of Children in Household Decisionmaking:
            Application of a New Taxonomy of Family Role Structure.
            In *Marketing to the Changing Household.* Edited
            by M. Roberts and L. Wortzel, pp. 41-52. Cambridge,
            Mass: Ballinger.

Oppong, C.
   1970   Conjugal Power and Resources: An Urban African Example. *Journal of Marriage and the Family* 32:676-680.

Orlove, Benjamin
   1980   Ecological Anthropology. *Annual Review of Anthropology* 9:235-273.

Painter, Michael
   1984   Changing Relations of Production and Rural Underdevelopment. *Journal of Anthropological Research* 40:271-292.

Palacio, J.
   1982   *Food and Social Relations in a Garifuna Village.* PhD. Dissertation, University Microfilms: Ann Arbor.

Park, C.
   1982   Joint Decisions in Home Purchasing: A Muddling-Through Process. *Journal of Consumer Research* 9:151-162.

Peters, Pauline
   1986   Household Management in Botswana: Cattle, Crops, and Wage Labor. In *Understanding Africa's Rural Households and Farming Systems.* Edited by J. Moock, pp. 133-154. Boulder: Westview.

Prais, S. J. and H. S. Houthakker
   1971   *The Analysis of Family Budgets.* Cambridge: Cambridge University Press.

Roberts, Mary and Lawrence Wortzel (eds.)
   1984   *Marketing to the Changing Household.* Cambridge, Mass: Ballinger.

Rodman, H.
   1972   Marital Power and the Theory of Resources in a Cross Cultural Context. *Journal of Comparative Family Studies* 1:50-67.

Ross, Harold
   1973   *Baegu: Social and Ecological Organization in Malaita, Solomon Islands.* Urbana: University of Illinois Press.

Rubenstein, H.
   1987   *Coping with Poverty: Adaptive Strategies in a Caribbean Village.* Boulder: Westview.

Rudie, Ingrid
    1970   Household Organization: Adaptive Process and
            Restrictive Form: A Viewpoint on Economic Change.
            Folk 12:185-200.
Rutz, Henry
    1989   Fijian Household Practices and the Reproduction of
            Class. *The Household Economy: Reconsidering the
            Domestic Mode of Production.* Edited by R. Wilk.
            Boulder: Westview Press.
Sahlins, M.
    1972   *Stone Age Economics.* Aldine: Chicago.
Scanzoni, J.
    1972   *Sexual Bargaining: Power Politics in the American
            Marriage.* Chicago: University of Chicago Press.
Schmink, M.
    1984   Household Economic Strategies: Review and Research
            Agenda. *Latin American Research Review*
            19(3):87-102.
Schultz, T.W.
    1981   *Economics of Population.* Reading, Mass:
            Addison-Wesley.
Segalen, Martine
    1986   *Historical Anthropology of the Family.*
            Cambridge: University of Cambridge Press.
Sirgy, M. J.
    1984   *Marketing as Social Behavior.* New York:
            Praeger.
Smith E.
    1984   Anthropology, Evolutionary Ecology, and the Explanatory
            Limits of the Ecosystem Concept. *The Ecosystem
            Concept in Anthropology.* Edited by Emilio Moran,
            pp. 51-86. Boulder: Westview Press.
Smith, Joan., I. Wallerstein and H. Evers eds.
    1984   *Households and the World Economy.* Beverley
            Hills: Sage Publications. pp.23-36.
Smith, Joan, I. Wallerstein and H. Evers
    1984   Introduction. In *Households and the World
            Economy.* Edited by J. Smith, I. Wallerstein and H.
            Evers. Beverley Hills: Sage Publications. pp.7-13.
Smith, T.C.
    1959   *The Agrarian Origins of Modern Japan.* Stanford:
            Stanford University Press.

Spiro, Rosann
    1983   Persuasion in Family Decision-Making. *Journal of Consumer Research.* 9:393-402.
Sternthal, B. and C. Craig
    1982   *Consumer Behavior: An Information-Processing Perspective.* Englewood Cliffs: Prentice-Hall.
Van Esterik, P.
    1985   Intra-Family Food Distribution: Its Relevance for Maternal and Child Nutrition. In *Determinants of Young Child Feeding and their Implications for Nutritional Surveillance.* Ithaca, NY: Cornell International Nutrition Monograph Series, Number 14. pp. 74-149.
Wallerstein, Immanuel
    1984   Household Structures and Labor-Force Formation in the Capitalist World-Economy. In *Households and the World Economy*, edited by J. Smith, I. Wallerstein and H. Evers, pp. 17-22. Beverley Hills: Sage.
Whitehead, Ann
    1981   'I'm Hungry, Mum': The Politics of Domestic Budgeting. In *Of Marriage and Market: Women's Subordination in International Perspective.* Edited by K. Young, C. Wolkowitz and R. McCullagh, pp. 88-111. London: CSE Books.
Wilk, Richard
    1981   *Agriculture, Ecology And Domestic Organization Among The Kekchi Maya.* Ph.D. Dissertation, University Microfilms, Ann Arbor.
    1983   Little House in the Jungle. *Journal of Anthropological Archaeology.* 2(2):99-116.
    1984   Households in Process: Agricultural Change and Domestic Transformation among the Kekchi Maya of Belize. In *Households: Comparative and Historical Studies of the Domestic Group.* Edited by Netting, Wilk and Arnould. Berkeley: University of California Press.
    1987   The Search for Tradition in Southern Belize: A Personal Narrative. *America Indigena* 47(2):77-95.
    1989   Houses as Consumer Goods: The Kekchi of Belize. In *The Political Economy of Consumption.* Edited by Ben Orlove and Henry Rutz. Latham, MD: University Press of America.

Wilk, Richard R. and Robert M. Netting
    1984   Households: Changing Form and Function. In
            *Households: Comparative and Historical Studies of
            the Domestic Group*, edited by R. Netting, R. Wilk
            and E. Arnould,, pp. 1-28. Berkeley: University of
            California Press.
Wolf, Eric
    1966   *Peasants*. Englewood Cliffs: Prentice Hall.
Woodford-Berger, Jane
    1981   Women in Houses: The Organization of Residence and Work
            in   Rural   Ghana.   *Antropologiska   Studier*
            30-31:3-35.
Yanagisako, Sylvia
    1984   Explicating Residence: A Cultural Analysis of Changing
            Households   Among   Japanese-Americans.   In
            *Households: Comparative and Historical Studies of
            the Domestic Group*, edited by R. Netting, R. Wilk
            and E. Arnould, pp. 330-352. Berkeley: University of
            California Press.
Young, K., C. Wolkowitz and R. McCullagh
    1981   *Of Marriage and Market: Women's Subordination in
            International Perspective*. London: CSE Books.

CHAPTER 13

1990 AND BEYOND: SATELLITE REMOTE SENSING AND ECOLOGICAL ANTHROPOLOGY

Francis Paine Conant

I. Introduction

Important changes — some of them good — have taken place in satellite remote sensing since 1984. The present review touches on the good news and the bad, and develops an argument on the relevance of scale and resolution that was understated in my earlier article (Conant 1984). Remote sensing as a tool for discovering unsuspected human and environmental interactions — the focus of the original article — gains additional force first by being flexible with regard to resolution levels, and second by georegistering the satellite data with other sources of information at a resolution level best suited to the investigation at hand. Finer resolution is not necessarily better.

A. The Good News and the Bad

The positive changes in remote sensing include the refinement and wider use of geographic information system (GIS) programs which greatly enhance the interpretation of satellite data by integrating multiple sources of information. The satellite technology itself has evolved so that finer spatial, spectral, and temporal resolution is available. The microcomputer hardware and software for image construction from the satellite data now exists in considerable variety. And perhaps most important, there is a (slowly) growing population of remote sensing users in anthropology, with significant discoveries continuing to be made in ecological anthropology and archaeology.

But there is a down side as well. The bad news is that there has been a 10-fold increase in the costs of satellite digital data and image products. Furthermore, archiving the early Landsat data has been seriously reduced by budget cuts made by recent administrations in Washington. Satellite data of important historical value has been, and continues to be, erased. And while the budget is now somewhat more secure, for a time in 1989 funding for Landsat operations was almost reduced to day-by-day operations.

In the decade of the 1990's it seems certain that policies will continue to be set which have as their goal commercialization or "privatization" of government services. Landsat and SPOT data pricing policies are examined in the February, 1989, issue (v. 10,no. 2) of the *International Journal of Remote Sensing.* One possibility is that Landsat data may be sold on a subscription basis (Williams 1989). While this might be feasible for very large users of the data, such as the US Department of Agriculture or the US Agency for International Development, the individual researcher, often academically based, will require some sort of subsidy in order to acquire the data.

B. A Look Back: Aerial Photography and Ecological Anthropology

While the earliest use of aerial photography in anthropology is put at 1907, when Stonehenge was photographed from the air (Ebert and Lyons 1983), it was not until after World War I that, as a result of the extensive use of aerial photography for intelligence purposes, applications were extended to ethnology and archaeology (see Vogt 1974; Schorr 1974). According to Schorr another relevant factor was:

the information requirements of expanding colonial empire administrations. It turned out that periodic aerial photographic reconnaissance, while satisfying the need for strategic military information, could also yield a wide variety of essential demographic, social, cultural and resource data with great efficiency and reliability over large territories. (p.164)

Schorr credits Marcel Griaule in 1937 as perhaps the first to use aerial photography in African ethnographic research. In what may have been the first use of aerial photography in American ethnography (for purposes other than simple illustration), Walter Goldschmidt used air photos and "a strong magnifying glass" in 1943 to predict the future pattern of rural settlement in irrigated and non-irrigated counties of Idaho, Oregon, and Washington (Goldschmidt 1943 and Goldschmidt 1983, personal communication).

The first use of Landsat data in anthropology was by Priscilla Reining (1973) and was done on General Electric's Image 100, a million dollar machine, with a GE engineer, Dwight Egbert, doing the actual classification. The task was to see if it was possible to identify individual Mali villages in the satellite data for West Africa. It was — but this could only be confirmed by fieldwork (what NASA calls "ground truthing") and by manually aligning existing maps (themselves based in part on aerial photography) with the Landsat imagery. In the present era of satellite remote sensing, aerial photography is by no means dead. As discussed below, air photos form an intermediate information plane between ground level observations and those recorded aboard the spacecraft. This synchronous use of aerial photography can be critical in the interpretation of the satellite data. Aerial photography also can help extend the time frame for perceiving ecological changes: in Kenya, for example, some aerial photographs from the 1940's and more extensive aerial surveys in the 1950's provide important historical extensions to the 1970's start of satellite remote sensing. And, of course, even a hand-held camera shooting obliquely from a light aircraft can provide immensely important information. But since it is still relatively new and often not well understood, the emphasis here will be on satellite remote sensing. Certain strictures apply, however.

C. Satellite Remote Sensing and the Public Domain

In what follows, the focus is on satellite remote sensing for ecological studies of living populations, especially those of the Third World. Only brief mention will be made of remote sensing in archaeology which has become a successful and relatively specialized field of its own (see, for example, Peregrine 1988, Sheets and Seever 1988, Ebert and Lyons 1983). This is so perhaps because remote sensing in archaeology is directed more at the revelation of previously unsuspected traces of past human settlements than with the discovery of processes relating to on-going, and changing human settlements.

The emphasis here also will be on what is called 'passive' remote sensing, which depends on detecting energy reflected from earth's surface. This is in contrast to 'active' remote sensing involving the transmission of radar-like waves and then analyzing the reflected signals. Finally, the emphasis is on the use of remote sensing data which are (a) repeatable and (b) in the public domain. With regard to repeatability polar orbiting satellites make it possible to accumulate data for different seasons as well as different years. While interesting data often result from episodic remote sensing, as from the space shuttles, they may well be impossible to replicate or verify since one space flight is unlikely to exactly repeat the orbit or the timing of a previous mission. By contrast satellite polar orbits are at regular intervals (16 days for Landsats 4 & 5, and every 5, 10, 15, 20, and 25 days, according to the look angle of the instruments aboard, for SPOT). Additionally, for there to be any chance of independent verification the satellite data must be available to *all* researchers, that is it must be in the public domain nationally and internationally. The studies described below are based on repeatable, public domain data.

Section (II) below considers some of the advantages and limitations of satellite remote sensing in human ecology studies. Section (III) is on some of the basic features of this new and hybrid science; (IV) is on the importance of geographic information systems (GIS) in the interpretation of remote sensing data; (V) presents briefly some recent studies which seem particularly significant in ecological anthropology, and (VI) is a brief discussion and conclusion on the future of remote sensing in human ecology.

II. Why remote sensing?

For somewhat different reasons, Arensberg (1981) and Bohannan (1980) have both called for a better integration of anthropology with other sciences. Bohannan sees anthropology as having fallen out of an earlier "econiche" (where it had been at work productively between the pressures of racism and colonialism and those of liberalism) into a cranny "between sociobiology on the one hand and the policy sciences on the other" (1980:519), and he argues for more integration, not less, if anthropology is to survive as an effective discipline. Remote sensing is of particular interest in the context of Bohannan's remarks. In terms of "levels" of investigation, remote sensing is more "elemental" or "basic" than either biology or sociobiology *and* it has become an important tool in the "policy sciences." Remote sensing is a hybrid field, based on the optical, physical, and computer sciences. It is equally capable of investigating non-living phenomena (atmospherics, for example) as well as organic processes such as plant phenology, estimating above-ground biomass, and the distribution and effect of human activities on the local surroundings. From the outset remote sensing dovetailed neatly with the geographical and geological sciences, in part because of the longstanding involvement of these fields with the earlier form of remote sensing, aerial photography, and cartography. As satellite remote sensing moves from the level of experimentation to that of application, it *and* the disciplines associated with it are becoming central to making policy — estimating crop yields, for example, or locating potential resources, as well as in projecting the results of massive development programs such as those involving the Senegal and Niger rivers of West Africa. Monitoring drought and desertification, giving early warning of famine in the African *Sahel* all have immense relevance for policy making. Geographers so far have been the ones mainly involved, not anthropologists despite their knowledge of indigenous subsistence practices and their fieldwork orientation that could contribute so much to "ground truthing". In urging new directions on our discipline, Bohannan (1980) warns that anthropology cannot afford to be out of touch with its times. And these times very much include remote sensing as a critical tool in ecological research and policy-making.

Arensberg argues that redefining domains and methodologies are critical steps in the histories of many sciences: "they have learned to model more and more complex organization linking elementary parts and greater wholes" (1981: 566). In many ways this is what has happened in the field of remote sensing: it started with photons, and now has, broadly, a capability of looking at cultural events and processes. In citing the use of Landsat satellite data in studying the Pokot of East Africa, Arensberg notes that what emerges is the centrality of the division of labor between women and men in effecting change in the local environment. Arensberg argues the need for a holistic approach in anthropology. This means in part avoiding single factor explanations (crop determinism, or the protein search, for example). It also means looking for complex interactions, as in the division of labor, to understand something as intricate as human ecology. Finally, holism also means integration of the sciences, "ours and the others'" (578). Thus Arensberg, like Bohannan, urges a wider horizon for anthropology. In this regard, at least our discipline got off to a good and early start in satellite remote sensing.

The initial Earth resources orbiting satellite, Landsat 1, was launched in 1972. Priscilla Reining (1973) was the first anthropologist to make use of the data, and in fact it was her study and enthusiasm which led to my own involvement in remote sensing. The National Science Foundation and the Wenner Gren Foundation were quick to see the potential value of the data and sponsored several workshops during those earlier years of remote sensing (Conant, Reining, and Lowes 1975, Conant 1978). Some of the earlier hopes proved too optimistic --- the limitations of cloud cover were insufficiently appreciated --- and it took some time to truly understand the additional limitations or advantages of relatively coarse resolution levels of the passively acquired data. The trade-off is important: coarse resolution allows inspection of a larger area; fine resolution means more detail but within a smaller area, given comparable limits on data capture. Anthropologists, myself very much included, are so conditioned to paying attention to minutiae that we sometimes find it difficult to shift focus to make broader observations on an entire series of villages or an entire population rather than on the one or two villages we are working in.

But while remote sensing allows us a broad look, the satellite data only become meaningful when they can be interpreted against what we have learned through fieldwork, on the ground. This is what the multi-stage approach, referred to above, allows us to do. Some anthropologists still resist the use of remote sensing as a result of the mistaken notion that the data will somehow "replace" fine-grained observations made in the course of traditional fieldwork. This error results partly from the oversell of Landsat data which was made by the National Aeronautical and Space Agency (NASA) in the 1970's: the satellite data, constructed as imagery, was constantly referred to in press releases as Landsat "photography" which of course it was not (and is not). No matter how much "like a photograph" (that is, how fine the resolution), on-ground field observations will remain critical for satellite imagery interpretation. The trick is to extrapolate from the minutiae of local observations so that these become relevant in the context of the satellite data. For example, a very small village may not be "seen" in the satellite data. The much larger scoured area around the hamlet may well be visible, however.

If on the ground we have noted larger details, such as the area of scouring, then we can use this to infer the presence of a hamlet in the satellite data. Thus *appropriate* ground observations made in the context of traditional field work remain central to the interpretation of the satellite data.

A. The Larger Surround & the "Bird's Eye View"

Satellite remote sensing offers the anthropologist an opportunity to study a fieldwork area within its much larger surroundings --- not just once, but repeatedly. The inter-relationship of a subsistence system and settlement pattern within an environment can be studied both in space and time, and lead to the discovery of unsuspected human and environmental interactions. The "bird's eye" view which satellite remote sensing affords makes possible quantitative measures of the areal extent and nature of biotic associations and subsistence resources, and to measure their changes through time. Identifying and quantifying transitional areas or ecotones is accomplished as readily as for major ecozones (Johnston and Bonde 1989). Partly because of the scales involved, entire ethnic areas, populations, communities (and settlements within them) can be studied from something approaching a holistic point of view.

B. Spatial Extensions

Of all the advantages which satellite remote sensing offers perhaps the most important is the possibility of testing the validity of locally-observed interrelationships ---for example, cropping systems and terrain characteristics — over very much wider areas. What appears to be a regular relationship locally may or may not prove to hold when extended over a larger area. At the point a swidden system, elevation, and a pattern of rainfall, for example, no longer coincide then some sort of a boundary marker has been reached; a series of such points or markers then make up a boundary — thus helping resolve the problem of boundary definition and recognition in human ecosystems (see Moran and Ellen, this volume). A time factor (seasonal and supraseasonal) is also involved.

C. Temporal Extensions

Perhaps because traditional fieldwork is so expensive in time and money it is rare for observations to extend for much more than a seasonal round or two. Re-studies are also expensive and sometimes can only be made for brief periods at a time or season when travel is possible. Thus the question of how permanent are the boundaries for a subsistence practice may be difficult to answer by means of traditional fieldwork, but an approximate answer may be obtainable through sequential use of remote sensing data representing different seasons and different years.

Sequential satellite data has proven itself of great value in mapping broad ecozones, and, more subtly, the ecotones or zones of transition between them. The "normalized difference vegetation index" (NDVI) derived from the NOAA weather satellites and the AVHRR ("advanced very high resolution radiometer") instrument aboard them has allowed the continent-wide *repeatable* mapping of deserts, semideserts, dry savannas and dry grasslands, forests, and wet savannas (see Szekielda 1988 for a general description of the techniques involved.) Note that the derivation of continental and even global vegetation indices began with the development of a hand-held radiometer for use *in the field* (Tucker, Jones, Kley, and Sundstrom 1981), leading eventually to land cover classifications for the entire African continent at different seasons (Tucker, Townshend and Goff 1985) and for estimating drought conditions (Tucker 1989)). Using many of the same techniques, but applying them to a much more local level, the *Centre de Suivi Ecologique* in Senegal is successfully integrating NDVI data with subsistence practices in the Ferlo region where cropping and open range livestock management subsistence systems have been modified by the development of a series of artesian wells (CSE 1988). The boundaries drawn between ecozones and the ecotones between them are dynamic and shifting, and reflect in part the nature of human activities taking place within them. Studies of deforestation, as on Madagascar, which utilize earlier aerial photographs and vegetation maps as well as satellite data, not only disclose past trends but predict forest survival rates into the future (Green and Sussman 1990). Thus the wide-area view, at low levels of resolution, offered by remote sensing allows testing the extent to which local observations hold good over larger areas, regionally, for an entire continent, or even globally. Sequential views make possible not only trend analysis through change detection but also are essential in determining the constancy of the association of environmental and cultural features thought to define an area and, hence, its boundary. The repetitive nature of remote sensing from orbiting space platforms allows one to test for the duration of a subsistence practice and its effects on the surrounding area.

D. The Multi-Stage Approach

Remote sensing data may be acquired on the ground, from conventional aircraft, and from satellites such as the Landsat and SPOT series of spacecraft. When working in an area which has not been closely mapped, and this is often the case in developing areas of the tropics and near tropics, it is necessary to integrate all levels of data acquisition in what is sometimes called a "multi-stage" approach. For such integration to take place some sort of data management system is needed to register one set of observations with another. A major development in recent years has been the increasing availability and sophistication of geographic information systems (GIS) for microcomputers. GIS allow not only georeferencing one data set with another but also statistical analyses based on the correlation of spatial distributions. One such system, *Idrisi*, is described below.

E. Limitations of Remote Sensing

A major limitation of passively-derived remote sensing data is cloud cover, which effectively blocks surface reflections from reaching the scanning devices aboard orbiting spacecraft. Some parts of the world such as New Guinea, highland Peru, and coastal West Africa are so seldom cloud free that passive remote sensing is rarely possible. Tropical forests in general are associated with persistent cloud cover. Active remote sensing utilizing radar can penetrate not only the cloud cover but the vegetation canopy as well and so reveal surface and near-surface features of great importance to ethnology as well as archaeology. So far, however, radar sensed data are accumulated on specific missions, that is, episodically, and not on the regular basis made possible by the satellites in polar orbits. From the foregoing it becomes obvious that the greater number of cloud-free days in desert and semi-desert regions offer better conditions for remote sensing, and partly as a result of this, the African *sahel* or "shore" between desert and forested areas has become a kind of test-bed for remote sensing applications.

III. Some Elements of Satellite Remote Sensing

A. Sources of Information

There are some aspects of remote sensing which are important in ecological anthropology but which are in danger of being lost sight of. The theory, practice and hardware of remote sensing are all well covered in recent texts. The second edition of Lillesand and Kiefer (1987) covers photographic, scanning, active and passive remote sensors, and the operational advantages and limitations of conventional aircraft and earth resource satellites such as Landsat and SPOT. Szekielda (1988) offers a compact review of basic principles and applications of remote sensing in terms of satellite monitoring Earth's oceanic and continental features. Elachi (1987) emphasizes the physical basis of remote sensing. The *Remote Sensing Yearbook* (Cracknell and Hayes 1988) is especially useful for summary accounts of remote sensing developments in particular countries, new analytic techniques and applications, the names, addresses, and telephone numbers of remote sensing centers in dozens of countries, a tabular listing of satellites launched, lost, and recovered, and an extensive bibliography based on the Canadian Remote Sensing Online Retrieval System (RESORS). In addition there is the comprehensive *Manual of Remote Sensing*, published by the American Society for Photogrammetry and Remote Sensing (ASPRS). The *Manual* (Colwell 1983) has particularly useful sections on the history of aerial photography and satellite remote sensing.

Some major journals in remote sensing include *International Journal of Remote Sensing, Photogrammetric Engineering and Remote Sensing,* and *Remote Sensing of Environment.* The *Proceedings* of conferences and symposia sponsored by the Environmental Research Institute of Michigan (ERIM, Ann Arbor) and by the Laboratory for the Applications of Remote Sensing (LARS, Purdue University) are invaluable sources of new methods and preliminary results in remote sensing. The EOSAT company issues *Landsat Data User Notes* and *Landsat Application Notes* which keep users informed of new applications and products. *Spotlight* provides much the same information for SPOT users.

B. Instruments

The main source of passive remote sensing satellite data is energy from the Sun which is differentially reflected back into and through the atmosphere surrounding the Earth's surface. Some of this reflected energy is in that very small part of the electromagnetic spectrum which is visible to the human eye. "Invisible" wavelengths of great importance in remote sensing are the near-, far- and thermal-infrared portions of the spectrum. The visible wavelengths are greatly affected by conditions in the atmospheric "pathway" between Earth's surface and the orbiting space platforms. The near- and mid-infrared reflections are particularly sensitive indicators of seasonal growth and the surface condition of Earth's natural and managed plant cover.

1. *Photo-optical Instruments* The instruments aboard the remote sensing space platforms are essentially of two types: photo-optical and scanning. In the optical type instrument the lens simultaneously focuses light reflected from everywhere within the camera's field of view onto a photosensitive screen, which is then scanned and recorded. The RBV (return beacon videcon) cameras aboard Landsat 3 provided 30 meter resolution but proved operationally unreliable. The scanning instruments proved to be far more reliable than originally predicted.

2. *Scanning Instruments* In the scanning type of instruments reflected light is directed by a rapidly moving mirror across a series of photoelectric sensors, each tuned to a different wavelength. The multi-spectral scanner (MSS) aboard the Landsat platforms responds to four different wavelengths at 80 m resolution. The "thematic mapper" (TM) aboard Landsat 5 responds to 7 different wavelengths at 30 m resolution.

C. Satellite Platforms

1. *SPOT and Landsat* The French SPOT (*Systeme Pour l'Observation de la Terre*) platform has pointable sensors which, on demand, allow them to scan away from nadir, the "straight down" angle of view. Another important potential of SPOT's "sideways look" is to provide overlapping data from which images are constructed that allow stereoscopic viewing. The sensors aboard SPOT-1 are two "high resolution visible" (HRV) systems. One system offers 10 meter resolution in the panchromatic mode and 20 meter resolution in the multispectral mode. The HRV systems do not use a

scanning mirror like the MSSs, and thus have fewer moving parts (for a comparison of SPOT and Landsat, see Chavezn and Bowell 1988).

Whether aboard SPOT as HRV instruments or on Landsat platforms as MSS systems, the sensors function in fundamentally the same way. They sample energy reflected from Earth, at the wavelengths known to be characteristic of different surface and atmospheric features: water, clouds, minerals and rock types, soils, plant and forest cover, and cultural features such as roads, settlement areas, crops and cropping systems. The reflected light is sampled by these scanning systems, converted from analogue to digital format, and [on Landsat 6] stored on tape aboard the spacecraft.

2. *Resolution: Spectral, Spatial, and Temporal*

It is necessary to keep in mind three kinds of resolution. *Spectral* resolution refers to how finely the electromagnetic spectrum is sliced by the instruments aboard the spacecraft. Spectral resolution is not necessarily related to spatial resolution. A photo-optical instrument such as the RBV camera aboard Landsat 3 was sensitive over a broad portion of the panchromatic part of the spectrum and thus was "coarse" spectrally. The optical quality of the lens system and the field of view, however, was such that it could resolve linear features on the ground down to 30 m nominal ground resolution. The latest MSS and TM instruments aboard Landsat 5 have relatively "fine" spectral resolution in that the visible and infrared portions of the spectrum are sampled through up to 7 quite narrow "windows". The instantaneous field of view and other constraints of the MSS and TM instruments allow a lineal or spatial resolution of 30 m, on the ground. SPOT instruments, in their multispectral mode, collect data through only 3 windows on the spectrum and yet also have a spatial resolution of 20 m. In its broad panchromatic operating mode, SPOT has a 10 m, resolution.

Spectral resolution is important because the finer it is the more specific assignments can be made of a pattern of reflectance to a particular feature or condition of Earth's surface. Active vegetation, for example, has a different spectral response pattern than senescent, dormant, or diseased vegetation. High spectral resolution allows identification of these different conditions of the plant cover, as well as differentiating between kinds of cover, including crop types.

Spatial resolution is important because if it is too coarse the picture elements ("pixels") to which the spectral data are assigned may blend the reflectance pattern of an important feature on the ground with its surroundings within the same pixel area. For example, anthropologists are often concerned with quite small field sizes in slash-and-burn farming systems. If a field is smaller than a pixel its reflectance pattern may be "lost" in the reflectance signal assigned to the particular pixel in which it lies. Fine spatial resolution, however, is self-limiting in that a satellite image with high resolution covers a smaller overall area than the scene produced by a similar system but one with coarser spatial resolution. Because there are limits to data storage technology there is a trade off between the fineness of detail one can pack within an imaged area and the size of that area. The argument that increased spatial resolution automatically yields more information is not always true. It can be as, or more, important to determine spectral patterns over larger areas than it is to "see" details within a smaller area. A prime example is the use of the NOAA/AVHRR data with a ground (at nadir) resolution of 1.1 km in order to map global vegetation (see Justice et al 1985).

Temporal resolution refers to the frequency with which the same area on Earth's surface may be scanned. Landsat 4 and 5 have a temporal resolution of once every 16 days. The SPOT system is unusual in that its temporal resolution is variable. The basic orbital period is 26 days. But by varying the angle of view of the HRV sensors (thereby varying the width of the scanned area) temporal resolution can be modified. This capability is particularly important in identifying crop development and distinguishing stages of crop development from that of the surrounding plant cover. Sequential satellite data, obtained at different seasons, may allow field or crop detection when "target" and "background" signal patterns are quite different, as they may well be when a field has been cleared, or a crop is mature.

D. Satellite Data Products

When close to a ground receiving station or a relay satellite the MSS data are transmitted by radio. A "tracking and data relay satellite" (TDRS) was designed to facilitate data transmission to the ground, but because of budget constraints and the higher priority assigned to defense related data, the TDRS system has not been as useful as hoped for the transmission of Landsat data. Ground processing of these transmissions may include expanding the data, correcting for operational anomalies of the space platforms and the

MSS or HRV systems, and reformatting so that the data when imaged conform to one or more of the current coordinate systems, including the Universal Transverse Mercator system (UTM). The satellite data when offered for sale are available as (a) digital tapes, (b) floppy disks, and (c) as already made-up imagery in the form of transparencies or prints (at scales from 1: 1,000,000 to 1:50,000 for TM data).

Table 1 lists the wavelengths sampled by the MSS instrument and the TM scanner aboard Landsat 4 and 5, the platforms in the Landsat series currently operational. Table 2 lists the wavelengths sensed by the two HRV systems aboard SPOT-1. Both tables roughly indicate the applications appropriate to each wavelength. In actual practice the power of remote sensing to discriminate one surface event from another is achieved by manipulating data not from one but several bands together, including ratioing one band to another and treating it as an additional data set.

No matter how fine the spectral, spatial or temporal resolution may be much information still will remain "locked up" unless it is matched against other sources of information. This is the job of the geographic information system, or GIS.

IV. Geographic Information Systems

The mushroom-like growth of GIS in the 1980's is only partly tied to the extraordinary power of today's micro- and minicomputers. Map overlays, a "central tool in current GIS" programs, were used over a century ago in studies of landscape architecture (Chrisman 1988: p.I-102, citing Steinitz, Parker and Jordan 1976). In anthropology, thematic maps of various culture traits such as weapons, house types, and clothing were being "onion-skinned" (overlain) before the turn of the 20th century (Mason 1894), but as compared to geography, mathematical modeling of spatial distributions and associations remained relatively undeveloped.

In effect, remote sensing data (by aircraft as well as by satellite), ground survey observations, maps, tabular or textual materials --- as long as the information from such sources can be spatially represented --- can be integrated by the "tools" provided by a GIS. The operations to be performed include data input, storage, and retrieval; data transformation such as fitting to a common scale, fitting data to new projections; data manipulation and measurement; testing for data integrity, and data presentation as a display on a

Table 13.1. Landsat MSS and TM Bands, Wavelengths, and Some Uses [1]

Band No.		Wave-lengths[2]		:	Some Uses[5]
MSS[3]	TM[4]	MSS	TM	:	
1 (4)	1	0.5-0.6 "green"	0.45-0.52 "blue"		MSS + TM: water; geological & cultural features; decid. vs. conif. forests
2 (5)	2	0.6-0.7 "red"	0.53-0.61 "green"		MSS: geological & cultural features TM: vegetation assessment
3 (6)	3	0.7-0.8 "near-infrared"	0.62-0.69 "red"		MSS: wetlands; plant vigor TM: Plant species identification; above ground plant biomass (AGPB)
4 (7)	4	0.8-1.1 "near-infrared"	0.78-0.91		MSS + TM: soil moisture, plant vigor, AGPB
	5		1.57-1.78 "mid-infrared"		TM: snow vs. clouds; plant stress
	7		2.08-2.35 "mid-infrared"		TM: hydrothermal mapping, mineral & rock identification
	6		10.42-11.66 "thermal infra-red"		TM: thermal mapping; plant stress; soil moisture

[1] Adapted from Lillesand and Kiefer (1987), pages 548-549 &   Table 9.4, and from Szekielda (1988) Table 2-3-1.

[2] In microns [3] Band numbers in parentheses apply to Landsats 1 - 3

[4] TM band numbers 6 & 7 are reversed because of late design changes

[5] Uses are highly summary. Bands are commonly used in various combinations and rarely singly.

Table 13.2. SPOT-1 HRV Systems, Modes, and Some Uses [1]

<table>
<tr><th colspan="4">Multispectral Mode</th></tr>
<tr><td>HRV Bands</td><td>Wavelengths[2]</td><td>Resolution</td><td>Some Uses</td></tr>
<tr><td>1</td><td>0.50-0.59</td><td>20m</td><td rowspan="3">Crop species & vegetation types. Cultural features & geomorphology</td></tr>
<tr><td>2</td><td>0.61-0.68</td><td>20m</td></tr>
<tr><td>3</td><td>0.79-0.89</td><td>20m</td></tr>
<tr><th colspan="4">Panchromatic Mode</th></tr>
<tr><td>1</td><td>0.51-0.73</td><td>10m</td><td>Mapping & georeferencing. May be used with SPOT or Landsat multispectral data</td></tr>
</table>

[1] Adapted from Szekielda 1988 pp. 26-27.

[2] in microns

computer screen, or as hard copy via a printer or plotter. According to Burrough, among the questions which can be asked of a GIS are:

(a) Where is object A?

(b) Where is A in relation to place B?

(c) How many occurrences of type A are there within distance D of B?

(d) What is the value of function Z at position X?

(e) How large is B (area, perimeter, count of inclusions?)

(f) What is the result of intersecting various kinds of spatial data?

(g) What is the path of least cost, resistance, or distance along the ground from X to Y along pathway P?

(h) What is at points X1, X2,...?

(i) What objects are next to objects having certain combinations of attributes?

(j) Reclassify objects having certain combinations of attributes

(k) Using the digital database as a model of the real world, simulate the effect of process P over time T for a given scenario S. (Burrough 1986: 9).

While the analytical portions of GIS are likened to the primitive operations of traditional mathematics "such as addition and subtraction," Berry warns that even such a straightforward measure such as distance between two points may be modified by "barriers" and "connectivity" affecting the reality of travel between the two points and which can be included in map algebra and mathematics (Berry 1987). Robinson and Lundberg (1987) show that some uncertainties can be resolved best by linking knowledge-based systems and semantic modeling to GIS.

Today, independent research involving both image analysis and alignment with other information sources via GIS is possible on "desktop" and even "laptop" microcomputers. The earlier monopoly of mainframe systems was challenged in the late 1970's by IMPAC, one of the first microcomputer programs for image analysis systems using "off the shelf" hardware (Egbert 1979). Throughout the 1980's many increasingly powerful and sophisticated image analysis systems were developed and became available at costs ranging from several thousands of dollars (ERDAS, for example) to only a few hundred dollars (see EIDETIC).

With a broadened base of satellite data users, both on campus and off, the need for layering Landsat data classifications over existing maps and other spatially arranged information sources has increased greatly. This need coincided with, and has been met (at least partially) by the parallel development of hardware and software for managing and graphically displaying digital data in the engineering, architectural, and cartographic professions. GIS today is a hybrid tool available on mainframes, mini-, and microcomputers. Like image analysis programs, GIS may cost several thousands of dollars (see ARC-INFO), or only $100 or so (see IDRISI, Eastman 1988). GIS capabilities are sometimes built into satellite image analysis programs or are available as stand-alone programs (for a review of likely future developments in GIS, see Dobson 1988, ACSM 1988).

The capability of GIS to reformat various information sources to a common scale raises important issues of quality control. The quality of information gathered at one scale may deteriorate at another: for example, the accuracy of observations made by an anthropologist on the subsistence system for one village is almost certainly to be affected by re-scaling those observations to include a much larger (or smaller) area. Thus the quality of the original information may suffer, and whatever analyses are performed in a GIS (for whatever purpose: epidemiological, for example) should be accompanied by a warning as to the effect on reliability of combining different information sources which have been reformatted to a common scale (Conant 1988; Moran, this volume chapter 10).

Some 30 years ago Porter (1960) gently spoofed too enthusiastic an application of spatial modelling to real-life situations, and Chrisman also voices a similar concern over whether small prototype research "will scale up to handle full-scale problems" (1988: I-105). These cautions take on added meaning as GIS applications continue to grow, and grow. For example, UNEP's Global Environmental Monitoring System (GEMS) has developed GRID (Global Resource Information Database) to the point where it now includes these datasets:

Parameter	Coverage
Political boundaries	Global (G)
Natural boundaries	G
Elevation	G
Soils	G
Vegetation	G
Vegetation index (weekly)	G
Vegetation index (seasonal)	Africa (A)
Cultivation intensity	G, A

Watershed	A
Rainfall, Wet days, windspeed	A
Precipitation anomalies	G
Temperature anomalies	G
Surface temperature	G
Albedo (four seasons)	G
Ozone distribution	G
Population density (1960)	A
Tse-tse-fly distribution	A
Elephant range and density	A
Endangered species	A
Protected areas	A

(UNEP 1988, *GRID News* 1,3:p.5)

In addition to the above, a variety of other kinds of information are being entered into the GRID system, such as HIV-1 and HIV-2 seropositivity levels as discovered locally and nationally by independent researchers with data accumulated in desktop GIS files. Programs for moving data between IDRISI and GRID are now being written to facilitate such exchanges (*GRID News* 1989, 2, 1: 11). Chrisman, almost uniquely in the current literature, has noted the need to include matters of equity and social and cultural goals in the application of GIS analyses (1987).

V. Resolution Levels and Discovery of Ecological Factors

Of the many applications of remote sensing which directly bear on human ecology the few mentioned here were chosen either because they utilize scales and resolution levels far different from those anthropologists are accustomed to, and yet achieve results of great interest, or because they point to previously unsuspected factors at work in subsistence systems and resource utilization in Africa.

Ernestine Cary (1985) has made one of the very few controlled studies on the effects of varying resolution levels on the correlation of diverse productive systems with elevation, soils, and population density. Her study area was the western highlands of Kenya. Figure 1A shows the outline of the Landsat scene enclosing the study area, which is characterized by multiple and radically different production systems (see Figure 1B). Figure 2 includes Cary's study area and Conant's (see below). The production systems in Cary's study area include plantation size fields of maize, wheat,

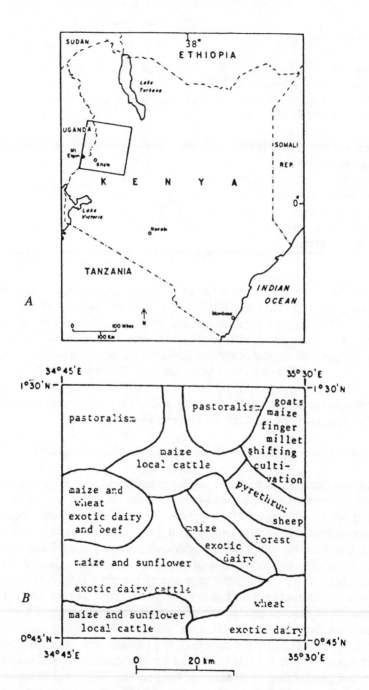

Figure 13.1. *A,* location of Landsat path 182, row 59 (after Cary 1985, p. 62). *B,* agricultural areas and their characteristics based on field observations; boundaries are approximate (after Cary 1985, p. 55).

pyrethrum, and sunflower; closed range management of exotic beef, dairy cattle, and sheep; indigenous pastoralism (open range management of Zebu cattle, goats); and indigenous shifting cultivation (maize, sorghum, finger millet) and goats.

Cary created a gridded geographic data base of 100-meter cells. The values for elevation and soil color were digitized from existing maps. A "greenness" index was calculated from MSS bands 5 and 7, and used as an indicator of agricultural patterns. On ground observations were made to verify the Landsat index. The values for elevation, soil, and agricultural pattern were assigned each of the 100-meter cells. Population data were gathered from maps and published tables and assigned to 1000-meter cells. Cell-by-cell comparisons were then made of agriculture to each of the encoded variables. The grid cells were then aggregated in successive steps, from 100 x 100m to 10,000 x 10,000m. At each step the strength of the association between agricultural pattern with the other variables was determined. Cary found that the use of cells of regular shape and equal size leads to the "same assessment of relative strength of relationship throughout a broad range of scales, for the particular variables included in this study" (p.146).

This is an important finding for several reasons. One is that significant information on human subsistence activities (as represented in the various agricultural patterns found in the area) and their relationships to prominent physical features (elevation, soils) and socio-economic factors (population) is retained through successive levels of aggregation. Gross levels of resolution are revealing of large-area relationships --- and these, it would seem, can be as important to investigate as the fine-grained phenomena observed in conventional anthropological fieldwork. Cary's study also shows the importance of careful quantification procedures. By finding suitable indicators of soil types she was able to avoid treating nominal classes along with interval data for the other variables. There are many possible extensions of Cary's work. One would be to encode estimated energy expenditures and yields characteristic of each of the agricultural subsistence systems; another would be to characterize these systems by the relative energy input of women, men, and children, as well as by oxen and tractor. Depending on the variables being used, a first question to be asked is *at what level of resolution or aggregation are the best correlations obtained*? And the worst? If adequate correlations of the chosen variables are obtained over a range of resolution or aggregation

Figure 13.2. Landsat scene 21107-06413, band 5 ("panchromatic"). The nominal scale is about 1 to 1 million. The image area is also shown in Fig. 13.1A. The Cary study area referred to in the text is approximately in the lower left quadrant of the image, not including the higher slopes of Mt. Elgon, which rises to some 14,000 feet above sea level. The dashed line encloses the area exploited by Pokot swidden farmers and open-range herding of indigenous livestock. The conversion from grassland to thorntree bushland noted in Conant's study took place in the southern portion of the Masol plains, at about 2,000 feet above sea level. The letters indicate approximate locations of the Pokot (P) and their neighbors: Sebei (SE), Karamojong (K), Jie (J), Turkana (T), Rendili (R), and Samburu (SA). The Kitale area is characterized by large-scale cultivation and closed-range management of exotic livestock.

levels, and if a concern is to include as large an area as possible (perhaps to make inter-regional comparisons), then the resolution level can be chosen which best represents the problem at hand. In other words, the finest level of resolution ("which looks most like a photograph") is not always the best. According to Collier:

"To generalize, the interrelations between any two given variables may be scale-related; in other words, they are best perceived when one, not another scale of analysis is selected. The scale of analysis that is optimal for the interrelation of any two variables may not be optimal between any other pair of variables" (1974:90).

Lambin (1988), concerned with agrarian systems in Burkina Faso, (West Africa) found good correspondence between 4 different field patterns as these were displayed in Landsat MSS imagery and the distribution of 4 ethnic groups as previously mapped. Lambin characterized the field patterns as nebulous, dispersed, forming a network (*resaux*), and concentrated. Landsat imagery for the dry and wet seasons in 1985 were utilized and compared with a series of 12 thematic maps covering such topics as ethnic distributions, agrarian landscapes, agro-ecological zones, lithology, and soils. Ethnic mappings best correlated with the Landsat imagery for the wet season, and agrarian landscape mapping had the second best correlation. For the dry season imagery, agro-ecological distributions had the best fit and ethnic distributions came a very close second.

Lambin finds that the determinants of the agrarian system (social and economic factors) are themselves inscribed in the agrarian landscape (*"les determinations des systeme agraire se retrouvent inscrites dans le paysage agraire"*, p. 341). This leads to the possibility that representations of the field patterns found in the satellite imagery may be useful in selecting a particular group or subsistence pattern for further field study. Changes in the field patterning as these might show up in future satellite imagery may also serve as indicators of socio-economic processes affecting particular ethnic groups.

Combining the satellite data with other sources, Lambin also attempts to read out of the imagery estimates of areas under cultivation, yields, and population size. Population estimates obtained by extrapolating from the imagery are within 13% of census figures, and a 17% average error of area under cultivation. Lambin (1988) feels that while such estimates may be improved with the data

provided by the Landsat TM system or SPOT HRV, the MSS data already
are sufficient to point the way for ground surveys and further
fieldwork research into suspected factors affecting agrarian systems.

In my own work (Conant 1982), while the various pixel counts
involved are likely to be improved by TM and/or SPOT HRV data, it is
the discovery of *process* rather than absolute values which
matters most, at least initially. The study concerns the changes in
vegetation cover from grassland to thorn bush on the plains of west
central Kenya following the *withdrawal* of Pokot herders from
the area. The Pokot farmers and herders exploit the slopes of the
Eastern Rift Valley escarpment and the lowland plains between Mt.
Elgon in the southwest and Lake Turkana to the north (see Figure 2).
Farmers outnumber herders about 3:1, and marriages between farming
and herding families facilitate the exchange of farm and herd
surpluses (Conant 1965). The usual complaint about Pokot herding
practices was that overgrazing by both cattle and goats had reduced a
once lush grassland to a near-barren condition. In the 1950's British
administrators tried to enforce a block-grazing scheme in order to
halt what they perceived as further degeneration. This was done on
the supposition that the Pokot either lacked understanding about the
relationship between the numbers of animals and environmental
degradation, or they simply did not care.

Neither supposition was true. Indigenous Pokot grazing strategies
were blocked by the imposition of boundaries imposed by the colonial
powers. The Pokot cared enough about what grazing lands were left to
them within their "reserve" to resist moving out of the area even
when, in the 1970s, *ngoroko* raiders, mainly from Uganda,
began slaughtering children, women, and men in their drive to capture
Pokot cattle and goats. Automatic weapons, grenade launchers, and
armored half-tracks devastated Pokot herding encampments. The Pokot
withdrew from *Simbol*, a prime grazing area, in 1974.

As part of a National Science Foundation grant (#BNS-77-15622, as
well as additional grants from the Wenner Gren Foundation and the
City University of New York) to investigate the use of remote sensing
and satellite data for studying subsistence systems in East Africa,
Landsat 3 data were secured for the years 1973, one year before
Simbol was abandoned, and for 1978, four years after abandonment.
Computer assisted analysis of the data for the two dates showed a
surprising result: following its abandonment, Simbol converted from
an area of mixed thornbush and grass to mostly thornbush. In the
absence of people and their livestock the grass almost

disappeared and Simbol became an impenetratable sea of thorntree. The Pokot who had fled the area claimed that Simbol was lost forever for grazing by cattle or browsing by goats. According to informants, including those from neighboring pastoral groups, once Acacia is established it dominates over grass. Only constant predation, primarily by goats, of the Acacia new growth, seed pods and sprouts keeps the thorntree in check and allows grassy areas to spread. Thus goats seem necessary to maintain the grassy cover needed by the cattle, and this explains, at least in part, the maintenance of mixed herds of grazers (cattle) and browsers (goats). A further finding was that, at least among the Pokot, management of goats is often the task of women. This division of labor, and the essential role of goats, persists despite the greater value and status assigned to cattle and the men who herd them.

VI. Summary and Conclusion

The low resolution Landsat MSS data (78 m.) was adequate to discover the ecological change in the Simbol area, and pointed the way toward a fruitful line of investigation on the ground. Would finer resolution have helped? Almost certainly it would have, in terms of the accuracy of pixel counts. But one can imagine arriving at the discovery of the processual inter-relationships between grass and Acacia, keeping goats as well as cattle, and the Pokot division of labor using an even *coarser* resolution than that inherent in the early Landsat data.

Cary's work on resolution levels shows a remarkably stable relationship between different production systems, elevation, soils, and population density across a wide range of resolution levels and grid sizes. Lambin's discovery of different field patterns and correspondence with diverse ethnic groups in Burkina Faso was made using 78 m. resolution data, as was my own discovery of grass to Acacia changes in Kenya. Several points follow from these experiences. One is that for purposes of initially discovering ecological change, especially of the kind in which humans are playing a part, the resolution levels of first generation satellite data are adequate. A second point is that archiving and retrieval of early Landsat data can prove a vital historical resource in present and future satellite studies. The third point is that inspection of satellite data at whatever levels of resolution can help formulate hypotheses for further testing on the ground--- that is, suggest particular directions which fieldwork might initially (or subsequently) take. Fourth, there is no cause for alarm: the use of satellite data, enhanced by integration via GIS with other information sources, is no threat to fieldwork in anthropology or other field sciences. In fact it seems the opposite may be true: the satellite data are raising questions which never before could be asked, and the answers to such questions are to be found on the ground, through fieldwork. Satellite data interpretation offers an opportunity to anthropology and anthropologists. Integrating our field with that of the others involved in remote sensing is a prescription for exciting research for the 1990s and beyond.

Abbreviations

AA	American Anthropologist
AAG	Association of American Geographers
AAAS	American Association for the Advancement of Science
ACSM	American Congress on Surveying and Mapping
ARC/INFO York	Environmental Systems Research Institute, Inc. 380 New Street, Redlands, CA 92373. Tel: (714) 793-2853.
ASPRS	American Society for Photogrammetry and Remote Sensing
AVHRR	Advanced Very High Resolution Radiometer
CSE	Centre de Suivi Ecologique, Dakar.
EIDETIC	Eidetic Digital Imaging Ltd. 1210 Marin Park Drive, Brentwood Bay, British Columbia, Canada V0S 1A0. Tel: (604) 652-9326
EOSAT	Earth Observation Satellite Company. Customer Services, 300 Forbes Blvd., Lanham, Maryland 20706. [Tel: (800) 344-9933; (301) 552-0537]
ERDAS	ERDAS, Inc., 2801 Buford Highway, Suite 300, Atlanta, Ga 30329. Tel: (404) 248-9000.
GEMS	Global Environment Monitoring System. See UNEP.
GIMMS	Global Inventory, Monitoring and Modelling Studies. Laboratory for Terrestrial Physics, Code 620, GSFC, Greenbelt, MD 20771.
GIS	Geographic Information Systems
GRID	Global Resource Information Database. PO Box 30552, Nairobi, Kenya. See UNEP.
GSFC	Goddard Space Flight Center. See GIMMS.
HERSL	Human Ecology and Remote Sensing Laboratory, Anthropology Department, Hunter College, CUNY.

IDRISI	GIS system developed by J. Ronald Eastman, Graduate School of Geography, Clark University, Worcester MA 01610
IGIS	International Geographic Information Systems
IJRS	International Journal of Remote Sensing.
LIS	Land Information Systems
MSS	Multi-spectral Scanner
NASA	National Aeronautics and Space Administration
NOAA	National Oceanographic and Atmospheric Agency
NDVI	Normalized Difference Vegetation Index
NTIS	National Technical Information Service
PE&RS	Photogrammetric Engineering and Remote Sensing. Monthly journal of the ASPRS.
S	Science. AAAS, Washington DC.
SMMR	Scanning Microwave Multichannel Radiometer (37 GHZ)
SPOT	System Probatoire d'Observation de la Terre
TM	Thematic Mapper. LANDSAT 5, 4 bands, 30m resolution
UNEP	United Nations Environment Programme, Nairobi.
URISA	Urban and Regional Information Systems Association
XS	SPOT multi-spectral, 3 band, 20m resolution scanner

References Cited

ACSM
   1988   *GIS/LIS'88 Proceedings: accessing the world.*
           Volumes 1 & 2. Third Annual Conference, San Antonio,
           November 30 - December 2, 1988. Sponsored by ACSM,
           ASPRS, AAG, URISA.
Arensberg, Conrad
   1981   Cultural Holism through Interactional Systems.
           American Anthropologist 83:562-581.
Berry, Joseph K.
   1987   Computer-assisted Map Analysis: Characterizing
           Proximity   and   Connectivity. *International
           Geographic Information Systems (IGIS) Symposium: The
           Research Agenda* NASA 1987, II, 11-21.
Bohannan, Paul
   1980   You can't do nothing. *American Anthropologist*
           82(3):508-524.
Burrough, P. A.
   1986   *Principles of Geographic Information Systems for*

Land Resources Assessments. London: Oxford
University Press.

Cary, Ernestine
1985   Spatial Patterns of Kenya Agriculture and Their
Relationship to Elevation, Soils, and Population: A
Study of Grid Cell Sizes. New York: Columbia
University, Doctoral Dissertation.

Chavez, Jr., Pat S. and Jo Ann Bowell
1988   Comparison of the Spectral Information Content of
Landsat Thematic Mapper and SPOT for Three Different
Sites   in   the   Phoenix,   Arizona   Region.
Photogrammetric Engineering and Remote Sensing,
LIV, 12:1699-1708.

Chrisman, Nicholas R.
1988   Challenges for Research in Geographic Information
Systems. *International Geographic Information
Systems (IGIS) Symposium: The Research Agenda*,
NASA v. I, pp.I/101-I/112

1987   Design of Geographic Information Systems based on
Social   and   Cultural   Goals.   *Photogrammetric
Engineering and Remote Sensing*, LIII,10:1367-1370.

Colwell, Robert N., editor in chief
1983   *Manual of Remote Sensing.* Volume I: Theory,
Instruments and Techniques. Volume II: Interpretation
and Applications. Falls Church, Va.: ASPRS

Collier, George A.
1974   The Impact of Airphoto Technology on the Study of
Demography and Ecology in the Highland Chiapas.
*Aerial   Photography   in   Anthropological   Field
Research.* Edited by E. Vogt. Cambridge, Mass:
Harvard University Press pp. 78-93.

Conant, Francis Paine
1988   Using and Rating Cultural Data on HIV Transmission in
Africa.   R.   Kulstad,   editor,   *AIDS   1988*.
Washington, DC: AAAS.

1984   Remote Sensing, Discovery and Generalizations in Human
Ecology.   *The   Ecosystem   Concept   in
Anthropology.* Edited by E.F. Moran. Washington
D.C. AAAS.

1982   Thorns Paired, Sharply Recurved: Cultural Controls and Rangeland Quality in East Africa. *Anthropology and Desertification.* Edited by Brian Spooner and T. Mann, London: Academic Press, pp. 111-122.

1978   The Use of Landsat Data in Studies of Human Ecology. *Current Anthropology* 19:382-4.

1965   Korok: A Variable Unit of Physical and Social Space among the Pokot of East Africa. *American Anthropologist* 67,2:429-434.

Conant, Francis P., Priscilla Reining and Susan Lowes

1975   Satellite Potentials for Anthropological Studies of Subsistence Activities and Population Change. Report of a NSF Workshop. Washington DC: mimeographed.

Cracknell, Arthur and Ladson Hayes, eds.

1988   *Remote Sensing Yearbook 1988/89.* London: Taylor & Francis

CSE

1988-   *Images du CSE. Bulletin trimestriel.* Dakar, CSE.

Dobson, Michael

1988   Trends in Hardware and software for GIS. *International Geographic Information Systems (IGIS) Symposium: The Research Agenda.* NASA 1988, v. I, pp. I/69-I/74.

Eastman, Ron

1988   *IDRISI, A Grid-based Geographic Analysis System.* Worcester, Massachusetts: Clark University, Graduate School of Geography.

Ebert, James Ian and Thomas R. Lyons, author-editors

1983   Archaeology, Anthropology, and Cultural Resources Management. *Manual of Remote Sensing.* Falls Chjurch, VA: ASPRS. R. Colwell, ed. V.II, Ch. 26.

Egbert, Dwight

1979   *IMPAC Image Analysis Package for Microcomputers.* Greenport, NY: Egbert Scientific Software.

EIDETIC

1990   RSVGA: A Digital Image Analysis System for Remotely Sensed Data. Available through EOSAT or EIDETIC Digital Imaging, Ltd.

Elachi, Charles
     1987   *Introduction to the Physics and Techniques of Remote Sensing.* New York: John Wiley.
EOSAT
              LANDSAT Data Users Notes. Quarterly Publication of the EOSAT Company. 4300 Forbes Boulevard, Lanham, Maryland 20706.
              LANDSAT Applications Notes. EOSAT Company.
Goldschmidt, Walter R.
     1983   Personal communication re the "sociological infiltration of remote sensing, an exercise I did while working for the USDA early in my career."
     1943   Some Evidence on the Future Pattern of Rural Settlement. *Rural Sociology*, 8,4:387-395.
GRID News
     1988   Published quarterly. Nairobi: UNEP/GEMS.
Green, Glen M. and Robert W. Sussman
     1990   Deforestation History of the Eastern Rain Forests of Madagascar from Satellite Images. *Science* 248:212-215.
Johnston, Carol A. and John Bonde
     1989   Quantitative Analysis of Ecotones using a Geographic Information System. *Photogrammetric Engineering and Remote Sensing* LV,11:1643-1647.
Justice, C. O., J. R. G. Townshend, B. N. Holben and C. J. Tucker
     1985   Analysis of the phenology of global vegetation using meteorological satellite data. *International Journal of Remote Sensing* 6 (8):1271-1318.
Lambin, Eric
     1988   L'Apport de la Teledetection dans l'Etude des Systemes Agraires d'Afrique: l'Exemple du Burkina Faso. *Africa* 58, 3:337-52.
Lillesand, Thomas M. and Ralph W. Kiefer
     1987   *Remote Sensing and Image Interpretation.* 2nd edition. New York: John Wiley.
Mason, Otis T.
     1894   Technogeography, or the Relation of the Earth to the Industries of Mankind. *American Anthropologist* 7 (2):137-161.
NASA
     1987   *International Geographic Information Systems (IGIS)*

Symposium: The Research Agenda. Proceedings. Volume I: Overview of Research Needs and the Research Agenda; Volume II: Technical Issues; Volume III: Applications and Implementation. NASA, Arlington, Virginia, November 15-18.

Peregrine, Peter
  1988   Geographic Information Systems in Archaeological Research: prospects and problems. *American Congress on Surveying and Mapping* 2, 873-879.

Porter, Philip W.
  1960   Earnest and the Orephagians--- A Fable for the Instruction of Young Geographers. *Annals of the Association of American Geographers*, 50(3):297-299.

Reining, Priscilla
  1973   ERTS Image Analysis. Preliminary Report on ID#1080-10163, Site North of Segou, Republic of Mali, West Africa. Springfield, Va., NTIS.

Robinson, Vincent B. and C. Gustav Lundberg
  1987   Organizational and Knowledge Base Considerations for the Design of Distributed Geographic Information Systems --- Lessons from Semantic Modeling. *International Geographic Information Systems (IGIS) Symposium: The Research Agenda.* V. II, pp.245-256.

Schorr, Thomas S.
  1974   A Bibliography with Historical Sketch. *Aerial Photography in Anthropological Field Research* Edited by E. Vogt. Cambridge: Harvard University Press, pp. 163-188.

Sheets, Payson and Tom Sever
  1988   High-Tech Wizardry. *Archaeology*, November/ December: 28-35.

Steinitz, C. F., P. Parker, and L. Jordan
  1976   Hand-drawn overlays: Their History and Prospective Uses. *Landscape Architecture*, 66,444-455.

Szekielda, Karl-Heinz
  1988   *Satellite Monitoring of the Earth.* New York, Wiley.

Tucker, Compton J.
  1989   Comparing SMMR and AVHRR Data for Drought Monitoring. *International Journal of Remote Sensing* 10:1663 -1672

Tucker, C. J., W. H. Jones, W. A. Kley, and G. J.Sundstrom
   1981   A Three-band Hand-Held Radiometer for Field Use.
           Science 211:281-283.
Tucker, Compton J., John R. G. Townshend, and Thomas E. Goff
   1985   African Land-Cover Classification Using Satellite Data.
           Science 227: 369-375.
UNEP
   1988-   *GRID News.* Quarterly. PO Box 30552, Nairobi,
           Kenya
Vogt, Evon Z., editor
   1974   *Aerial Photography in Anthropological Field
           Research.* Cambridge, Mass.: Harvard University
           Press.
Williams, C. P.
   1989   Landsat Commercialization. *International Journal of
           Remote Sensing* 10,2:265-274.

CHAPTER 14

ECOSYSTEM APPROACHES IN HUMAN BIOLOGY: THEIR HISTORY AND A CASE STUDY OF THE SOUTH TURKANA ECOSYSTEM PROJECT

Michael A. Little, Neville Dyson-Hudson,
Rada Dyson-Hudson, James E. Ellis,
Kathleen A. Galvin, Paul W. Leslie,
and David M. Swift

Ecological Studies in Human Biology

The disciplinary origins of human biology or biological anthropology are quite diverse and cross-cut the social and biological sciences. Today, the field is unified by the theoretical framework of human evolution and the concept that behavior and biology interact within human cultures and societies to facilitate *adaptation to the environment.*

"Adaptation" as a concept in human population biology preceded the interest in ecological systems as the context of adaptation. Research on human adaptation to the environment in the early 1950's centered on narrowly defined climatic characteristics (Coon, Garn and Birdsell 1950; Newman and Munro 1955; Roberts 1953). During the same period, the ecologist and epidemiologist Marston Bates argued persuasively for the development of a field of "human ecology" in which ecological principles were to be applied to the study of human populations (1953, 1960). Bates played an influential role because he was one of the few scientists who was willing to consider humans as simply another species that was subject to evolutionary and ecological processes, while at the same time, recognizing some of the unique properties of our species.

Despite Bates' arguments for the development of a "human ecology," much of the research conducted throughout the 1950's was biogeographical and emphasized adaptation of human morphological

389

features to climatic extremes. However, other trends were under way at this time that were to strengthen the research foundations of human biology. First, demographic studies on tribal societies (Birdsell 1953; Roberts 1956; Spuhler 1959) underlined the need to consider populations and population variation rather than the individual as the unit of investigation. Second, work by environmental physiologists (Brown and Page 1952; Scholander *et al.* 1957, 1958), with their interests in human physiological responses to environmental stress, stimulated research by human biologists in areas other than just morphology (Baker 1958; Roberts 1952). Third, in addition to climate, other environmental elements were seen as imposing stress on humans (Newman 1960).

By the 1960's pioneering human biologists on both sides of the Atlantic were developing ecological approaches to the study of human adaptation (Baker 1962; Newman 1962; Weiner 1964). Such ecological approaches, or perspectives, arose from the evolutionary traditions of physical anthropology, which were centered firmly on the concept of "adaptation to the environment" (Little 1982). These trends were reinforced by the initiation of the *International Biological Program* and its *Human Adaptability* component during the mid-1960's (Hanna *et al.* 1972).

International Biological Program

Planning for the International Biological Program (IBP) began in 1961, with the program being initiated by the International Council of Scientific Unions in July 1964. The IBP extended over a period of ten years and was divided into the three phases of planning, research, and synthesis. In the United States, the IBP became operational in 1967, and was administered through the National Academy of Sciences.

The objectives of the IBP were clearly ecological in scope with stated goals of furthering basic scientific research on biological productivity, natural resource use and management, biological systems, and human adaptation (Golley 1984). IBP research in the U.S. was divided into two components: 1) *Environmental Management*, which was dominated by several large "analysis of ecosystems" projects (grassland, desert, tundra, coniferous forest and deciduous forest) and 2) *Human Adaptability*, which was constituted of three major and numerous minor projects (National Academy of Sciences 1974). Although the U.S. involvement in the IBP was limited at the beginning, by the end of the synthesis period in 1974 its contributions had been substantial. The program on *Analysis of Ecosystems* was instrumental in bringing together multidisciplinary groups to develop sophisticated computer-based models characterizing the properties and processes of *ecosystems* as integrated units. Such comprehensive tasks required many scientists working at various levels of research design, field organization, technical assistance, data collection, analysis, modeling, and management of the research organization. It is certain that a major contribution to the IBP by the U.S. was the development of the ecosystem paradigm as an operational concept (Johnson 1977), synthesizing population and evolutionary biology with systems and ecological theory. As a result, the ecosystem paradigm and ecosystem analyses have emerged as one of the most powerful conceptual frameworks in the environmental sciences.

Many of the human adaptability projects in the U.S. and abroad were initiated at a conference in Austria in July, 1964 (The Wenner-Gren Foundation's Burg Wartenstein Conference; see Baker and Weiner 1966). At this conference, papers were presented that dealt with planned research and research synopses of human adaptation on all of the major continents of the earth. Although the human adaptability projects were much more modest than the analysis of ecosystem projects, there were, however, marked similarities. The approach taken to investigate the patterns of adaptation of a single population was multidisciplinary; that is, investigators from many different sciences worked together on an integrated project designed to solve a series of scientific problems and to gain an understanding of the whole human-environment system. Projects were also multinational with scientists drawn from host nations and the United States. There were strong ecological interests in several of the human adaptability projects, but for a variety of reasons, including lack of involvement of ecologists and lack of funds, ecological research never became well-integrated within these major projects. In fact, among the human adaptability projects, "ecology" was really translated to mean "environment," and the emphasis was placed on human adaptation to the environment or to environmental stress.

Attempts were made to bring about further integration between ecological research and human adaptability research. A Conference was held on "Man in the Ecosystem" (Little and Friedman 1973) in which the theme was "to incorporate humans as an integral component of ecosystem studies." Following three days of close contact between ecologists and human biobehavioral scientists, it became clear that collaborative ecosystem research incorporating humans was most readily accepted when technologically simple populations within relatively intact or only slightly modified natural ecosystems were considered for investigation. It was observed, further, that in more technologically advanced societies, political, trade and communication networks transcended ecosystem boundaries and magnified the complexity of the analysis.

The conceptual differences among social scientists, human biologists, and ecologists on how they view ecosystems, i.e., as relatively "open" or "closed" ecosystems, was a serious barrier to exchange of ideas at this conference (Little and Friedman 1973). Definition of boundaries, to many, is an exercise in taxonomy, and it continues to constrain interdisciplinary communication (see discussion in Ellen, Netting and Moran, this volume). Ethnic (Barth 1969; Ross 1975), human population (Brues 1972), and ecosystem (Terborgh 1971) boundaries can each be delineated by different criteria. Ethnic boundaries may be defined by non-transfer of shared values or certain cultural practices and limited interaction (Barth 1969:11); population boundaries may be defined by mating practices and social barriers (Harrison and Boyce 1972); ecosystem boundaries may be defined by marked change in the abiotic environment (steep environmental gradients), by species competition, or by the identification of ecotones (Terborgh 1971). Ecologists define an ecotone as a transition zone between plant communities (Hansen et al. 1988). Interest in ecotones by ecologists has increased in recent times because of parallel interests in landscape ecology, patch dynamics, and biodiversity. The latter tends to be greater in ecotones than in established ecosystems or plant communities (di Castri et al. 1988). Collaborative efforts that involve the definition of intersecting boundaries of different systems introduce even more complex criteria for boundary definition. Finally, temporal boundaries are less often considered but add a further dimension of sublime complexity (Shugart 1978; Ellen, Fish and Fish, Butzer, Hastorf, and Netting, this volume).

One area identified by ecologists and biosocial scientists alike as fertile ground for collaborative work was energy and nutrient flow studies (Little and Friedman 1973). Accordingly, a workshop on "Energy Flow in Human Communities" (Jamison and Friedman 1974) was organized as a "means of integrating the study of human communities with the study of the surrounding ecosystems, ... and to understand the implications of different patterns of energy usage for human populations." The overall recommendation from the workshop was that the most useful techniques for analyzing energy flow data were systems analysis and simulation models.

Although these and other recommendations concerning integration of humans into ecosystem studies came too late to affect the IBP human adaptability research, some progress was made toward a "human ecology" in population biology. Much of this progress occurred through human adaptability projects conducted within tundra, desert, tropical forest, coniferous forest and mountain ecosystems. A discussion of this research follows.

Multidisciplinary Studies in Human Adaptability

More than 200 human adaptability projects were completed throughout the world by the end of the IBP, although most were quite modest in what they achieved, with one or two investigators working on a narrowly-defined problem (Weiner 1977). Perhaps 10 percent or fewer of the projects were multidisciplinary, involved research teams with an integrated approach to learning how one or more populations had adapted to a specific environment, and had an organizational network designed for rapid information exchange. *High mountain projects* were conducted in: 1) the Pamirs of the Soviet Union, 2) the Simien mountains of Ethiopia, 3) the Nepalese Himalayas, and 4) the Peruvian Andes. *Circumpolar and boreal forest studies* were done of: 1) reindeer-herding Lapps in a joint Scandinavian project, 2) Eskimos from Alaska, Canada and Greenland in a four-nation cooperative venture, 3) Aleutian Islanders in Alaska, and 4) Ainu natives on the island of Hokkaido in Japan. *Tropical humid forest projects* included: 1) collaborative work on New Guinea natives by the United Kingdom and Australia, 2) studies of Congo Pygmies in the Central African Republic, Cameroon and Zaire, and 3) a massive effort to study the Yanomama Indians of the Brazilian and Venezuelan Amazon. In addition to these single-ecosystem multidisciplinary projects, others were designed to explore the *effects of environmental change through migration.* Studies were carried out on migrants to Israel, Japanese migrants to Hawaii and California, migration of Venda Bantu to cities in South Africa, and migration of Andean Indians in Peru to the lowland tropical forest and to the semi-arid Pacific coast. Three of these projects will be discussed briefly to outline the extent to which ecosystem studies played a role in their formulation and outcome.

There were many human adaptability projects completed by U.S. scientists, but several large, multidisciplinary research programs exemplified the U.S. contribution to the IBP. Each project reflected a single frame of reference that was environmentally (but *not* ecosystematically) oriented: that of investigating the adaptive mechanisms of populations living under some form of environmental stress (National Academy of Sciences 1974:69). The multidisciplinary projects were: (1) *International Study of Circumpolar Peoples* with Inuit (Eskimo) (Jamison *et al.* 1978) and Aleut (Laughlin 1970; Laughlin and Harper 1979) subprojects; (2) *Population genetics of Native Americans* (largely Amazon Yanomama) (Neel *et al.* 1977); (3) *Biology of Human Populations at High Altitude* (Baker and Little 1976). These projects will be discussed briefly to outline the extent to which ecosystem studies played a role in their formulation and outcome.

Alaskan Eskimo (Inuit) and Aleut Population Biology

The northern Alaskan Inuit and Aleut studies were independent sub-projects on peoples who were living within quite different ecosystems and exploiting different kinds of food resources. Coastal Inuit depended on sea mammals and caribou in the frigid northern tundra and sea-ice, while Aleut were dependent on shellfish, fish, birds, and sea mammals that flourish as the result of the nutrient-rich upwelling systems around the more temperate Aleutian Islands (Laughlin 1970). Both sub-projects were interested in population biology, including food resource exploitation, nutrition, general health, and ecosystem constraints on population growth.

As noted above, the Alaskan Inuit project was part of a four-nation effort to study Eskimos around the Arctic Circle. U.S. investigators were responsible for the north coast and inland areas of Alaska; Canadians worked at Igloolik; the French studied Inuit on the east coast of Greenland; and Danish scientists worked on the west coast of Greenland. Since comparable methods were employed in studies of health, disease, growth, demography, nutrition, genetics, and acculturation, it was possible to compare Inuit from widely diverse environments (Milan 1980).

Very little modeling was done on either the tundra ecosystems with Inuit as a component or the island ecosystem including Aleuts, with the notable exception of Hett and O'Neill's (1974) model of carbon flow. This model is illustrated in Figure 14.1.

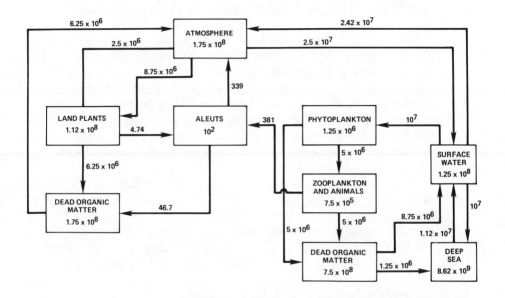

Figure 14.1. An Aleutian ecosystem model of carbon flow. Terrestrial components are to the left in the diagram and marine components are to the right. Values are in kilograms of carbon. (After Hett and O'Neill 1974.)

The model was based on data from the literature on the amount of carbon in each state variable and estimates of early Aleut population size (16,000), human biomass (mean weight by sex and age classes), and food inputs. An assumption in the model was that Aleuts derived 95 percent of their food from sea animals and five percent from land plants (a 19:1 ratio of sea to land food). However, analysis indicated a sensitivity to changes in the marine subsystem of 10,000:1, rather than 19:1, as expected. The marine subsystem was much less subject to seasonal change than the terrestrial subsystem, and thus the "sea appears to offer a more 'dependable' source of food." Also, the model suggested that the rate of ecosystem recovery from disturbance showed little dependence on Aleut activities. A number of other interesting hypotheses concerning Aleut origins and the human carrying capacity of the Aleutian ecosystem were generated by the carbon-flow model.

Some modeling of energy flow through an Inuit system was done by Kemp (1971) based on data from two hunting families from Baffin Island, Canada. There were, however, no successful attempts to coordinate the extensive IBP tundra ecosystem project at Barrow, Alaska (Brown *et al.* 1980) with any of the Alaskan Inuit human adaptability studies (Jamison *et al.* 1978; Milan 1980).

Yanomama Indian Population Biology

The Yanomama project was an integrated research program with a focus on population genetics, tribal structure, environmental pressures, and human evolution in a population of Amazon lowland natives (Neel *et al.* 1977). The research team was constituted of a social anthropologist (Chagnon 1968) and numerous population geneticists and biomedical scientists from the United States, Venezuela and Brazil. Although the investigators were very much interested in patterns of exploitation of the wet tropical forest environment and Yanomama techniques of survival and resource exploitation, the project was not oriented toward ecosystem ecology *per se*. However, considerable effort was devoted to innovative genetic and demographic modeling of the Yanomama (MacCluer *et at.* 1971). Indeed, there are no comparably detailed demographic and genetic studies of a tribal population anywhere in the world as those conducted among the Yanomama.

Andean Indian Population Biology

Investigations of the biological and cultural patterns of adaptation of a high-altitude Quechua population were centered in Nuñoa in southern Peru at elevations ranging from 3800 to 5000 meters above sea level (Baker and Little 1976). Abiotic characteristics of this *puna* ecosystem included low atmospheric and oxygen pressure, broad diurnal temperature variation with cold nights, low humidity and seasonal aridity, and high levels of solar radiation. In addition to these severe climatic conditions, highland peoples were faced with limited nutritional and energy resources as well as problems associated with altitude-aggravated and altitude-induced disease states.

The approach of the project was both ecological and evolutionary, with the major goal being to define how Andean natives had adapted genetically, developmentally, physiologically, and behaviorally to life under these environmental stresses. Research included comparative laboratory tests, field studies, population comparisons in relation to migration, studies of the relationships between sociocultural practices and biological attributes, and energy flow analysis. It was in the area of energy flow analysis that some very important insights were gained about the human population and the managed *puna* system. Figure 14.2 is an energy-flow diagram developed by Thomas (1976) based on H.T. Odum's (1971) modeling framework.

Energy values in the diagram represent thousands of kilocalories (kcal) utilized annually by a Quechua Indian family of two adults and four children. This approach led Thomas to several conclusions about the energetic efficiency of the mixed subsistence system of herding and cultivation. Comparing food energy production (outputs) with labor expenditures (inputs), cultivation of plant foods yielded more than a 10:1 return. Livestock herding *in the absence of trade* provided only a 2:1 return. However, because animal products were highly valued at lower elevations, where sheep and camelid livestock could not be kept easily, the net gain when trade of animal products for plant foods is included gave a ratio of more than 7:1.

These relationships demonstrated the utility of a mixed subsistence pattern in this area of the Andes. Cultivation yields alone were insufficient to meet all of the human food requirements under the present subsistence system and demographic structure. Livestock provided a relatively stable protein resource, dung for

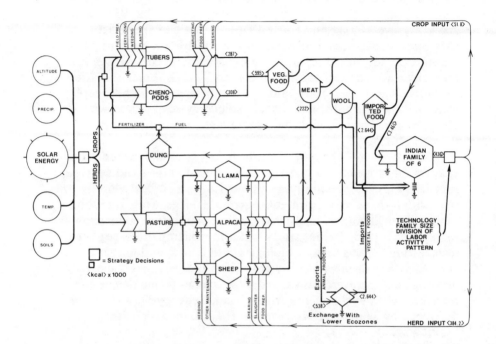

Figure 14.2. Energy flow in a highland Andean *puna* ecosystem. Values are in thousands of kilocalories (kcal x 1000) annual flow with reference to a Quechua Indian family of two adults and two children. (Adapted from Thomas 1976.)

cultivation and for cooking fuel and facilitated the acquisition of valuable trade commodities.

The human adaptability studies reviewed here demonstrated the value of investigating human populations within an ecosystem framework. However, these projects focused largely on human adaptive responses to the extant ecosystem. There was little consideration of the influence of the human population on ecosystem dynamics and development. Only patterns of energy flow were followed, whereas some ecosystem studies emphasized the influential role of, and limitations imposed by, nutrients such as nitrogen and phosphorus (Gorham et al. 1979; Odum 1969; Woodmansee 1978).

Man and the Biosphere Program

In the late 1960's, while the International Biological Program was in its data gathering phase, planning began within UNESCO to develop an international program of research called the *Man and the Biosphere Program* (MAB) (di Castri 1976). One of the main perspectives of the MAB program was spelled out in the 1971 MAB Paris report as its first objective: "to identify and assess the changes in the biosphere resulting from man's activities and the effects of these changes on man" (Unesco 1972). This underlined the central role of humans in the biosphere and supported the contention that human populations were integral parts of the ecosystems within which they lived. The MAB program, as originally formulated, promoted interdisciplinary science designed to encourage research that would provide long- term benefits to the human residents of participant nations. Implied in these early MAB perspectives was the notion that an understanding of the ways that the biosphere will continue to provide resources for people is through an interactive view of humans and the environment (see Bennett, this volume). That is, not only was a knowledge needed of human impact on the environment, but also needed was detailed information on the impact of human-induced environmental changes on humans in terms of their status and welfare.

As the MAB program developed, a number of ecosystem-oriented subprograms were defined for research activities. These included the following ecosystems: Arctic, Temperate, Mediterranean, Humid Tropical, Arid Tropical, and Mountain. Other subprograms included: Urban Ecosystems, Grazing Lands, Environmental Perception and Education, Conservation, and Environmental Pollution. In 1986, a reorientation of the MAB program was initiated in an attempt to focus research and encourage more of an emphasis on "scientific problems." The reorientation also resulted from an increasing awareness of the role of humans in producing changes in the biosphere (UNESCO 1987). Four cross-cutting themes were identified as part of this reorientation. The first three themes dealt with the influence of humans on the biosphere: 1) Ecosystem functioning under different intensities of human impact; 2) Management and restoration of human impacted resources; 3) Human investment and resource use. The last theme was identified as 4) Human response to environmental stress. This fourth theme is an excellent focus for studies in human biology which are directed to an understanding of human adaptation within an ecological context.

Stress can be viewed as a perturbing force arising from the environment that has a disruptive effect on humans. New diseases, toxic pollutants, or modified social systems can act as specific stresses, or the act of migration to a new area can expose the newcomer to multiple stresses in the novel environment. Stressful conditions can arise as the result of the introduction of a new food that can not be tolerated by members of a population (e.g., lactose intolerance), or by the cultural disintegration of traditional social values that effectively had dealt with common problems experienced by the family or by the community. Stress and its impact on health and well-being can be assessed by measures of: nutritional status, physical fitness, morbidity and mortality, child growth, reproductive function, social integration, affective function and psychological status, and even degree of substance abuse (Baker 1977, Mazess 1975, 1978, Little and Haas 1989). The *means of adjustment* to environmental stress, whether it originates from the sociocultural or the natural environment can be thought of as *adaptation*. Patterns of adaptation in human populations require some time to develop and to become refined. Environmental stress, as is likely to be defined for human populations, is largely a function of environmental change, and this phenomenon of dramatic and rapid change is what characterizes human existence today.

There are two projects that serve as models for MAB research, not only with regard to work that applies to Theme 4, but to the other themes as well. Both projects were conducted in Kenya. The first is the Integrated Project on Arid Lands (IPAL), which was a comparative study of "ways in which tropical savannas respond to natural and man-made stresses and disturbances" and the effects of these disturbances on the human populations (Lusigi 1981). This international MAB affiliated program of research compared several pastoral populations. In addition, IPAL dealt with varied systems of subsistence and resource extraction by the populations, and range savanna ecosystem conditions were studied using the concepts both of resilience and stability. Although emphasis was placed on human impact to the environment, human adaptation to the environment constituted a relatively small component of this project. Despite this limitation, IPAL was interdisciplinary and international in scope, took place at a regional scale of analysis, and fell well within the objectives of MAB projects.

The second project is also international and interdisciplinary, but not directly MAB-affiliated. It is the South Turkana Ecosystem Project, an ongoing, collaborative research program employing ecologists, social scientists, and specialists in biomedicine and human biology. The objectives are to understand and explain (1) how the Turkana are able to extract resources and survive in an arid and stressful environment by nomadic pastoral subsistence, (2) how these extractive techniques modify the dry savanna environment, and (3) what are the effects of the environment and Turkana cultural practices on the adaptability and health of the people. Of particular interest to MAB Theme 4 are the behavioral responses to stress through cooperation, reciprocity, exchange of livestock and labor, and nomadic movement. Of equivalent interest are the measures taken of biological adaptability to the environment, among them: reproduction, child growth, physical work capacity, nutritional status, disease prevalence, and mortality (Little 1980, 1989, Little *et al.* 1988a). Additional studies are being conducted of human responses to changes from a nomadic to a settled life style, and there will be valuable baseline data on traditional means of adaptation in order to assess the effects of future economic development on the people. Preliminary results suggest that Turkana nomadism is an effective means of extracting resources from the environment without major perturbation, while at the same time providing an effective means to buffer the population from periodic food shortages produced by drought (Coughenour *et al.* 1985, Ellis *et al.* 1987, Ellis and Swift 1988).

The remainder of this paper will outline the conceptual framework and some research results from this multidisciplinary ecosystem-human ecology project, initiated in 1980 and still underway in northwest Kenya.

The South Turkana Ecosystem Project

The South Turkana Ecosystem Project has as its major objective the analysis of the role of human populations in a dry savanna ecosystem. Ecological anthropology and plant, animal and soil ecology have been incorporated within an ecosystems framework in which the research design focuses: 1) on the influence of pastoralists and their livestock on the structure and productivity of an arid savanna ecosystem and 2) on the patterns of adaptation of the human populations enabling them to survive and persist in this ecosystem.

South Turkana is a region of northwest Kenya characterized as arid to semi-arid or at the xeric end of the East African savanna moisture gradient. The topography of the region is rugged and soils are largely coarse sandy alluvium, thin stony lithosols, or degrading basalts. Ambient temperatures are uniformly high throughout the year which, when combined with a low annual rainfall (100 to 600 mm/yr) (Little and Johnson 1985, Coughenour *et al.* in press) that is concentrated into only a few months (March to May) and a high evapotranspiration rate, leads to a marked moisture deficit. Figure 14.3 illustrates the high degree of variation in rainfall that is found seasonally and from year-to-year at one weather station. Seasonal drought occurs every year, whereas longer-term droughts (e.g., June 1979 to March 1981), which can have disastrous effects on pastoralists, occur, on average, once per decade (Ellis *et al.* 1987). Soil and rainfall patterns contribute to a vegetation that is xerophytic and dominated by *Acacia* spp. trees, a variety of shrubs, and grasses and forbs. The South Turkana Ecosystem Project has concentrated on a region inhabited by a southern branch of the Turkana tribe, a group of migratory or nomadic pastoralists who herd cattle, camels, goats, sheep and donkeys, and who live under relatively traditional conditions. They call themselves the Ngisonyoka Turkana.

Figure 14.4 is a simplified model of the state variables, material flows, and controls in the Turkana ecosystem as originally conceptualized when the project was designed in the early 1980's. It remains a useful representation of some of the relationships in the system. Basically, plant biomass is consumed by livestock (cattle, camels, goats, sheep and donkeys), and the products of the livestock (milk, blood, meat, hides) are either consumed by the Turkana or

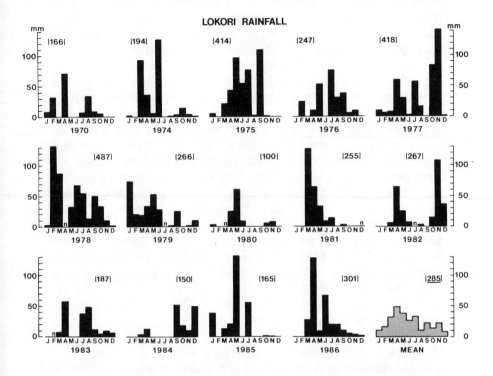

Figure 14.3 Rainfall at the Lokori weather station in South Turkana (1 57'N, 36 02'E, 762 m elevation).

traded for food and other items. In order to maintain the output of livestock products, the Turkana must manage the livestock in specific ways. The impact of their management has the potential to alter photosynthesis rates, respiration rates, plant species composition, total plant biomass, and soil nutrient conditions.

The project has several principal and many lesser components that reflect both the project objectives and the capabilities of the research personnel. There is considerable complementarity in these components. For example, livestock are of interest to the ecologists in terms of their impact on the ecosystem and their consumption of plant productivity (Coppock *et al.* 1985, 1986a), while sociocultural anthropologists are concerned with livestock management strategies and how the cultural system facilitates or constrains the implementation of these strategies (N. Dyson-Hudson 1984, McCabe 1988, McCabe *et al.* 1988). Human biologists, on the other hand, are concerned with the physical effort required to manage the livestock, the dietary products derived from the animals, and the adequacy of these products as they contribute to child growth, nutritional status and health status of the Turkana (Little *et al.* 1988a, 1988b).

The People and the Study Area

Work has been conducted largely with a "neighborhood" Turkana community of the Ngisonyoka tribal subsection. The whole community, which may congregate during the wet season, numbers up to 50 family production units or more than 1000 persons. Of these, we have long-term, close contact and excellent rapport with 10 to 20 family units. Census and demographic data, information on family unit movements and social relations of herd owners and members of their families, and data on individual health and biology has been gathered on more than a thousand individuals.

The social organization of the Turkana has several structural levels and is characterized by some flexibility and individual variation. The largest unit with which we are working is the Ngisonyoka subsection (subtribe) or *socio-territorial group*. This group has a spatially demarcated home range which is an area where Turkana expect to encounter the people so named and where they have priority of resource use. The smallest level of social organization is the *family production unit* represented by the family *awi* or settlement and satellite herding camps.

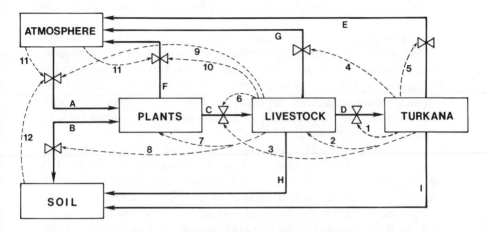

Figure 14.4. A control diagram representing state variables (atmosphere, soil, plants, livestock, Turkana people), energy and material flows (A through I), and some system controls (dashed lines 1 through 12 with "bow tie" control junctures) for the South Turkana ecosystem. Energy and material flows are: A-photosynthesis, B-nutrient uptake and decomposition, C- livestock consumption, D-Turkana consumption, E-Turkana energy expenditure, F-plant respiration, G-livestock energy expenditure, H-livestock nutrient decomposition and release, I-Turkana nutrient decomposition and release.

Typically a production unit consists of an extended family, sometimes with other dependents, but in its basic form, including a family head/herd owner, his wives and their children. On occasion, the awi will include two or more herd owners tied by bonds of kinship or friendship. Intermediate between the socio-territorial and family production groups is the transient set of awis that form *neighborhood associations* or *networks*, called *adakars*. They constitute temporary common interest groups that form under a variety of conditions for defense as well as for social living, act as information networks, and provide a context of general insurance against emergencies in the small settlement groups of which they are constituted (Dyson-Hudson and Dyson-Hudson 1982, R. Dyson-Hudson 1989).

The project study area is defined by the region inhabited by members of the Ngisonyoka subsection (see Figure 14.5). It is approximately 9000 km^2 in size and consists of a central north- south massif with elevations up to 2500 m, a broad alluvial dry- season valley in the west adjacent to the Turkwel River (*Naroo*), an alluvial wet-season plain in the central area (*Toma*), lava hills in the east, a rocky pediment in the southeast (*Nadikam*), and lava plains north of the Kalapata River. Vegetation characteristics of each of these areas are described by Ellis and Coppock (1985) and Coughenour and his colleagues (in press). There is a north-south rainfall gradient, with wetter areas in the south and very dry conditions toward Lodwar in the north (Little and Johnson 1985). Vegetation and water availability are the major ecological factors determining livestock and human rangelands exploitation patterns. A general vegetation map of the ecosystem was developed from on-ground surveys, aerial photographs, and LANDSAT imagery (Ellis and Dick 1985). In addition, water-holes were mapped in a large part of South Turkana, and their permanence, quality for different livestock species, and constraints on their use were recorded (Dyson-Hudson and McCabe 1985).

Human Resource Extraction

How the Turkana wrest a living from an environment that has limited resources is one of the fundamental questions of the project. This is best addressed by reference to the energy flow model or diagram of Coughenour and colleagues (1985) that represents, synoptically, the pattern of Ngisonyoka resource exploitation in

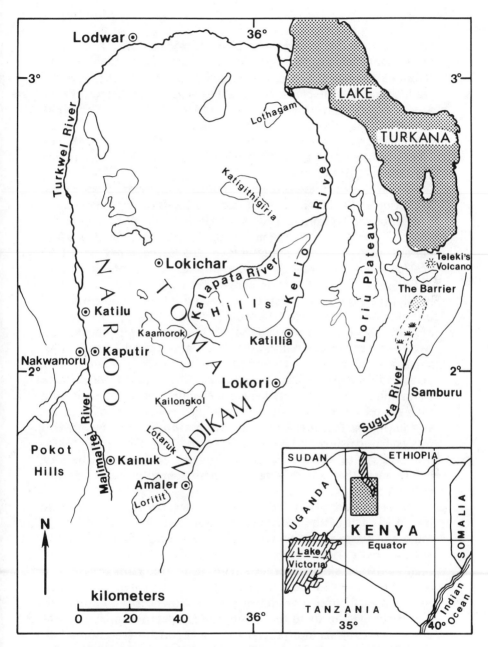

Figure 14.5. Map of the South Turkana region. The research area of the South Turkana Ecosystem Project (and Ngisonyoka territory) is roughly bounded by the Turkwel and Kerio Rivers, although Ngisonyoka Turkana may move beyond these borders from time to time, particularly during drought.

South Turkana (see Figure 14.6). The left side of the diagram traces
the flows of energy from plant products (forage) consumed by
livestock (donkeys, cattle, sheep, goats, camels) to the livestock
products (milk, meat, blood) consumed as food by the people. Values
are in GJ/person/year (1 GJ = 2.389 x 10^5 kcal). The right
side of the diagram represents non-pastoral sources of food that are
either gathered and hunted or traded. A central flow item, derived
from shrubs and trees, leads to fuel use and settlement construction
materials (Ellis *et al.* 1984).

A number of interesting relationships are apparent from a close
inspection of the diagram. First, pastoral foods constitute about 75
percent of the Ngisonyoka dietary calories, and all foods provide
about 1275 kcal/person/day (an average for all ages), which is a low,
but adequate per capita energy intake (1.95 GJ/person/year = 1275
kcal/person/day). These data are in general agreement with results
of Galvin's (1985, 1988, Galvin and Waweru 1987) dietary work, and
predicted energy requirements from activity levels, body size, and
population structure (Leslie *et al.* 1984, Little *et
al.* 1988b). Four categories of plants serve as food for the
livestock: herbaceous plants (including grasses and forbs); dwarf
shrubs (principally *Indigofera* spp.); *Acacia* seed
pods; and foliage from shrubs and trees. The Turkana utilize these
plants and capture the nutrients for their livestock by maintaining a
high degree of mobility and by moving livestock quickly to tap
seasonal grass flushes. Turkana also maintain separate grazing
orbits for different livestock species (McCabe *et al.* 1988),
and are quite skilled at exploiting an ecologically patchy
environment that is highly variable both spatially and temporally
(Ellis and Swift 1988).

If the pathways from the four plant categories in Figure 14.6 are
traced to each of the livestock species, it is clear that donkeys,
cattle, and sheep are principally herbaceous grazers, whereas camels,
and to a lesser degree, goats, depend heavily on browse vegetation
such as dwarf shrubs, larger shrubs, and tree foliage for food
(Coppock *et al.* 1986a). These graze/browse needs of given
species are of particular concern to herd owners who must lead
species-specific herds to the appropriate vegetation mix, and who
must take into account the seasonal variation (see below) in
availability of categories of vegetation.

Each of the livestock species makes a different contribution to
pastoral food production on an annual basis. Camels are the

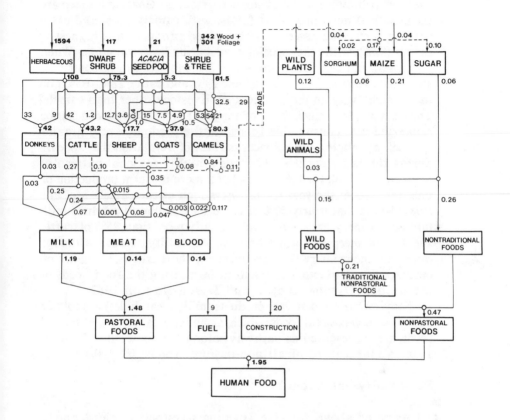

Figure 14.6. Some of the pathways for the flow of energy through the Ngisonyoka
Turkana pastoral system. Values are in gigajoules of energy (GJ) per person
per year (1 GJ = 2.389 x 105 kcal of energy). (Adapted from Coughenour *et
al.* 1985.)

principal milk producers (56%), with cattle (21%) and goats and sheep
(20%) contributing equal, but smaller, amounts. Goats and sheep are
the principal meat producers (56%), with camels (33%) and cattle
(10%), contributing lesser amounts. Blood production is highest in
camels (82%), with other species contributing lesser amounts. When
milk, meat, and blood food production is combined, camels contribute
about 56%, goats and sheep contribute 23%, and cattle, 18%. Based on
these values, why do the people keep livestock other than camels?
The answer is that much more of the plant biomass produced can be
converted into consumable foods via the predominantly grazing
livestock by tapping the herbaceous flush of vegetation that appears
during the rainy periods. There is a highly seasonal pattern of
production of milk for the grazers, but a more uniform production of
milk throughout the year for browsers such as the camel. McCabe
(1984) found that nearly 50% of camels were able to produce milk
throughout the year, whereas only slightly more than 25% of cattle
and 16% of sheep and goats had this capability. Hence, although
camels are the primary food producer for the Turkana, a balance of
other species of livestock is essential in meeting the food needs of
the people (Dyson- Hudson and Dyson-Hudson 1982).

Some animals and animal products (milk, meat, hides) are sold or
traded for non-pastoral foods, which, along with gathered and hunted
foods, are represented as pathways along the right side of Figure
14.6. About a third of all non-pastoral foods are wild foods.

Seasonality and Drought

As noted above, the five Turkana livestock species (camels,
cattle, goats, sheep, donkeys) feed on different forage components,
their diets being separated on the basis of plant species, plant
parts, vegetation height, and habitat use (Coppock *et al.*
1986a). It is clear, also, that different vegetation canopy layers
(herbaceous, dwarf shrub, tall shrub, tree) undergo quite different
phenological sequences due, most likely, to variations in rooting
depth and ability to gain access to water during rainy and dry
seasons. For example, large acacia trees with deep root systems are
nearly always found along borders of dry stream beds where
below-ground water reserves persist throughout most dry seasons.
Thus, diet quality, nutritional condition, and production of human
food (including milk, blood and meat) vary greatly throughout the
year among the livestock species (Coppock *et al.* 1986b,

1986c, Galvin 1985, McCabe 1984). Turkana patterns of exploitation are sensitive to these variations in food production, most of which are a function of seasonal variation.

In any given year, good vegetation conditions for livestock feeding (plenty of green herbaceous, shrub and arboreal growth) persist only for a few months during and after the period of rains. A sequence of annual events can be described, keeping in mind that the Turkana system is highly variable through time and "an average year" is really an illusion. First, a period of several months of seasonal drought is followed by a rainy season, beginning in March or April (see Figure 14.3). Within a few days, grass and forb shoots begin to arise from the sandy soil, and, if the rains continue, sandy plains are transformed into grassy, flowered meadows, and other vegetation, including shrubs and trees, becomes green. At this time, livestock begin to feed actively on new vegetation, regain some of their lost body weight, and, if they give birth to young, begin to lactate and produce milk. Milk yields tend to peak about two or three months after the peak rainfall (Galvin 1985, McCabe 1984). These peak milk yields result from livestock having had this period of heavy feeding to regain lost body weight. In parallel with livestock, humans also lose body weight during the dry season, although the percent weight loss is considerably less for Turkana than for their livestock. For example, livestock may lose up to 20 percent of body weight (Wilson *et al.* 1985, Coppock *et al.* 1986c), while human loss of body weight seldom exceeds 12 percent. Weight gain in humans follows weight gain in livestock by several weeks (Galvin 1985). Basically, the seasonal pattern is: (1) structured by rainfall, (2) followed by a vegetation flush, (3) followed by improved weight of livestock, (4) leading to increased milk production, and (5) leading to increased food intake by the Turkana and by a regaining of lost body weight. It is a typical seasonal pattern of exploitation of successive trophic levels with time lags in the sequence.

Figure 14.7 illustrates the remarkable variation by season of human milk consumption as a percentage of total caloric intake (Galvin 1985). Toward the end of the rainy season in 1982 when livestock were maximally productive, nearly 90 percent of caloric intake was in the form of milk or milk products. During seasons in which livestock milk production and human milk consumption are low, other foods (such as blood, meat, maizemeal, wild foods) make up some of the human energy needs, but some of the energy needs are derived from human body stores. This leads to weight loss in adults that may amount to several kilograms during particularly dry years.

Weight losses and gains in small samples of Turkana women and men are shown in Figure 14.8 (Galvin 1985). Fluctuations in body weight of these magnitudes indicate a negative energy balance between caloric intake and expenditure (through activity) that principally occurs during the late dry season. In most Western individuals with adequate reserves of body fat, weight loss of even up to ten percent of total weight would not be a health threat. However, the Turkana have limited reserves of fat and have a small muscle mass, so very

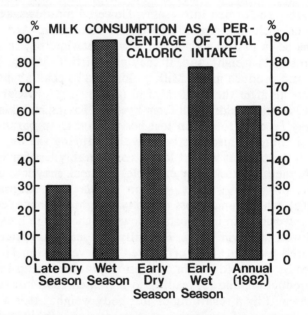

Figure 14.7. Milk consumption as a percentage of total caloric intake by season for Ngisonyoka Turkana. The year 1982 was marked by two distinct rainy seasons with a four-month dry season between the two. (After Little 1989, data from Galvin 1985.)

slight losses of body weight may impair the health of both women and men. Another factor that contributes to weight loss in adults is a value that Ngisonyoka Turkana adults will undergo voluntary fasting in order to ensure that children are well fed. Hence, although the adults are buffered from major weight loss (from starvation) by other sources of food, children have an additional level of buffering through greater access to food. The effects of limited food energy during the dry season on children and on their growth processes is not yet known, although some disruption of growth processes would be expected with seasonal hunger (Little 1989).

One of the best documented effects of seasonality on Ngisonyoka Turkana is in numbers of births (Leslie and Fry 1989). Month and year of 452 Ngisonyoka Turkana births over a 25-year period were determined, and they showed a highly seasonal distribution with a birth peak in April and May. However, as noted by the authors, the timing of conceptions is likely to be of greater significance than

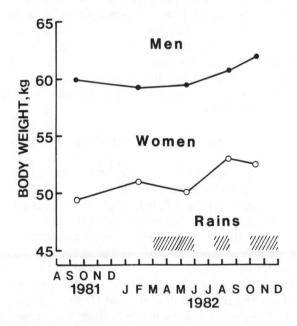

Figure 14.8 Body weight changes by season in Ngisonyoka Turkana. (Data from Galvin 1985.)

the timing of births. With this in mind, births were projected backward nine months to be converted to month of conception. The distribution of estimated conceptions by month is superimposed on average rainfall from Lokori in Figure 14.9. The degree of seasonality found for the Turkana is one of the most extreme reported in the literature and is likely to be linked to the sequence of events initiated by seasonal rainfall, and following through the vegetation flush, livestock recovery and milk production, human health and nutritional status.

In addition to seasonality, periodic droughts of varying duration influence the Ngisonyoka Turkana and their ecosystem. While single-year droughts are, in effect, similar to a very long dry season, multi-year droughts have devastating effects on livestock populations and their productivity. For example, livestock populations may drop by 50 percent or more during severe droughts (McCabe 1988). Such devastating droughts induce shifts in human food acquisition, stimulate human emigration from the pastoral sector, and cause far-flung movements in search of livestock forage. The relatively high frequency of these multi-year droughts (about one in ten years) appears to play an important role in limiting livestock numbers and in inducing non-equilibrium interactions between livestock and the plant community (Ellis *et al.* 1987, Ellis and Swift 1988). Figure 14.10 illustrates relationships between livestock numbers and forage availability (plants) during single-year and multi-year droughts.

Health and Adaptability

The nutritional status, child growth, body composition, fitness level, reproduction, disease status, and population dynamics of the Turkana people are largely a function of behavioral patterns of pastoral resource exploitation within an arid, highly seasonal, stressful ecosystem (Little 1980, 1989). The stresses imposed on the Ngisonyoka Turkana include: (1) high temperatures and intense solar radiation, (2) limited moisture and low, unpredictable rainfall, (3) limited resources and highly seasonal vegetation availability, (4) endemic human and livestock diseases, and (5) social pressures that arise from within and outside of the immediate human population. Approaches to the study of Turkana health and adaptability have focused on the six topics discussed below. Each topic can be thought of as a measure of adaptation to the arid, pastoral, savanna environment.

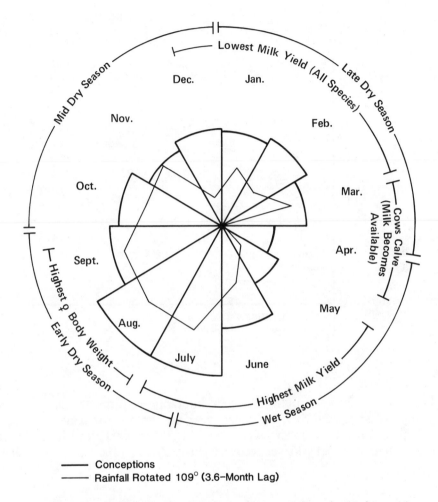

—— Conceptions
—— Rainfall Rotated 109° (3.6–Month Lag)

Figure 14.9. Seasonal distribution of conceptions in Ngisonyoka Turkana in conjunction with seasonal rainfall and activities of people and livestock. The rainfall distribution in the figure has been rotated (advanced) to its best fit with the conception distribution, to provide an estimate of the total lag in the chain of responses described in the text. (After Leslie and Fry 1989.)

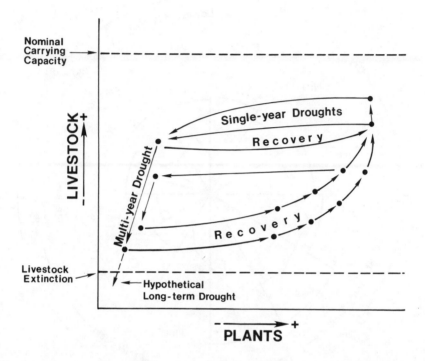

Figure 14.10. Turkana plant-livestock interactions under the influence of frequent drought perturbations. Single-year droughts allow relatively rapid recovery in livestock numbers, whereas multi-year droughts lead to dramatic losses of livestock and a longer recovery period. (After Ellis and Swift 1988.)

Diet and nutrition

The Turkana have a diet which consists of pastoral foods (milk, meat, and blood); traded foods (largely maize meal, but with some sugar included); and wild foods, through gathering and hunting (Galvin 1985, 1988). Protein intake is very high at between 200 and 400 percent of recommended daily intakes (FAO/WHO 1973). Energy intake, however, is very low, particularly in adults and some adolescents. During the dry season, some adults may fast every few days, and even go without food for several days. These practices lead to losses in body weight that are common during the several months of the late dry season, as noted above (Galvin 1985).

Food intake varies considerably by season, where, although milk is a staple, it only constitutes a small portion of the diet during prolonged dry seasons or under drought conditions. During periods when milk production is low, animals are tapped for blood, small stock (goats and sheep) will be slaughtered for meat, and live animals will be sold to purchase maize meal. Under extreme conditions, such as the drought from mid-1979 to early 1981, meat consumption was very high because animals were dying and were providing large amounts of food (McCabe 1984, Ellis *et al.* 1987). There were later, harmful effects of this drought. Following the onset of rains in 1981, most livestock had very low milk production because they had aborted or lost their young animals during the drought (Dyson-Hudson 1989). Hence, the year following the drought was a difficult one because herds were depleted and milk production was low. All of these seasonal and longer-term events underline how difficult it is to characterize the diet of the Turkana in any "average" or typical manner.

Infant, child and adolescent growth

The growth patterns of infants, children, and adolescents are good indicators of health status at different ages. Growth of more than 800 Turkana youths and young adults has been assessed by anthropometric measurements since 1981. Most of the data are cross-sectional (i.e. measurements taken at one time) (Little *et al.* 1983), although some children have been measured longitudinally (more than one measurement) (Little and Johnson 1987). Figure 14.11 shows growth in height and weight of Turkana

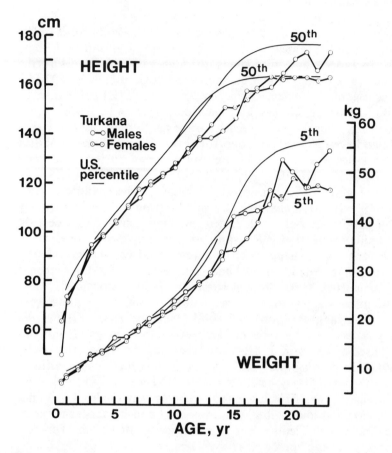

Figure 14.11. Growth in height and weight of Ngisonyoka children, adolescents, and young adults. (After Little and Johnson 1987.)

boys and girls. Linear growth (height) of Turkana children lags behind United States children throughout the growth period, but both males and females approach very closely the 50th percentile rankings in early adulthood. Shortened stature or "stunting" is not characteristic of the Turkana in adulthood. Growth in weight, however, presents a different picture. Here Turkana show substantial retardation in weight throughout the growth period, such that as young adults, they still fall at or below the 5th percentile for U.S. standards.

It been suggested that Turkana children are able to show adequate linear growth despite limited energy intake because of their high protein intake (see above) (Little and Johnson 1987). This protein hyper-sufficiency may drive linear growth processes and act on Turkana children as protein supplements have acted on children from other populations (Lampl 1978, Frisancho 1980).

Adult size and body composition

Adult Turkana men and women are tall and lean by world standards. When compared with U.S. norms, muscle mass tends to be very small in men and relatively small in women, and fat stores are low in both sexes (Little and Johnson 1986). These limited tissue reserves place Turkana adults at risk during periods of food scarcity during dry seasons and drought. Fat stores, which are depleted when energy intake is less than output, tend to decline in women from a peak at age 18 throughout the woman's reproductive years (Little *et al.* 1983). This probably results from the increased energy needs of pregnancy and breastfeeding, but also is probably due to the increased physical activity associated with marriage and managing the home settlement.

Activity and physical fitness

The Turkana are quite active in the management of their livestock and their own lives. Day-to-day tasks include: walking herds to distant grazing and browsing areas; milking, bleeding, and watering animals; collecting firewood and fetching water; visiting other settlements; moving and constructing new settlements; and other, more sedentary, activities. These activities are divided by age and sex, such that boys and young men carry out most of the herding tasks, whereas girls and women are responsible for managing the settlement, providing firewood and water, and milking the livestock. Watering the livestock, a strenuous activity during the dry season, is shared by young men and women.

Of particular interest is the balance between activities which expend energy and food energy intake by season, since weight loss occurs during the late dry season. A preliminary estimate of seasonal variation in activities (by timed activity observation) suggested that there are major differences in activity levels by season (Little *et al.* 1988b), which verifies the role of limited food intake on seasonal weight loss. Although not fully verified, there is some evidence that people cut back on activities when food is in short supply, presumably to conserve energy.

The relationships between some of the Turkana activities and physical fitness were tested by Curran-Everett (1990), who measured aerobic capacity (i.e. maximal oxygen consumption, or max VO2) in a sample of Turkana men. Previous work (diPrampero and Cerretelli 1969) had found an extremely high aerobic capacity in a small sample of Turkana men. Curran-Everett's (1990) study suggested a lower aerobic capacity more in line with expected fitness for an active, but not trained, people.

Reproduction

Based on earlier work by Brainard (1981) on recently settled Turkana who had been nomadic during their reproductive years, fertility was shown to be relatively high, and, in fact, higher than fertility of Turkana who had been sedentary for many years. These results were at variance with the general literature in which settled or sedentary existence was always accompanied by greater fertility than in nomads (Henin 1968, Ganon 1975). The results of the recently-settled Turkana study were verified when practicing Turkana nomads were found to have high fertility as well (Leslie *et al.* 1988). In this study, Turkana were found to have a completed fertility (i.e. total live births per post-menopausal woman) of about 7 births. Since the time of Brainard's study, high fertility has also been observed in Malian Fulani nomads (Hill 1985).

In the context of birth seasonality, these results lead to a series of interesting questions about Turkana reproduction that we are currently exploring. For example, marked seasonality of births in Turkana women suggests that there is some periodic environmental stress that limits fertility. Under these circumstances, one would expect Turkana fertility to be relatively low because of this pattern of periodic environmental stress. This is not the case: fertility is high. Another interesting question is: what is the basis for the depressed birth rate during the late dry season and early wet season? It is inappropriate to explain this as simply a function of poor nutritional status in women, when many other variables, such as intrauterine and infant mortality, breastfeeding patterns, male fecundity, and frequency of coitus, can influence female fertility (Leslie and Fry 1989). Current investigations are involved in dealing with some of these variables and measuring more directly characteristics of the female ovulatory cycle.

Disease and mortality

The health picture of Ngisonyoka Turkana is less clear than other aspects of their human biology. We know that malaria is a serious problem based on clinical data, spleen or liver enlargement, anemia, and symptom reporting, and that prevalence of the disease seems to average about 10 percent of the population (Shelley 1985). Analysis of stool samples indicated that other parasites were in low prevalence and not a serious threat to child and adult health (Shelley 1985). Nomadic children seem in good health, and, despite the annual period of hunger, do not show any symptoms of marasmus or kwashiorkor. Despite the appearance of relatively good health, infant mortality appears to be higher in nomads than in settled Turkana farmers (Brainard 1986), so much more research is needed to fully define disease and mortality patterns in Turkana youths and adults. Current research focuses on collecting prospective data on infant mortality and immunocompetence assessment of adults and infants.

Integration of Human Biology and Ecosystem Studies

Past studies of human biological adaptation to the environment suffered from two major shortcomings. The first is that very little information was gathered on the sociocultural patterns of the people under investigation. These data are important because in traditional subsistence societies the structure and organization of the society is at the same time (1) a reflection of behavioral adaptations to the ecosystem and (2) a sociocultural system with rules of behavior that regulate, at least in part, the exploitation of the ecosystem. Among the United States IBP human adaptability projects, sociocultural data links were weak in the Andean Indian, Alaskan Eskimo, and Aleut

(Laughlin and Harper 1979) projects. The Amazon Basin Yanomama Indian project was a notable exception in its comprehensive cultural studies but was deficient in examining environmental variables.

The second shortcoming of human adaptability studies has been in the limited amount of understanding sought about the ecosystem within which human adaptation takes place. If an environmental stress design is applied in biological adaptation research, then the dynamic attributes of the stresses arising from the ecosystem must be defined. In the case of the Turkana, their limited dietary energy intake could result, among other things, from low milk production due to inadequate plant forage quality, or social factors associated with herd management practices. The conditions surrounding the low human energy intake can only be defined by detailed work on the trophic relationships between primary productivity, the livestock, and their human managers. Modeling and simulation can be used to solve scientific problems with complex sets of relationships such as these, but extensive fieldwork is necessary for validation of the models, as in the case of the South Turkana Ecosystem Project.

It is our belief that the structure, dynamics, maintenance and evolution of human populations are controlled by the action of sociocultural, ecological, and human biological factors. These controls are highly complex and interactive. Accordingly, an integrated, multidisciplinary, ecologically-based research (i.e. ecosystemic) approach is likely to be most productive in understanding these relationships.

Acknowledgments

This research was conducted as a part of the South Turkana Ecosystem Project. Research clearance was provided by the Government of Kenya, Office of the President, under No. OP/13/001/7C 120/54. Funding was provided by the U.S. National Science Foundation under NSF Grants BNS 78-15923, BNS 80-107800, DEB 80-04182, DEB-82-06864, BSR-86-04774, and BNS-87-18477. We acknowledge, with appreciation, our close associations with the Kenya Medical Research Institute (KEMRI), the University of Nairobi, and the Kenya National Museum. Our thanks are due to our co-workers who contributed to the project at all levels. They are: Drs. D. Layne Coppock, Michael B. Coughenour, Linda Curran-Everett, Brooke R. Johnson, Jr., J. Terrence McCabe, Geoffrey M.O. Maloiy, Mutuma Mugambi, Mworia Mugambi, Karen Shelley, and Jan Wienpahl. We particularly thank Eliud A. Lowoto, whose field expertise was invaluable, Lopeyon, for his knowledge of the South Turkana environment, Mohammed Bashir, for his companionship and hard work, and our Turkana hosts for their friendship and laughter.

References Cited

Baker, P.T.
    1958   Racial differences in heat tolerance. *American Journal of Physical Anthropology* 16:287-305.

Baker, P.T.
    1962   The application of ecological theory to anthropology. *American Anthropologist* 64:15-22.

Baker, P.T.
    1977   Problems and strategies. *Human Population Problems in the Biosphere: Some Research Strategies and Designs*, Edited by P.T. Baker. MAB Technical Notes 3. Paris: Unesco. Pp. 11-32.

Baker, P.T. and M.A. Little eds.
    1976   *Man in the Andes: A Multidisciplinary Study of High Altitude Quechua.* Stroudsburg, Pa.: Dowden, Hutchinson, and Ross.

Baker, P.T. and J.S. Weiner eds.
    1966   *The Biology of Human Adaptability.* Oxford: Clarendon Press.

Barth, F.
    1969   *Ethnic Groups and Boundaries: The Social Organizations of Culture Difference.* Boston: Little, Brown and Co.

Bates, M.
    1953   Human ecology. *Anthropology Today: An Encyclopedic Inventory.* Edited by A.L. Kroeber. Chicago: University of Chicago Press. Pp. 700- 713.

Bates, M.
    1960   Ecology and evolution. *Evolution After Darwin, Vol. I, The Evolution of Life.* Edited by S. Tax. Chicago: University of Chicago Press. Pp. 547- 568.

Birdsell, J.B.
    1953   Some environmental and cultural factors influencing the structuring of Australian aboriginal populations. *American Naturalist* 87:171-207.

Brainard, J.M.
  1981   Herders to Farmers: The Effects of Settlement on the
          Demography of the Turkana Population of Kenya. Ph.D.
          Dissertation in Anthropology. Binghamton: State
          University of New York.
Brainard, J.M.
  1986   Differential mortality in Turkana agriculturalists and
          pastoralists. *American Journal of Physical
          Anthropology* 70(4):525-536.
Brown, G.M. and J. Page
  1952   The effect of chronic exposure to cold on temperature
          and blood flow to the hand. *Journal of Applied
          Physiology* 5:221-227.
Brown, J., P.C. Miller, L.L. Tieszen, and F.L. Bunnell
  1980   *An Arctic Ecosystem: The Coastal Tundra at Barrow,
          Alaska.* Stroudsburg, Pa.: Dowden, Hutchinson, and
          Ross.
Brues, A.M.
  1972   Models of race and cline. *American Journal of
          Physical Anthropology* 37:389-400.
Chagnon, N.A.
  1968   *Yanomamo: The Fierce People.*   New York:
          Holt, Rinehart, and Winston.
Coon, C.S., S.M. Garn, and J.B. Birdsell
  1950   *Races: A Study of the Problems of Race Formation in
          Man.* Springfield, Illinois: C.C. Thomas.
Coppock, D.L., J.T. McCabe, J.E. Ellis, K.A. Galvin, and D.M. Swift
  1985   Traditional tactics of resource exploitation and
          allocation among nomads in an arid African environment.
          *Proceedings of the International Rangelands
          Resources Development Symposium, Society for Range
          Management, Salt Lake City.* Edited by L.D. White
          and J.A. Tiedeman. Pullman: Department of Forestry and
          Range Management, Washington State University. Pp.
          87-96.
Coppock, D.L., J.E. Ellis, and D.M. Swift
  1986a   Livestock feeding ecology and resource utilization in a
          nomadic pastoral ecosystem. *Journal of Applied
          Ecology* 23:573-583.
Coppock, D.L., D.M. Swift, and J.E. Ellis
  1986b   Seasonal nutritional characteristics of livestock diets

in a nomadic pastoral ecosystem. *Journal of Applied Ecology* 23:585-595.

Coppock, D.L., D.M. Swift, J.E. Ellis, and K. Galvin
  1986c   Seasonal patterns of energy allocation to basal metabolism, activity and production for livestock in a nomadic pastoral ecosystem. *Journal of Agricultural Science* (Cambridge) 107:357-365.

Coughenour, M.B., J.E. Ellis, D.M. Swift, D.L. Coppock, K. Galvin, J.T. McCabe, and T.C. Hart
  1985   Energy extraction and use in a nomadic pastoral ecosystem. *Science* 230(4726):619-225.

Coughenour, M.B., D.L. Coppock, J.E. Ellis, and M.Rowland
  In press Herbaceous forage variability in an arid pastoral region of Kenya: Importance of topographic and rainfall gradients. *Journal of Arid Environments.*

Curran-Everett, L
  1990   Age, Sex, and Seasonal Differences in Work Capacity of Ngisonyoka Turkana Pastoralists. Ph.D. Dissertation in Anthropology. Binghamton: State University of New York.

di Castri, F.
  1976   International, interdisciplinary research in ecology: Some problems of organization and execution. The case of the Man and the Biosphere (MAB) Programme. *Human Ecology* 4(3):235-246.

di Castri, F., A.J. Hansen, and M.M. Holland (eds.)
  1988   A New Look at Ecotones: Emerging International Projects on Landscape Boundaries.*Biology International*, Special Issue 17. Paris: International Union of Biological Sciences.

diPrampero, P.E. and P. Cerretelli
  1969   Maximal muscular power (aerobic and anaerobic) in African natives. *Ergonomics* 12:51-59.

Dyson-Hudson, N.
  1980   Strategies of resource exploitation among east African savanna pastoralists. *Human Ecology in Savanna Environments.* Edited by D.R. Harris. London: Academic Press. Pp. 171-184.

Dyson-Hudson, N.
  1984   Adaptive resource use strategies of African pastoralists. *Ecology in Practice, Part I:*

Ecosystem Management. Edited by F. di Castri,
F.W.G. Baker, and M. Hadley, eds. Dublin: Tycooly
International Publishing. Pp. 262-273.

Dyson-Hudson, N. and R. Dyson-Hudson
1982   The structure of East African herds and the future of
East African herders. *Development and Change*
13:213-238.

Dyson-Hudson, R.
1989   Ecological influences on systems of food production and
social organization of South Turkana pastoralists.
*Comparative Socioecology: The Behavioral Ecology of
Humans and Other Mammals.* Edited by V. Standen and
R.A. Foley. Oxford: Blackwell. Pp. 165-193.

Dyson-Hudson, R. and N. Dyson-Hudson
1980   Nomadic   pastoralism.   *Annual   Review   of
Anthropology* 9:15-61.

Dyson-Hudson, R., and J.T. McCabe
1985   *South   Turkana   Nomadism: Coping   with   an
Unpredictably Varying Environment.* New Haven: Human
Relations Area Files. Hraflex Books.

Ellis, J.E. and D.L. Coppock
1985   Vegetation patterns in Ngisonyoka Turkana, Appendix
II. *South Turkana Nomadism: Coping with an
unpredictably Varying Environment.* New Haven:
Human Relations Area Files. Pp. 315- 330.

Ellis, J.E. and O. Dick
1985   *Landsat vegetation map of Turkana District,
Kenya.* Oslo: Norwegian Agency for International
Development.

Ellis, J.E. and D.M. Swift
1988   Stability of African pastoral ecosystems: alternate
paradigms and implications for development.
Journal of Range Management 41:450- 459.

Ellis, J.E., D.L. Coppock, J.T. McCabe, K. Galvin, and J. Wienpahl
1984   Aspects of energy consumption in a pastoral ecosystem:
Wood use by the South Turkana. *Wood, Energy and
Households, Perspectives on Rural Kenya.* Edited by
C. Barnes, J. Ensminger, and P. O'Keefe. Stockholm and
Uppsala: The Beijer Institute and the Scandinavian
Institute of African Studies. Pp. 164-187.

Ellis, J.E., K. Galvin, J.T. McCabe, and D.M. Swift
1987 *Pastoralism and Drought in Turkana District,Kenya.* Bellevue, Colorado: Development Systems Consultants, Inc. A report to the Norwegian Aid Agency (NORAD).

Frisancho A.R.
1980 Role of calorie and protein reserves on human growth during childhood and adolescence in a Mestizo population. *Social and Biological Predictors of Nutritional Status, Physical Growth, and Neurological Development.* Edited by L.S. Green and F.E. Johnston. New York: Academic Press. Pp 49- 58.

Galvin, K.
1985 Food Procurement, Diet, Activities and Nutrition of Ngisonyoka, Turkana Pastoralists in an Ecological and Social Context. Ph.D. Dissertation in Anthropology. Binghamton: State University of New York.

Galvin, K.
1988 Nutritional status as an indicator of impending food stress. *Disasters* 12(2):147-156.

Galvin, K. and S.K. Waweru
1987 Variation in the energy and protein content of milk consumed by nomadic pastoralists of northwest Kenya. *Food and Nutrition in Kenya: A Historical Review.* Edited by A.A.J. Jansen, H.T. Horelli and V.J. Quinn. Nairobi: UNICEF and University of Nairobi. Pp. 129-138.

Ganon M.F.
1975 The nomads of Niger. *Population Growth and Socioeconomic Change in West Africa.* Edited by J.C. Caldwell, N.O. Addo, S.K. Gaisie, A. Igun, and P.O. Olusanya. New York: Columbia University Press. Pp. 697-700.

Golley, F.B.
1984 Historical origins of the ecosystem concept in Biology. *The Ecosystem Concept in Anthropology.* Edited by E.F. Moran. Boulder: Westview Press. Pp. 33-49.

Gorham, E., P.M. Vitousek and W.A. Reiners
1979 The regulation of chemical budgets over the course of terrestrial ecosystem succession. *Annual Review of Ecology and Systematics* 10:53-84.

Gulliver, P.H.
1951   *A preliminary survey of the Turkana: a report compiled for the government of Kenya.* Communication from the School of African Studies, No. 26 (n.s.). University of Cape Town, South Africa.

Gulliver, P.H.
1955   *The Family Herds.* London: Routledge and Kegan Paul.

Hanna, J.M., S.M. Friedman, and P.T. Baker
1972   The status and future of the U.S. Human Adaptability Research in the International Biological Program. *Human Biology* 44:381-398.

Hansen, A.J., F. di Castri, and M.M. Holland
1988   Ecotones: What and why? *Biology International* 17:9-46 Paris: International Union of Biological Sciences.

Harrison, G.A. and A.J. Boyce
1972   Migration, exchange and the genetic structure of populations. *The Structure of Human Populations.* Edited by G.A. Harrison and A.J. Boyce. Oxford: Clarendon Press. Pp. 128-145.

Henin R.A.
1968   Fertility differentials in the Sudan. *Population Studies* 22:147-164.

Hett, J.M. and R.V. O'Neill
1974   Systems analysis of the Aleut ecosystem. *Arctic Anthropology* 11(1):31-40.

Hill A.G.
1985   The recent demographic surveys in Mali and their main findings. *Population, Health and Nutrition in the Sahel: Issues in the Welfare of Selected West African Communities.* Edited by A.G Hill. London: Kegan Paul International. Pp. 41-63.

Jamison, P.L. and S.M. Friedman
1974   *Energy flow in human communities: proceedings of a workshop.* University Park, Pa.: Human Adaptability Coordinating Office of the U.S./IBP.

Jamison, P.L., S.L. Zegura, and F.A. Milan eds.
1978   *Eskimos of Northwestern Alaska: A Biological Perspective.* Stroudsburg, Pa.: Dowden, Hutchinson, and Ross.

Johnson, P.L.
1977   *An Ecosystem Paradigm for Ecology*. Oak Ridge,
Tennessee: Oak Ridge Associated Universities.
Kemp, W.B.
1971   The flow of energy in a hunting society.
Scientific American 225(3):104-115.
Lampl M., F.E. Johnston, and L.A. Malcolm
1978   The effects of protein supplementation of the growth
and skeletal maturation of New Guinean schoolchildren.
Annals of Human Biology 5:219- 227.
Laughlin, W.S.
1970   Aleutian ecosystem. *Science* 169:1107-1108.
Laughlin, W.S. and A.B. Harper eds.
1979   *The First Americans: Origins, Affinities, and
Adaptations*. New York: Gustav Fischer.
Leslie, P.W. and P.H. Fry
1989   Extreme seasonality of births among nomadic Turkana
pastoralists. *American Journal of Physical
Anthropology* 79:103-115.
Leslie, P.W., J.R. Bindon, and P.T. Baker
1984   Caloric requirements of human populations: a model.
Human Ecology 12:137-162.
Leslie, P.W., P.H. Fry, K. Galvin and J.T. McCabe
1988   Biological, behavioral and ecological influences on
fertility in Turkana. *Arid Lands Today and
Tomorrow: Proceedings of An International Research and
Development Conference*. Edited by E.E. Whitehead,
C.F. Hutchinson, B.N. Timmermann, and R.C. Varady.
Boulder: Westview Press. Pp. 705-712.
Little, M.A.
1980   Designs for human-biological research among savanna
pastoralists. *Human Ecology in Savanna
Environments*. Edited by D.R. Harris. London:
Academic Press. Pp. 479-503.
Little, M.A.
1982   The development of ideas on human ecology and
adaptation. *A History of American Physical
Anthropology, 1930-1980*, Edited by F. Spencer. New
York: Academic Press. Pp. 405-433.
Little, M.A.
1989   Human biology of African pastoralists. *Yearbook of
Physical Anthropology* 32:215-247.

Little, M.A. and S.M. Friedman
   1973   *Man in the ecosystem: proceedings of a
           conference.*   University Park, Pa.: Human
           Adaptability Coordinating Office of the U.S./IBP.
Little, M.A. and B.R. Johnson, Jr.
   1985   Weather conditions in South Turkana, Kenya. *South
           Turkana Nomadism: Coping with an Unpredictably Varying
           Environment.* Edited by R. Dyson- Hudson and J.T.
           McCabe. New Haven: Human Relations Area Files. Pp.
           298-314.
Little, M.A. and B.R. Johnson, Jr.
   1986   Grip Strength, muscle fatigue and body composition in
           Nomadic Turkana pastoralists. *American Journal of
           Physical Anthropology* 69(3):335-344.
Little, M.A. and B.R. Johnson, Jr.
   1987   Mixed longitudinal growth of Turkana pastoralists.
           Human Biology 59(4):695-707.
Little, M.A. and J.D. Haas
   1989   Introduction: human population biology and the concept
           of transdisciplinarity. *Human Population Biology:
           A Transdisciplinary Approach.* Edited by M.A.
           Little and J.D. Haas. New York: Oxford University
           Press. Pp. 3-12.
Little, M.A., K. Galvin and M. Mugambi
   1983   Cross-sectional growth of nomadic Turkana pastoralists.
           Human Biology 55(4):811-830.
Little, M.A., K. Galvin, K. Shelley, B.R. Johnson, Jr., and M.
Mugambi
   1988a   Resources, biology and health of pastoralists.
           *Arid Lands Today and Tomorrow: Proceedings of an
           International Research and Development Conference.*
           Edited by E.E. Whitehead, C.F. Hutchinson, B.N.
           Timmermann, and R.C. Varady. Boulder: Westview Press.
           Pp. 713-726.
Little , M.A., K. Galvin and P.W. Leslie
   1988b   Human growth, health and energy requirements in nomadic
           Turkana pastoralists. *Coping with Uncertainty in
           Food Supply.* Edited by I. de Garine and G.A.
           Harrison. Oxford: Oxford University Press. Pp.
           288-315.
Lusigi, W.J.
   1981   *Combatting   desertification   and   rehabilitating*

degraded population systems in northern Kenya. Technical Report A-4, UNESCO/UNEP Integrated Project in Arid Lands. Nairobi: UNESCO/UNEP.

MacCluer, J.W., J.V. Neel, and N.A. Chagnon
1971   Demographic structure of a primitive population: a simulation. *American Journal of Physical Anthropology* 35:193-207.

Mazess, R.B.
1975   Biological adaptation: aptitudes and acclimatization. *Biosocial Interrelations in Population Adaptation.* Edited by E.S. Watts, F.E. Johnston, and G.W. Lasker. The Hague: Mouton. Pp. 9-18

Mazess, R.B.
1978   Adaptation: a conceptual framework. *Evolutionary Models and Studies in Human Diversity.* Edited by R.J. Meier, C.M. Otten, and F. Abdel- Hameed. The Hague: Mouton. Pp. 9-15.

McCabe, J.T.
1984   Livestock Management among the Turkana: A Social and Ecological Analysis of Herding in an East African Population. Ph.D. Dissertation in Anthropology. Binghamton: State University of New York.

McCabe, J.T.
1988   Drought and recovery: livestock dynamics among the Ngisonyoka Turkana of Kenya. *Human Ecology* 15(4):371-389.

McCabe, J.T., R. Dyson-Hudson, P.W. Leslie, P.H. Fry, N. Dyson-Hudson and J. Wienpahl
1988   Movement and migration as pastoral responses to limited and unpredictable resources. *Arid Lands Today and Tomorrow: Proceedings of An International Research and Development Conference.* Edited by E.E. Whitehead, C.F. Hutchinson, B.N. Timmermann, and R.C. Varady. Boulder: Westview Press. Pp. 727-734.

National Academy of Sciences
1974   *U.S. Participation in the International Biological Program: Report No. 6 of the U.S. National Committee for the IBP.* Washington, D.C.: National Academy of Sciences.

Neel, J.V., M. Layrisse, and F.M. Salzano
1977   Man in the tropics: the Yanomama Indians.

Population   Structure   and   Human   Variation.
Edited by G.A. Harrison. London: Cambridge University
Press. Pp. 109-142.

Newman, M.T.
  1960   Adaptations in the physique of American aborigines to
           nutritional factors. *Human Biology* 32:288-313.

Newman, M.T.
  1962   Ecology and nutritional stress in man. *American
           Anthropologist* 64:22-33.

Newman, R.W. and E.H. Munro
  1955   The relation of climate and body size in U.S. males.
           American   Journal   of   Physical   Anthropology
           13:1-17.

Odum, H.T.
  1971   *Environment, Power and Society.*   New York:
           Wiley- Interscience.

Odum, E.P.
  1969   The   strategy   of   ecosystem   development.
           *Science*164:262-270.

Roberts, D.F.
  1952   Basal metabolism, race and climate. *Journal of the
           Royal Anthropological Institute* 82:169-183.

Roberts, D.F
  1953   Body weight, race and climate. *American Journal of
           Physical Anthropology* 11:533-558.

Roberts, D.F.
  1956   A demographic study of a Dinka village. *Human
           Biology* 28:323-349.

Ross, J.K.
  1975   Social borders: definitions of diversity. *Current
           Anthropology* 16:53-72.

Shelley, K.
  1985   Medicines for Misfortune: Diagnosis and Health Care
           Among Southern Turkana Pastoralists of Kenya. Ph.D.
           Dissertation in Anthropology. Chapel Hill: University
           of North Carolina.

Scholander, P.F., K.L. Anderson, J. Krog, F.V. Lorentzen, and J.
Steen
  1957   Critical Temperature in Lapps. *Journal of Applied
           Physiology* 10:231-234.

Scholander, P.F., H.T. Hammel, S.J. Hart, D.H. LeMessurier, and J. Steen
    1958   Cold adaptation in Australian aborigines. *Journal of Applied Physiology* 13:211-218.

Shugart, H.H., Jr. ed.
    1978   *Time Series and Ecological Processes.* Philadelphia: Society for Industrial and Applied Mathematics.

Spuhler, J.N.
    1959   Physical anthropology and demography. *The Study of Population.* Edited by P.M. Hauser and O.D. Duncan. Chicago: University of Chicago Press. Pp. 728-758.

Terborgh, J.
    1971   Distribution on environmental gradients: theory and a preliminary interpretation of distributional patterns in the avifauna of the Cordillera Vilcabamba, Peru. *Ecology* 52:23-40.

Thomas, R.B.
    1976   Energy flow at high altitude. *Man in the Andes: A Multidisciplinary Study of High-Altitude Quechua.* Edited by P.T. Baker and M.A. Little. Stroudsburg, Pa.: Dowden, Hutchinson, and Ross. Pp. 379-404.

Weiner, J.S.
    1964   Part V. Human ecology. *Human Biology* Edited by G.A. Harrison, J.S. Weiner, J.M. Tanner and N.A. Barnicot. Oxford: Oxford University Press. Pp. 399-508.

Weiner, J.S.
    1977   The history of the human adaptability section. *Human Adaptability: A History and Compendium of Research.* Edited by K.J. Collins and J.S. Weiner. London: Taylor and Francis. Pp. 1-23.

Wilson R.T., A. Diallo, and K. Wagenaar
    1985   Mixed herding and the demographic parameters of domestic animals in arid and semi-arid zones of tropical Africa. *Population, Health and Nutrition in the Sahel: Issues in the Welfare of Selected West African Communities.* Edited by A.G. Hill. London: Kegan Paul International. Pp 116-138.

Woodmansee,R.G.
    1978   Additions and losses of nitrogen in grasslands ecosystems. *Bioscience* 24:81-87.

UNESCO
1972   International Co-Ordinating Council of the Programme on
        Man and the Biosphere (MAB). First session. Held in
        Paris, 9-19 November 1971. Paris: UNESCO/MAB.
UNESCO
1987   International Co-Ordinating Council of the Programme on
        Man and the Biosphere (MAB). Ninth Session. Held in
        Paris 20-25 October 1986. Paris: UNESCO/MAB.

CHAPTER 15

ECOSYSTEMS, ENVIRONMENTALISM, RESOURCE CONSERVATION, AND ANTHROPOLOGICAL RESEARCH

John W. Bennett

Preface

This paper—really an essay, or a philosophical rumination--has three main objectives: it is, first, a kind of polemic against the tendency to dichotomize human culture and behavior from the physical environment which sustains them. Secondly, it attempts to describe, by means of three case studies, the complex ways modern humans use and degrade the environment, and to suggest the kind of research activities which anthropologists--and other social scientists--must engage in to make a contribution to the unravelling of these complex processes. Third, it speaks for the need for a normative perspective--a viewpoint which takes reasoned, biased positions on the problem of resource use and abuse and proceeds toward scholarly attempts to examine them empirically.

There is a hypnotic appeal, especially to academics, in the scientizing mode of thought and rhetoric, of splitting hairs and avoiding political entanglements. Concepts like that of the *ecosystem* give us a comfortable sense of precise and rigorous knowledge of the way the environment works. Less often acknowledged is that if we neglect the major influence on the environment--humans and their institutions--this knowledge is of little meaning. While anthropologists perform archaistic studies of odds and ends of humanity, municipal authorities struggle with the chemical, geological, economic, and political problems of toxic wastes--with little help from social scientists. Agricultural research provides clever answers to the problem of increasing yields without ruining the soil and water, but rarely concerns itself with the problems of decision-making among farmers who must make a living and function as members of communities.

We talk about "sustained yield" of resources: a fine conservationist principle that has attracted little work by the social sciences on how to put it into practice in a competitive and individualistic economy, with the partial exception of economics. Entrepreneurial production--rapidly becoming the dominant form of production in the world--is rarely compatible with ecosystemic principles, and hence one wonders, again, if ecosystem and other concepts like it have any relevance to the real world. This world includes humans who struggle to survive and to realize their goals in a human society and culture. If these factors are not somehow built into our ecological science, we are all monuments to futility and elitist pomposity. This view has an acknowledged bias: it is scholarship and science from a jaundiced perspective, and let the chips fall where they may.

Formal Introduction

The paper contains two general and familiar ideas about the relationship of the concept of ecosystem to human affairs: first, if the human use of the physical environment is to be brought into some kind of balance, both human and physical factors must be conceived as a single system, i.e., a system in which human needs are satisfied *and* the yield of the resource is maintained. As things stand now, human needs come first, and only then adjustments are made in resource practices which may reduce exploitative use. Most of these adjustments, I suspect, simply displace exploitation or destructive use into some other system. The second, and perhaps not so familiar, idea is that the most important human factor in ecosystems is not some exotic force or process embodied in biological theory, but represents the purposes and actions of humans in real social contexts, i.e., human ecology *is* human behavior (Bennett 1980).

Politics, social change, greed, profit, self-actualization, ethics, and philosophy are all aspects of the human engagement with the physical environment, and these factors must be incorporated into our understanding of human ecology if we are to achieve a more sustainable use of the world. Human ends must be related to environmental ends; the quality of life must be synthesized with--and perhaps politically subordinated to--the quality of the environment.

If ecosystem means dynamic balance between resource and sustenance, this requires a restructuring of human purpose and cultural, political, and moral problems.

While these ideas were articulated in the idealistic ecology movement of the 1960's and '70's, its successor, the organizational environmental movement of the later '70's and '80's, has largely surrendered them in favor of an acceptance of the institutionalization of regulation of resource practices. Such regulations have failed, in large part, to attack the fundamental issue, which is the persistent priority of human demands on the environment.

The role of anthropology--or any other social discipline--in this situation must be dual: (1) to conduct research on the way physical phenomena become absorbed into human systems of needs, wants, and profit-seeking: a process I have called the "ecological transition" (Bennett 1976:2-5). This effort should not be simply an analysis of how humans use physical or natural phenomena to survive and realize their aims (the transformation of Nature into "natural resources"), but a normative inquiry obsessed with the question of whether the transformation process is generating environmental costs which future generations must somehow pay. That is, the human-centredness of our social disciplines must give way to a concept of "socionatural systems"--systems of effort and impact of humans on Nature, in which humans are part of the larger whole. The politics of need and want satisfactions must be included in the data protocols.

(2) The second role of these disciplines is even more difficult: the need to raise serious questions about fundamental social and ethical values of the 20th century--in particular, the dominant theme of self-gratification. These values need to be viewed with reference to the costs generated by efforts to realize such goals. Social scientists should take the leadership in documenting these costs, but even more important is the necessity of finding alternatives. Reduced expectations is the pathway of the rest of the 20th century: how can this pathway be reconciled with the dominant ethos? How can a culture of mass indulgence be realigned toward a culture of austerity and more modest expectations? Above all, how can we create a mass culture more concerned with posterity than with self-gratification in the here and now?

Ecosystems and Human Systems

The concept of *system*, originating in mechanical and biological investigations in the 1940's, was originally an arcane notion attracting workers in interstitial scientific and humanistic fields. By the 1980's, the concept has become commonplace in many fields, including the social sciences, and is extensively used as an analytic concept in applied fields like management and communications. The original philosophical tradition survives in General Systems Theory, a highly theoretical inquiry not to be confused with studies of empirical systems (for a typical treatment, see Laszlo 1972).

Ecosystem is really one of the specialized concepts pertaining to empirical systems and consists of a set of generalizations about the interdependent nutritional and demographic processes of plant and animal species living in defined physical environments (Tansley 1935 is the pioneer statement). However, during the 1960's, under the stimulus of the idealistic ecology movement, the concept began to be used by non-biological scientists and commentators in new ways. It was proposed that humans, who disturb natural ecosystems, should model their own uses of the physical environment on those of non-human components of ecosystems and should adjust their resource practices so as to insert human activities into ecosystems without strain to the biotic and abiotic components. Likewise, studies have been made in which human institutional processes, like economic behavior, have been proposed as models for ecosystem processes. These various attempts on both sides of the fence have not borne much policy or theoretical fruit, and most efforts seem to be mainly intellectual exercises in analogy (for extensive discussion of this point, see Bennett 1976, Chapter 6).

The basic ideas associated with the ecosystem concept are expressed by the tendency for natural species to exchange energy in such a manner as to create cyclical movement. For example, consumption of natural substances and energy conversions among species can result in a dominance of one species or the overconsumption of given substances, but this imbalance gives way, sooner or later, to compensatory phenomena, such as a rise in a predatory population which reduces the numbers of the dominant form; or the bloom of a new sessile plant form which shifts the dominance to a new consuming species. Such cyclical movements, with maintenance of approximately stable or average energy budgets, are sometimes approximated among human populations, especially low-energy-using and isolated tribal groups, but they do not begin to describe the major course of human ecological history (for further discussion, see Bennett 1976, Chapter 5).

This history was concisely synthesized by Eugene Odum in 1969, in one of the classic papers associated with the idealistic ecology movement of the 1960's and early '70's. In essence, Odum noted that natural systems tend to approach unity or stability in their relations of production to output, subsequent to youthful states where the ratio of production to output was greater than unity. Human systems, on the other hand, increasingly seek maximum output at the lowest possible energy expenditure. Therefore, since human purposes, like profit and gratification, intervene in and influence the process, unity or stability is never reached; the tendency is for demand for yield to increase exponentially and to require ever increasing amounts of energy. These rising costs of production are concealed or charged to other institutional systems. The classic example is, of course, the cost of fossil fuels consumed in crop production, which are charged off to national energy budgets or shunted off to the consumer. Thus, the full cost of food production is disguised. While improvements in fertilization, tillage, and other agronomic techniques may appear to lower the environmental costs of agricultural production by conserving soil or water quality, these procedures may result in increased environmental costs felt in some other sphere. Segmental approximation of ecosystemic balances in human activities thus might be illusory when all of the larger systemic processes are considered.

The crucial issue remains how to maintain high levels of production in order to meet social demands. While the idealistic ecology movement led to a series of environmental strategies which represent "improvements" in accordance with ecosystemic principles, there has been no fundamental change in social goals -- since demand remains at high levels although some erosion has taken place due to scarcities or rising costs (Hirsch 1976). The question therefore is whether these high levels of demand exerted by human systems on the physical environment can be attained or maintained without irreversible environmental damage or degradation. That is, can we continue to produce at a high want-satisfaction level and safeguard the environment while doing so?

The theoretical issue, then, of the relationship of ecosystem to the human sphere is simply whether or not human or social factors can be incorporated in balanced, sustained-yield ecosystems. That is, are humans part of ecosystems in the sense that their needs can be satisfied without running down the system? Is there a direct or linear relationship between the magnitude of human demands on the environment and the degradation of this environment? Or is the relationship curvilinear; i.e., can degradational processes be modified by superior strategies or technologies and still maintain high yields?

Several considerations arise at this point. First, there is the question of *time*. Obviously, degradation is a temporal process, taking place at different rates and depending on many factors. If you replace one-fourth of the forests you cut with seedlings, you delay the time it takes to achieve a state of denuded and eroded land; but if the replacement rate is only one-fourth, ultimately the denuded state becomes visible. Yet it is possible to claim, on the basis of replanting activity, that you are meeting some of the conditions of sustained yield. And you can always promise Society (or Nature) that when capital permits you can step up the rate of replanting from one-fourth to 100%. This is a noble promise, but one rarely kept.

A second factor, the *circumstances of regeneration* or recovery of an altered natural substance or resource, is, of course, also a temporal as well as a material process. No unmodified landscape, once transformed by humans (or any other species), ever returns to its original state because in a sense there is no such "original state." However, one can obtain *similar* states: equivalent biomass although with new and different species; similar energy potentials and changes, though not identical to the original; and so on. This constant change and recovery of physical phenomena is itself an aspect of evolutionary change; the human contribution is intrinsically no different from that of other species. It is simply quantitatively much greater. Moreover, the recovery rates of natural systems altered by humans tend to be slower and the return to earlier states tends to be less promising or certain. In addition, the chance of further intervention during the process of regeneration is always high. Thus the tendency is toward progressive degradation—or exponential curves of increasing output and exploitation. Eastern Mediterranean forests were slowly destroyed over a period of 2000 years by shipbuilding, goat grazing, firewood, and other uses; they have never recovered. This type of process can be duplicated for countless habitats or ecosystems throughout the world. [1]

But cyclical processes similar to those in natural ecosystems appear segmentally in the human domain. Grazing lands offer examples. Here regenerative capacities are often substantial, and what appears to be abusive or degradational usage may turn out, even in a few years' time, to be a cycle of use and recovery. The concept of "overgrazing," formulated originally by conservationists with little understanding of the practices of either nomadic herders or sedentary ranchers, ignored the tendency of these people to engage in cyclical resource use strategies (see Little *et al.* this volume). However, secular degradational and erosional patterns exist nonetheless. The conversion of large portions of the North American range to heavy brush cover, often too difficult or costly to remove and to replace with grasses, is a case in point. This example is especially interesting since the brush cover, caused by human abuse of grass, actually represents an increase in biomass over the grass stage. Humans consider it to be a deleterious change for *economic* not ecological reasons. The case also illustrates the tendency for judgments about ecological matters to become intertwined with human purpose and value.

All modern uses of the physical environment are mediated by *institutions*. The concept of institution is absolutely crucial to the problem of environment in the contemporary world. However, it is one concept that remains largely unfamiliar to anthropologists; indeed, most anthropologists this writer has talked to recently consider the term to be "sociological." This is due to the fact that anthropological theory was formulated on the basis of studies of societies "without institutions"--in a manner of speaking. That is, the formal-legal constituents of rules and purposes segregated by function, which characterize the institutions of civilization, is a process which was only weakly developed in (formerly) isolated tribal societies. Instead of institutions, anthropologists of the 1920's and '30's believed they had only "culture."

If anthropologists wish to adequately deal with contemporary environmental problems, they will be required to use the concept of institution--or to develop their own version of this fundamental concept. Such a task includes an objective look at the process of change and reform--very much a topic immersed in the "historical present." For example, the history of use of the Great Plains by

Americans in the past century offers impressive opportunities for the analysis of how institutions shaped environmental successes and failures: the land survey and its irrelevance for natural topography and resource placement; the institution of freehold land tenure in a region demanding collective use and sharing; high-yield cash-crop agriculture on marginal soils; and so on. Even the rehabilitation measures promulgated by the Roosevelt Administration as a result of the Dust bowl have been recently reanalyzed as an example of how the best of intentions in environmental management, if not balanced with an appropriate understanding of human behavior, can make the situation worse. Donald Worster (1979) has recently proposed that soil conservation and other subsidized programs cushioned the fears of farmers so that they assumed that soil abuse would be compensated by federal benefits; consequently, they were free to continue abuse management procedures in pursuit of high yields and profits, although the market institutions of the nation forced the smaller farmers to opt for such intensive cultivation.

Transient Herding and Rangeland Conservation

Now for some case studies. Tribal herding societies in Africa constitute an unusually complete case of changing resource practices, illustrating shifts from ecologically benign to abusive systems. Sometimes lumped together in the category of "nomadism," these groups in reality offer a wide variety of range utilization patterns involving varying types of settlement, movement, and sequential use of pasturage and water. Their transiency represents an adaptation to marginal resources in dryland environments where rainfall is not sufficient to provide year-round forage at any single locations. Hence, movement between pasture areas affording adequate grazing and water at different times of the year evolved over the centuries. Recurrent droughts encouraged the development of a series of organizational devices like the "herd-friend": each herd owner has one or more associates in other regions who are prepared to take part of his herd during periods of local drought, returning them, with suitable compensation, when the drought period is at an end. By moving the animals—and sometimes the residences or camps of the families—reasonable assurance of forage for a herd of reasonably stable size was obtained. The human populations of these groups remained static, grew very slowly, or fluctuated with conditions--as has been the case for most tribal subsistence adaptations where growth or profit was not the main objective (for discussions of the pastoral ecological and development problem, see Oxby 1975; Bennett, Lawry, and Riddell 1986; Galaty *et al.* 1981).

Beginning in the colonial period in the 19th century, these ecologically-adapted strategies of livestock and human production began to change under the stimulus of new conceptions of economic effort, and also as a result of altered access to pasturage, due to new political boundaries or new uses for the rangeland. These changes were greatly accelerated in the post-World War II period of independence when the colonies became nation-states, and the tribal people were gradually transformed into citizens who were expected to adhere to new concepts of tenure rights and to fulfill national development objectives. In the past 20 years, the effort to "develop" transient pastoral societies into sedentary livestock ranchers has constituted one of the principal targets of agrarian development in the dryland regions of sub-Saharan, East, and Southern Africa. The U.S. Agency for International Development committed $618 million to such projects over an approximately 12-year period in the 1960's and '70's. The projects were encouraged or initiated by country governments not only in order to increase the offtake of animals and diversify the product (much of the traditional industry was concerned more with blood and milk rather than beef) but also to find substitute occupations for people pushed out of rangelands targeted for crops, plantations, game parks and other uses.

Many of these projects happened to reach a peak of involvement during a period of recurrent drought of which the Sahelian drought is the best known segment (Dalby and Church 1973). The forced modification of transient pasture usage plus drought, *plus* the frequent failure of the country governments to follow through with promised pasture development and well-drilling, resulted in a deterioration of the dynamic resource-populations system. Constrained to limited pasturage, herders followed a variety of strategies: overgrazing, concentration around water sources, ignoring prescribed range areas, migrating across national boundaries, or selling herds and moving to the cities.

These processes, of course, actually began much earlier, under colonial governments, though the period of independence witnessed an acceleration. Rangelands in nearly all of these countries were in a state of deterioration and measures to reclaim pasture and hasten sedentarization were in operation. One approach to the problem of maintaining pastoralist production, while at the same time avoiding pasture abuse and herd fluctuation, involved the allocation by governments of large tracts of rangeland for which no immediate alternative use was foreseen, to herding groups in the form of cooperative or "group ranches." The herders were expected, and sometimes aided, in adapting to these large but much more restricted grazing areas, but on the whole the schemes worked poorly. Large as they were, they were not sufficient in time of drought, and the herd owners felt free to move their animals off them. More fundamentally, the combination of restricted collective pasture tenure and household herd ownership resulted in a "tragedy of the commons" of the classic type. During droughts, or in response to other factors modifying herd size, sales, or production, the owners simply attempted to maximize their production and thus competed for pasture; the end result being badly overgrazed group ranches (for a summary of the group ranch situation, see Oxby 1982). In any case, nowhere in Africa has a balance between resources, people, and animals been established--at least as it existed in the pre-colonial period (for a broad survey of all the issues, see Galaty et al. 1981).

It is perhaps too easy to attribute this collapse of a human-managed ecosystem to the nation-state and its development initiatives. There is, of course, evidence that the traditional system also went out of balance in the past, due to tribal wars, invasions, population increases and other factors. The rinderpest epidemics appear to have been the result of changes introduced by colonial governments (Kjekshus 1977), yet the diseases were present for centuries and native cattle breeds had developed degrees of immunity to the organisms. New, introduced breeds and new strains of the organism resulted in the devastation of herds in the 19th century. There probably were earlier episodes, now lost to history. A sober scientific perspective on the entire problem of African pastoralism would probably acknowledge a fluctuating interaction over centuries between people, animals, plants, disease organisms and other factors; the current troubles may represent the shift to a new ecosystem--or socionatural system--in African drylands. It is too early to say if the present situation represents a progressive and non-regenerative deterioration. The issue here is the fact that human populations have always displayed these cyclical patterns of use and abuse of resources; in some cases the degradation is irreversible; in others, regeneration is possible.

The book by Helge Kjekshus (1977) cited above in connection with *rinderpest* has a significant general theme: it proposes to demonstrate that contemporary management of herds and rangelands is not the first control system imposed on East Africa. Kjekshus shows that "control" of an ecosystem is a relative matter: beginning in the colonial era, attempts to dominate Nature in East Africa meant increasing production of many commodities but also a series of calamities, from the rinderpest epidemic and the continued difficulties created by the tsetse fly, to the groundnut fiasco, to increasing range degradation and soil erosion. In contrast, the relationship between humans and the physical environment in precolonial times was much more benign, although this can be exaggerated, as previously suggested, since there are indications of cyclical epidemics and environmental abuse. However, a recent increase in abusive trends seems undeniable; and for the first time, perhaps one is entitled to fear that range recovery may become progressively more difficult and pastoralism increasingly less viable. This is a matter of some moment since it threatens the contribution of animals to the diet and the productivity of otherwise marginal lands.

So far as anthropological research is concerned, the burst of effort in the past decade has been of inestimable value not only to the theory of cultural ecology--an intra-disciplinary concern--but also to the policy and practice of livestock management and economic development. In fact, the anthropological effort was largely paid for by the development agencies in their desperate attempt to determine why their carefully-conceived schemes went astray. In other words, the impetus came not from the discipline itself but from the participation of anthropologists in real-life issues. For the first time in anthropological research in African pastoralism, basic questions of policy, and of practical effort, were made the primary focus of research. The values involved were openly expressed: the welfare of the herders, as well as of the range and animals, must be a concern of the developers. The writer does not feel he is exaggerating when he says that we now know more about pastoralist societies and ecology than we know about the cultural ecology and livelihood of any other tribal-peasant part subsistence-part commercial production system in the Third World and that this knowledge was gained by making research topics sensitive to real human needs (see Lees and Bates, this volume).

Sustained Yield in Coastal Fisheries

The concept of *sustained yield* is usually offered as the major general objective of resource management systems designed to achieve something like ecosystemic continuity. This concept is ambiguous since it is often not clear whether it refers to the economic product derived from the physical resources or the resources themselves. In addition, every modern resource practice implicates many, not just one species. Sustained-yield management of forests may assure a continuing supply of trees, but with a diminished or vanished supply of other plants and animals living in the natural stand previous to intensive cutting and replanting. But a certain degree of regeneration is even possible here, providing that intact older stands are preserved as a source of other organisms to replenish the cutover tracts. But all this is very expensive in terms of contemporary economic arrangements. Labor, time, and other costs must be added to the product price, something the producers are not usually inclined to do. Tax laws and other instrumentalities may also penalize such management schemes. Sustained-yield versions of ecosystems become a complex matter of costs, prices, profits, laws, and management personnel. The history of forest regeneration on a world-wide basis gives no room for optimism.

In recent years strenuous attempts were made by the U.S. Government to establish sustained-yield management regimes for East Coast fisheries. Overfishing by the multitude of small private fishing boats and companies, particularly in New England, were beginning, in the 1960's, to develop into a classic case of abuse of a "commons": the Atlantic fish school. The evidence took the form of drastic changes in the numbers of fish of various species or in the virtual disappearance of some. Of course, pollution and fluctuating ocean temperatures also play a role, and the evidence is not as decisive as one would wish. But in any case, overpredation by humans was considered to exist. In 1976 the U.S. Government promulgated its Fisheries Conservation and Management Act, considered at the time a model of sustained yield management regulation. Coastal fisheries were extended out to 200 miles, and the general coastlines divided into 8 regions. Each region is supervised by a Regional Council, whose membership includes Federal, State, and industry representatives. The law requires that fisheries be managed for attainment of what is called *Optimum Sustainable Yield* (OSY); that is, the most you can take and still maintain the fish stocks. This combines both biological and natural factors, plus socioeconomic ones; that is, human activities and needs are supposed to be inserted into the "ecosystem." In essence, this requires some extremely complex trade-offs between fish populations, fish catches, economic costs, and needs of coastal human populations.

The new law and regulatory devices were greeted with optimism by environmentalists, but with skepticism by fisheries people. However benign the goals, the law itself made no attempt to define *Optimum Sustainable Yield* for the simple reason that no one knew how to define it or what types of data to include in the definition. Nor was there any systematic attempt to define the human institutions and activity patterns which modify the physical circumstances. No precedents for sustained yield existed save in the form of seat-of-the-pants management by the fisherman themselves. However, these strategies were site-specific: adjusted to microenvironments and changing physical factors, like wind or water temperature, as well as to social phenomena, markets, and consumption standards. No set of criteria applicable everywhere were visible. In short, the basic information necessary to sustain or manage an ecosystem including human activities in accordance with the law was simply not available in any precise form (see Wilson and Acheson 1981 and other reports of the study for a detailed history and analysis).

Most of the Regional Councils did make an effort to set quotas on total catches, these quotas adjusted to what little is known about the impact of fishing on stocks and the economic demands of the fishing industry. None of the parties have shown satisfaction with the results: the scientists feel the fish are being depleted; the fishermen resent regulation of any kind; and the Council members acknowledge they are doing their best but basically "pinning tails on donkeys." The issue is simply that the optimal fish catches in terms of conservationist standards do not necessarily correlate with acceptable economic returns to the fishermen. That is, a large catch can mean only modest income if the catch is made up of species which bring low prices on the market. Taste and preference standards influence the marketability of fish as they do other foodstuffs. Moreover, regulation assumes that the major objective of fishermen is to control and even reduce his own catch.

The study done by Acheson and Wilson shows that the current procedures followed by the Councils are uninformed; they reflect the lack of hard data on the components of the system mentioned earlier. There has been no inquiry by the Councils as to what is gained or lost by fishermen when certain regulations or strategies are used. The concept of a trade-off between the demands of the fishermen for income and the species survival of given fish stocks have not formed the basis of the legislation or the makeshift procedures followed thus far. In other words, if human activities are to be inserted into ecosystems, full consideration of the role of various behavioral and socioeconomic factors on the human side need to be integrated with the environmental data. As it stands at present, conservation was conceived largely as a matter of curtailing overfishing, not weaving human interests and needs within the socionatural system. (For comparable discussions of fisheries problems in Asian contexts, see Emmerson 1981).

While conservationists were impressed with the law, the fishermen's attitude might be summarized as a mixture of anger and contempt. Many, if not most, simply ignored the law since it was virtually impossible to enforce due to the lack of precise definitions. Moreover, the concept of OSY, while important, somehow has not attracted the kind of research effort from biologists and resources management specialists it deserves, and the available information by 1981 was no more abundant than in previous years and decades. To establish a sustained yield regime in a complex socionatural system like coastal fisheries would require an extremely expensive multidisciplinary research program--not to mention innovative thinking and experimentation.

However, the Wilson and Acheson research program was a least a first step—an economic-anthropologic partnership between several East Coast universities designed to construct a series of models of local fishing strategies. These models were of two main types: an econometric model of fishing as small entrepreneurial firms with an emphasis on risk and decision-making, and an anthropological model based on concepts of adaptive strategy and cultural factors of influence. Some crossover between the models was worked out, but on the whole, this integration remains to be accomplished. The chief contribution of the research to our understanding of fisheries as a cultural-ecological problem is the *ethnographic* information on how small fishermen combine information from social sources, their own competitive interactions, practical knowledge of fish behavior, and so on, to forge their own "seat of the pants" strategies. This is certainly an important first step in understanding the larger system of fishing and the conservation problem, but much work remains to be done.

Once again: we are dealing here with the concept of "socionatural system" (Bennett 1980). If human activities are to be inserted into ecosystems, the system itself has to be re-conceptualized: it is not a matter of a "natural" system being invaded by humans, but a complex whole system involving an interaction between the physical resources, animal species, and the human activities. This requires a shift in values as well: human components must be viewed as *analytically* equal to environmental components. The tendency to take human needs and interests as separate and with an always-higher priority from the environmental is what vitiates attempts to develop true conservationist and ecosystemic-management schemes. There is, of course, considerable room for pessimism as to whether these can be implemented given the deeply-institutionalized human-first consciousness of contemporary institutions.

Surface Mining and Land Reclamation

Another example of an attempt to utilize ecosystemic ideas in resource management is that of surface mining of minerals, particularly coal (see National Research Council 1981). The decision to exploit shallow deposits of coal in several Western states received predictable opposition from conservationists, preservationists, sportsmen, and ranchers, whose economic interests, values, and communities would be disrupted by the mining operations. That is, issues of concern were almost equally divided between physical and social aspects. The central theme of protest over surface mining for a generation has been the visual appearance of the spoils banks following extraction. This concern has taken the form of persistent attempts to obtain legislation designed to require companies to restore the landscape *to its original condition.*

The Surface Mining and Reclamation Act of 1977 was the culmination of this campaign; and while the Act did not completely satisfy its conservationist proponents, it represented a major step in establishing a degree of environmental responsibility in the industry (the Reagan administration issued a revised and weakened set of implementing regulations in 1983). The original law required that a minimum performance standard be adhered to in mining operations which transform the land surface and soils. The term "reclamation" is used in the Act throughout, although the word in the past has been associated more familiarly with natural "wastelands" to be developed for agricultural or recreation uses. "Restoration" is a more accurate term since the law provided for measures which would "restore the land affected to a condition capable of supporting the uses which it was capable of supporting prior to any mining, or higher or better uses of which there is reasonable likelihood. . ." (SMRA, Sect. 515(b)).[2]

The case also illustrates another important issue: the importance of aesthetic interests in the management of resources and their restoration or reclamation. This factor became one of the most difficult issues in the attempts to adhere to the provisions of the Act. Restoration of a mined surface for agricultural uses, industrial or residential development, or recreational purposes is found to be in conflict with aesthetic interests in a great many instances. The preservationist goal is to make the lands look like they did before mining; the value is placed on the scenic appearance or on the "native" contours and vegetation. Since in a majority of cases the vegetation, and in many cases the very contours of the land were previously modified by human activities, these goals are somewhat ambiguous. More important is the fact that reclamation of spoils and land levelling for agricultural purposes in many or most regions is more successful if the underburden soil material is left exposed, after a degree of levelling and recontouring, since this material has been found to be more fertile than the original topsoil. If recreational uses focus on water facilities, like sloughs or ponds, the rugged spoils contours may be preferable to levelling. In desert regions where the original vegetation was extremely sparse and heavily modified by grazing, restoration to this condition is easily done but this simply returns the land to an extremely low-productivity state. And so on.

Analogous to the fisheries legislation, the Surface Mining Act did not define the varying definitions of restoration or reclamation, nor did it base its provisions on research in different environments and biomes. Above all, the trade-offs and benefits of different uses of mined land before or after reclamation were not described. Nor are there provisions for involving local people in decisions on the pattern of reclamation. As in the fisheries case, the human activity factor was not included systematically in the specifications. This was to some extent deliberate since the issue was known to be extremely complex, and it was expected that application of the Act would develop and vary as experience accumulated. However, insufficient provision for such experimental modification was included. The Reagan Administration's modified regulations permit these flexibilities but also leave a series of loopholes which strongly suggest that the real intention is to return to the pre-1977 situation which absolved companies from responsibility. This may be contrasted to the situation in European countries, for example, Germany, where restoration of mined lands, in accordance with wishes and needs of local populations, have been a standard provision of law and a budgetary item in mining economics for many years.

The Act generated widespread controversy as soon as it went into effect. Questions about its provisions and basic philosophy were asked not only by mining companies, with their obvious vested interests in non-enforcement, but also by environmental groups concerned with the problem of alternative strategies of treatment of the land for different purposes following the mining activity. Community-oriented groups were concerned with the failure of the Act to clearly specify that local people should be consulted about the nature of the reclamation or restoration instead of making this a mandatory matter defined by Federal regulation. This concern generated enough political steam to encourage various groups in the Carter Administration to request the National Academy of Sciences-National Research Council to undertake a research project aimed at clarifying the whole issue.

Ultimately two NRC committees were appointed, one of them concerned with technical matters and the other with a fully multi-disciplinary inquiry into the role of land, soil, economics, and cultural interests involved in surface mining activity in American society. This committee worked for three years to collect information from published and field sources on the various conditions of surface mining and land restoration in various parts of the country where the activity was taking place--in particular, Appalachia and the northern Great Plains. The result was a report (NRC 1981) which while suffering from the usual ills of research-by-committee nevertheless managed--for the first time--to officially define the problems of surface mining both in terms of sociocultural factors and purely economic and technological factors. That is, the report is a kind of "cultural ecology" of surface mining. It was not reviewed as such by the *American Anthropologist* because anthropologists rarely identify cultural ecology as anything that exists outside of their own specialized arena of research subjects (tribals, peasants), but it is a document in this field, nevertheless. As such, it indicates the magnitude of the tasks before us: the sheer cost of assembling a thoroughly multi-disciplinary team of expensive experts to work for three years to produce something like a general-systemic model of a complex instrumental activity and that such cost can only be borne by the Federal government.

As one who participated in the work of this committee, I can testify to the fact that anthropologists were listened to carefully and that the anthropologists listened carefully to the economists and technical people. The work of such groups as the Anthropology Resource Center, which has shown a commendable interest in the impacts of energy development on Amerind reservation groups (Jorgensen et al. 1978), was taken seriously, and representatives of that research effort were asked to make presentations to the committee. All of this is to the good; if it does not generate immediate relief or implement social change, one must be patient. It is too easy to be cynical; anthropologists must work hard to be heard; but their message, when delivered intelligently and not truculently, begins to register.

This case illustrates the way *ecosystem* becomes absorbed into a matrix of human intentions, values, and economic activities. The goal of simply returning a landscape to a *natural* state is largely meaningless since it is difficult to define the nature of the previous state and its value to humans. Once a physical environment has been altered, questions of relative costs and benefits immediately arise. If, in a pluralistic society, there is no general consensus on the array of costs and benefits, then these must be determined through a complex social process involving dialogue between various interests. Each of these interests, or uses of the land, may have a different definition of what constitutes a balance between human interests and those of the environment. Biological standards of ecosystemic functioning are thus not sufficient to decide the case. Once again, the system must be re-conceptualized as a socionatural entity in which humans are defined as responsible components with other components of the system. This means subjecting human interests to a dispassionate scrutiny and making hard choices between these interests. Thus, systemic thinking in the resources field becomes a political process as well as a sociopsychological one.

Concluding Remarks

These considerations raise some important issues concerning how ecological science is to be used in practical environmental affairs. Ecosystem is one of those concepts that seems to acquire a life of its own: once the concept proved its usefulness as a way of analyzing interdependencies and processes among living species and physical phenomena, ecosystem became a reality: ecosystems existed. Hence we are persuaded to search for and to *find* ecosystems; and their properties are likely to be reified. In the early idealistic ecology movement, the fact that some subsistence-tribal societies were represented as approximating ecosystemic homeostatic properties was the occasion to recommend similar behavior for modern society, ignoring the fact that modern society operates on entirely different principles of resource use and with differing social and economic power magnitudes. While the exegesis based on the ecosystem concept formed a significant critique of these destructive practices, it could not supply remedies.

The remedies, as the later organizational environmental movement and its experiments in regulation and legislation have found, lie in everyday institutional forms and behavior. Ultimately the meaning of a total ecosystem is to be found in ourselves, not in Nature. If we are to insert ourselves into Nature in some sophisticated and constructive way, we must study ourselves as much as we study Nature, perhaps more.

Several research tasks have high priorities. In my opinion, one of the highest concerns the way different organizational forms and institutions establish high probabilities for specific resource practices. Are cooperative forms more conducive than competitive ones to resource conservation? Are collective property institutions like those of the Hutterites more congenial to a sense of guarding resources for posterity than individualistic frames? Are market pressures and profit motivations more exploitative than centrally-planned systems? A carefully-conceived and planned comparative study of these institutional forms, with systematic variation in other variables like population, climate, crops, and extractive modes of utilization, might begin to give us the information we need. The need for this type of information is critical in Third World countries where development programs in agriculture and industry have wrought havoc on the physical environment.

A second major task concerns the way societies allocate and control their use of resources. This task must begin with research on the institutions and organizations that any society has available to make resource decisions and organize the transformation of these resources into products or energy. How resources are used and distributed is governed, as noted earlier, by sets of rules which are as yet only generally understood. Intensive research needs to be done on particular systems of resource utilization and transformation: mining, petro-chemicals, cropping, fisheries, timberlands, and so on. Such studies must include the politics and economics of the institutional systems, as well as cultural values and habits which influence the power structure. Anthropologists have a vital role here since their ethnographic case-study method can be easily adapted to such research. Prototypes already exist: the recently- accumulating new material on African pastoralists in the development process is an example.

However, the problem is not one a single discipline can solve. A concept like socionatural system is the key, and this concept would require intimate collaboration among several disciplines in order to be effective as a research frame. So the main problem of defining cultural ecosystems becomes one of overcoming the social and cognitive barriers to collaborative work by people from different branches of scholarship. A culturally-oriented ecosystem or socionatural system concept is largely a structure of cognitive interdependence among people who at the present time are required to find their rewards in life largely in the form of personal intellectual satisfaction and by the prestige gained by impressing their disciplinary colleagues. The reinforcement of this segregative social process by the structure of the modern university is one of the more depressing aspects of the problem.

As things stand now, intimate relationships among disciplines are largely interdicted by the structure of the professions and the universities (for a discussion, see Bennett 1986, and other papers in the same volume). This sort of work can occur only in organizational settings outside universities or in sheltered zones inside the institutions, like research institutes enjoying outside funding. Government agencies are responsible for much constructive work of this kind, but this era may be drawing to a close due to financial constraints. Integrative research and scholarship, related to the pressing environmental problems of the age, will have to depend largely on enterprising and enthusiastic individuals who band together and attempt to overcome the early-20th century institutions that imprison them. Interdisciplinary research on socionatural systems is the main survival task of science; yet it is the one thing our vaunted establishments of learning and knowledge find it most difficult to sponsor. To realize the goals of scientific integration and system management of resources, we must change the arrangement of cognitive categories in the human mind and the social forms these create. This is a formidable task indeed.

Afterword: 1990

The Editor of this volume offered the writers an opportunity to revise their contributions for this second edition. I decided not to accept the invitation since a revision would in no substantial way change or improve the main points of the paper. I intend to rewrite and extend the paper in a volume of retrospective essays in human and cultural ecology I hope to complete sometime in 1991.

The plans for a completely new version include an extension and documentation of the arguments introduced in the this paper's Introduction. The key item I will elaborate is the moral issue of human greed and its control. Another task is a more detailed analysis of precisely how ecosystems become incorporated into human social systems, and thus come to be defined not by physical or biological principles, but by economic and political values. The task also involves a more comprehensive examination of the concept of *institution*, as the principal social vehicle through which natural phenomena are inserted into human effort and organization.

A second major step will be the addition of several case studies, in order to better show the great variety of ways human systems incorporate natural phenomena as "resources". Some of these studies will, I hope, present a somewhat more optimistic picture of conservationist postures and procedures than those selected for the paper in this volume. Related to this updating will be a consideration of several developments in social science and allied disciplines which also offer some hope for progress along the interdisciplinary lines discussed in the last part of the paper. Noteworthy among these is the "common property" movement which seeks to examine and promote cooperative and collective ownership and management of resources as a means to enhance social responsibility towards the environment (cf. Netting, this volume). This need is alluded to in the present paper in my remark about Hutterite "collective property" institutions. I shall also deal with some new research in the discipline of economics which may provide better tools for sociologists and anthropologists interested in analyzing socionatural systems.

Footnotes

[1]Incidentally, it is in historical contexts of resource utilization like this where anthropologists and archeologists can make salient contributions to our understanding of how humans use Nature (e.g., Cernea 1981).

[2]My brief excursion here into the semantics of the problem is deliberate in order to illustrate another important facet of the attempt to bring humans into relationship with ecosystems. Since government becomes the court of last resort, one must be attentive to the nuances of legal and legislative language because regulation requires enforcement, which becomes a legal process.

References Cited

Bennett, John W.
   1976   *The Ecological Transition.*   New York and
            London: Pergamon Press.
   1980   *Human Ecology as Human Behavior.* I. Altman, A.
            Rappaport and J. Wohlwill, *Human Behavior and
            Environment*, vol. 4. New York:
   1986   Interdisciplinary Research on People - Resources
            Relations.   *Natural Resources and People.*
            Edited by K. Dahlberg and J.W. Bennett. Boulder and
            London, Westview Press.
Bennett, John W., Steven Lawry, and James Riddell
   1986   Land Tenure and Livestock Development in Sub-Saharan
            Africa. U.S. Agency for International Development, AID
            Special Evaluation Study #39, Washington D.C.
Hirsch, Fred
   1976   *Social Limits to Growth.* Cambridge: Harvard
            University Press.
Jorgensen, Joseph G. *et al.*
   1978   *Native Americans and Energy Development.*
            Cambridge, Mass.: Anthropology Resource Center.

Kjekshus, Helge
    1977   *Ecology Control and Economic Development in East African History.* London and Nairobi: Heinemann.
Lazlo, Ervin
    1972   *Introduction to Systems Philosophy.*   N.Y.: Gordon and Breach Science Publishers.
National Research Council.
    1981   *Surface Mining: Soil, Coal, and Society. A Report Prepared by the Committee on Soil as a Resource in Relation to Surface Mining for Coal.* Washington, D.C.: National Academy Press.
Odum, Eugene
    1969   The Strategy of Ecosystem Development. *Science* 164:262-269.
Oxby, Clare
    1975   *Pastoral Nomads and Development.* London: International African Institute.
    1982   *Group Ranches in Africa.* Pastoral Network Paper 13D. London: Overseas Development Institute.
Tansley, A.G.
    1935   The Use and Abuse of Vegetational Concepts and Terms. *Ecology* 16:284-307
Wilson, James A. and James M. Acheson
    1981   A Model of Adaptive Behavior in the New England Fishing Industry. University of Rhode Island and University of Maine Study of Social and Cultural Aspects of Fisheries Management in New England. Report to the National Science Foundation, Vol. III.
Worster, Donald
    1979   *Dust Bowl: The Southern Plains in the 1930's.* New York: Oxford University Press.

NAME INDEX

SUBJECT INDEX